PRACTICAL PHARMACEUTICAL
LABORATORY AUTOMATION

PRACTICAL PHARMACEUTICAL LABORATORY AUTOMATION

Brian Bissett

CRC PRESS

Boca Raton   London   New York   Washington, D.C.

Library of Congress Cataloging-in-Publication Data

Bissett, Brian.
    Practical pharmaceutical laboratory automation / Brian Bissett.
        p. cm.
    Includes bibliographical references and index.
    ISBN 0-8493-1814-9 (alk. paper)
        1. Pharmaceutical technology—Automation. I. Title.

    RS192.B55 2003
    615'.19'00685—dc21

                                                                    2003040910

Visit the CRC Press Web site at www.crcpress.com

Dedication

For my grandfather, John Vasilchik

This next homework problem will make you think, so let's make this one extra-credit.

Dr. Ted Anderson
Former Electrical Engineering Professor
Rensselaer Polytechnic Institute

Foreword

You are about to enter the world of electronics, computers, interfacing, data acquisition and analysis, robotics and automation. An exciting world in which a computer can be programmed to control a robot, capture and process data from an instrument, measure temperature, or control a pump. But, how do you even begin to learn all this?

THE LATE 1970s

A pharmaceutical company hired a young electrical engineer to help design laboratory instruments and configure automation equipment. His friends asked why an electrical engineer was needed in a company working on chemistry and biology. Little did they know that the pharmaceutical world was moving from an era of beakers and test tubes to a new era of applied laboratory technology.

The engineer started his first day of work and brought along his sophisticated Hewlett Packard HP-35 scientific calculator and his trusty slide rule as a backup. He felt a little uneasy because he had no experience with circuit design, had never worked with a computer, and knew nothing about laboratory instrumentation or interfacing. However, the engineer had a mentor, his supervisor, who had over 45 years of experience in laboratory instrumentation design. He also had at his fingertips a complete machine shop with two mechanical designers/machinists.

A/D INTERFACE

The engineer's first assignment was the development of a flow-through spectrophotometer to analyze fermentation broths automatically. He worked with his supervisor and assembled a custom spectrophotometer from an ultraviolet (UV) light source, photomultipliers, and log amplifiers that provided an analog output voltage proportional to the optical absorbance. For this project, the analog data were acquired by a computer rather than a traditional strip-chart recorder. Management approved the purchase of a Digital Equipment Corporation PDP-11 computer. Interfacing the analog output to the input of the computer was a challenge. The young engineer had never worked directly with a computer; his closest encounter had been handing a stack of punch cards to a computer operator to run a college FORTRAN program assignment. General purpose analog-to-digital (A/D) cards were not available. How would he implement the interface?

Analog Devices, an integrated circuit manufacturer, had just developed its first digital panel meter (DPM) with an integral A/D converter and light emitting diode (LED) digital readouts. The engineer's solution to the interface problem was to tap into the binary coded decimal (BCD) wires of the DPM and connect them into the digital inputs on the PDP-11 computer. The interface had the added benefit of a digital readout for the operator. Fortunately, the pharmaceutical company had just hired a computer programmer whose assignment was to write the data acquisition software. The programmer wrote an assembly language program to read the digital inputs and to print the data to a dot matrix printer. The engineer was impressed by such a sophisticated program.

THE 1980s

PROGRAMMABLE PUMPS

The engineer's next assignment was the development of an automated enzyme reactor system. A scientist needed a method to add reagents and enzymes automatically into a spectrophotometer. Computer-controlled pumps were not available, so the engineer decided to connect a peristaltic pump head to a variable speed stepper motor. Superior Electric had just come out with a stepper motor controller. The controller was $19 \times 10 \times 12$ inches and weighed approximately 40 pounds; it was perfect for driving the 2-in. diameter motor.

Apple Computer, Inc. had just released the Apple II+ personal computer. The scientist was anxious to apply this computer to the application. Once again the engineer was confronted with a computer interface problem about how to control the stepper motor via the Apple II+ computer. The computer had an optional interface called RS-232. The engineer soon learned that this interface would have multiple applications in the future. However, the motor controller required transistor–transistor logic (TTL) pulses to control motor speed and direction. Intel Corporation had come out with the 8749H single chip microcontroller and a peripheral chip called a universal asynchronous receiver-transmitter (UART) that, when configured, would allow RS-232 commands to be captured and processed by the microcontroller.

The engineer dug through pages of documentation, learned assembly language, and months later finally had written an assembly language program to control the stepper motor via RS-232 commands. The scientist wrote a program in a language called BASIC and soon the Apple computer was controlling the motor-driven pumps.

ROBOT

One day the director of the biochemistry department received a call from a salesman who wanted to discuss a robotic system that might be useful in the laboratory. A few days later the owners of the robotic system gave a presentation and showed drawings of a small laboratory robot that had an arm that could rotate 360°, go up and down, rotate its wrist, and open or close its fingers.

The director was hesitant to purchase the robot without more details, so the company sent the engineer to Hopkinton, Massachusetts, to take a look. In a small garage, six engineers were working on the world's first commercial laboratory robot. Excited about the robot, the owner reviewed the electromechanical details and control circuitry. The engineer was amazed about how the electronic and mechanical parts were configured and controlled by a computer. He rattled off question after question and before he knew it, it was 2:00 in the morning.

The pharmaceutical company purchased the Zymark robot (serial no. 2) and applied it to a second-generation automated enzyme reactor. By then the scientist had replaced the Apple II+ computer with a large MassComp computer. The computer had a multichannel RS-232 interface. The engineer and the scientist thought that RS-232 was the way to go. Unfortunately, the Zymark robot controller had no RS-232 interface. So the engineer and the scientist called Zymark and within a few months received a custom-made interface box that allowed any computer with an RS-232 port to control the robot.

The scientist wrote software drivers in a language called "C". This program controlled the robotic system via RS-232 and captured the spectrophotometric data via an A/D converter. Scientists and management watched in awe as the robot went through its motions. Enzymes and reagents were dispensed into the cuvette, and the computer captured the spectrophotometer data in real time. The engineer was impressed when the scientist pressed a button and the computer crunched the numbers, performed a curve-fitting algorithm, and plotted results on a plotter.

THE MID 1980s

Years passed. The engineer's mentor and fellow machinists had retired and the engineer was now managing the group. One by one he hired a replacement team and was back in business ready to tackle the next instrumentation challenges.

THE 1990s

AUTOMATED CHEMICAL SYNTHESIZER

An energetic chemist came down to the engineer's lab to discuss building an automated parallel solid phase synthesizer. The idea was to use a number of pumps to flow reagents through columns containing resins. The chemical compounds would form on the resin in a process called solid phase chemistry. Kloehn Ltd. had recently introduced a programmable syringe pump with the now familiar RS-232 interface; pumps could be wired together into an addressable network. The engineer purchased a number of pumps and convinced the manufacturer to sell some of the pump's motor control cards. The engineer's design team configured the syringe pumps and designed stepper motor-driven positioners to move test tubes around.

IBM Corporation was marketing a computer called the IBM PC XT and Microsoft Corporation had released QuickBasic for DOS. This was the engineer's opportunity to write an instrument control program using a high-level language. Soon the computer was controlling the pumps and valves, reagents were flowing, and the chemist had synthesized his first compound.

WINDOWS OPEN

The engineer and his staff used QuickBasic for a few years. Then Microsoft Corporation released the Visual Basic (VB) program. Developing computer-controlled instruments running under Microsoft Windows was now possible. The engineer decided to teach a lunchtime course on VB and he and his staff started learning the new programming language. A young biochemist heard about the class and asked if he could join.

Everyone was amazed how easy it was to develop a graphical user interface (GUI). Digging through the material, they soon learned Microsoft had incorporated a VB control that allowed an application to send and receive commands via the PC's RS-232 port.

The young biochemist was now chapters ahead of the class. There was excitement about what could be done with VB in the laboratory. A year later, the biochemist was configuring the company's most sophisticated high-throughput-screening (HTS) system. Software drivers were written for numerous robots and instruments and the engineer's staff developed custom interfaces and specialized instrumentation. The biochemist saw his dreams materialize.

TODAY

The once-young engineer is still working at the pharmaceutical company; he and his staff are amazed by the rapid advances in technology. The PDP-11 computers have been replaced with high-speed laptops and personal digital assistants (PDAs). Motors with integral controllers have replaced the large stepper motor drivers. The RS-232 interface is still in use but is slowly being replaced with the universal serial bus (USB). Multichannel high-speed A/D converters that fit onto a PCMCIA card are available and the 8749H microcontroller has been replaced by hundreds of new high-speed microcontrollers. Software tools are available to stream data directly into spreadsheets and perform analysis. Internet-enabled instruments and sensors are appearing. The world of wireless has arrived. Microsoft has released Visual Basic .NET, the next generation VB product. The excitement is even more contagious: imagine what can be done with all this advanced technology.

As I (the engineer of the preceding account) look over the chapters within Brian Bissett's book, *Practical Pharmaceutical Laboratory Automation*, I shake my head and wish that I had had this book — a textbook to guide someone quickly into the world of computers, interfacing, automation, data acquisition, and analysis — then. You are part of a new era of applied technology in which the possibilities are endless. It is now your turn to feel the excitement and to meet future challenges. My wish is that these possibilities keep your attention as long as they have kept mine. Have fun.

Gary S. Kath
Senior Manager
Bioelectronics Laboratory Merck & Co., Inc.
Rahway, New Jersey

The Author

Brian Bissett holds graduate degrees in business and engineering from Rensselaer Polytechnic Institute in Troy, New York, and is employed by Pfizer Global Research and Development in Groton, Connecticut. He started working at Pfizer nearly 10 years ago in what was then the physical measurements laboratory. Presently, he works in the molecular properties group (MPG) in exploratory medicinal services (EMS), constructing assays and automating measurements for parameters such as ElogD, pKa, solubility, and stability.

Mr. Bissett's undergraduate work was done at The University of Rhode Island, where he received a degree in electrical engineering and worked as a co-op student at the Naval Undersea Warfare Center (NUWC). Prior to working at Pfizer, he worked onsite at NUWC on projects ranging from digital design on the new sonar intercept system (NSIS) to analog preamplifier testing and fabrication for towed arrays.

Preface

TARGET AUDIENCE

Who will get the most out of this book? The "ideal" person is someone who (1) works in any type of laboratory where the automation of processes is desired, (2) has a background in computer programming, (3) has a background in electronics (even at the hobbyist level), (4) has a background in doing basic mathematical analysis with spreadsheets, and (5) has a background in chemistry or biology.

I work with someone who is responsible for the development and construction of robotic assays to measure various physical/chemical properties of compounds stored in our sample repository. Out of necessity, he had to learn the fundamentals of Visual Basic programming and electronics in order to be able to construct some basic automated systems. He has constructed some pretty impressive automated systems, but every once in a while he will hit upon a stumbling block because he lacks an engineering/programming background. At times this can be an insurmountable disadvantage for someone with a chemistry or biology background who is capable of automating 80 to 90% of a very useful process, but does not have the background to make one last thing work correctly. This text was written to provide to people like my co-worker the resources necessary to aid them in completing their projects with a minimum of outside help.

Anyone planning on using Visual Basic for any purpose can expect to learn a great deal from this book. For those expecting to use Visual Basic in the near future, this book would be a great *second* read on the subject. Having taken an introductory course on Visual Basic or having read and understood an introductory book on Visual Basic will enable someone to get more out of this book. (I highly recommend any book published by QUE on Visual Basic.) Anyone with some knowledge of computer programming will come away from reading this book having learned new methods and algorithms, which will make life easier.

This book does not attempt to teach users the fundamentals of the topics it covers. It is intended for people who know the fundamentals of Visual Basic programming, electronics, macro writing, algorithm development, and spreadsheets. It teaches sophisticated methods to deal with real-world automation problems that occur time and time again in the workplace. The book is broken up into seven chapters; even if only one chapter is pertinent to a project, it will probably save a great deal of time to study that section.

There are two kinds of technical books. The first is the college textbook type; every scientist and engineer has shelves full of them. Some keep them on the shelf to remind themselves how much they suffered in college. Others keep them on the shelves above their desk to impress people. Although full of accurate technical information, many contain academic exercises wonderful for teaching a person the methodology to think like a scientist or engineer, but with little application to tasks faced on the job.

The second kind of technical book is the tried and true reference. It contains a lot of information that can be used on a day-to-day basis. These books can be readily identified because they sit right on a person's desk. Usually such books have many Post-it® notes stuck to various pages marking useful subjects that are utilized day in and day out on the job. It is my sincerest hope that this book will be the type of technical book that does not have a thin layer of dust on top of it or a spine that still cracks when it is opened. This book was written to share useful information of a nonproprietary nature that I have acquired and that has proven to be so useful to me on the job.

Many times in my professional life I have come across a problem that took a lot of persistence, searching, or experimentation to solve. Usually, upon discovering the solution, I would think, why is this not written down somewhere? Many things that should be in online help files are not. Most references are exactly that — references that provide a few sample applications and common algorithms which lack enough sophistication to be useful. This is a collection of the most useful tasks and their solutions that I have developed in my 8 years of experience in working in laboratory automation.

OVERVIEW

Chapter 1 and Chapter 2 cover using Microsoft Visual Basic for Applications in Excel to create robust macros to analyze data from laboratory instruments. Chapter 1 has a sample application that takes in hypothetical raw data from an instrument and performs a series of automated calculations and manipulations on the data. Chapter 2 builds upon the concepts learned in Chapter 1. It covers the best ways of implementing Excel macros, such as having all macros reside in "hidden" workbooks that can be run by menu items, all of which are loaded when Excel starts. Furthermore, it demonstrates the utility of using templates rather than hardcoding formatting changes to be performed in macros. Most useful, it shows how to have Excel access other independent applications to perform tasks that may not be possible within Excel.

Chapter 3 and Chapter 4 cover using Microsoft Visual Basic in ways that other books do not. Chapter 3 shows how to write a device driver to control an instrument with a given protocol. An example application is written to control a Kloehn laboratory syringe. Chapter 4 shows how to control commercially available hardware cards such as relay boards, digital I/O boards, and analog I/O cards. Examples are given using the shareware DLL "portio.dll" from Scientific Software Tools to control boards made by Keithley Metrabyte.

Chapter 5 shows how to use electronics to interface with any instrument that does not have a built-in interface, or has a built-in interface too limited for the process to be automated. Although these techniques are guaranteed to void the warranty on any instrument, having the ability to control an instrument remotely that was designed only to be used by hand provides a tremendous competitive advantage. It allows the automation of processes where previously it was impossible.

Chapter 6 is for people who realize that Excel is ubiquitous, but that it is severely lacking in its ability to solve certain types of mathematical problems. Because nearly everyone has the ability to open an Excel worksheet, it is nice to be able to distribute data in this manner. However, sometimes Excel is not robust enough to perform certain kinds of calculations. For such situations Microcal Origin is the perfect substitute, allowing the results to be saved as Excel worksheets. Origin also has its own built-in scripting language called Labtalk, which is very similar to "C" and can be utilized to automate data analysis tasks.

Chapter 7 is for anyone who has tried to write macros for any Agilent (formerly HP) HPLC, GC, or CE Chemstation. I know only one book on writing HP Chemstation macros, and if the information sought is not in it, then the searcher is out of luck. Some user-contributed software libraries exist, but information overall for writing macros is severely lacking. It is impossible to cut and paste information into Chemstation sequences; however, there is a way to write a macro to perform this task. This is a collection of tasks and documentation that should have been included with the Chemstation software package.

I have often wondered how many capable people have been hindered at successfully performing a task only because the information they seek is unavailable or severely limited. This book is for all those capable people looking to perform a specific task like the preceding one and needing some examples to help them understand the process and get going. Good luck!

NOTES ON ENGINEERING DESIGN

This is essentially an engineering book for chemists and biologists (nonengineers), so I would like to share two of what I consider the greatest *faux pas* of engineering. The first is a process that seems innocent. A person wants to accomplish some task and has numerous ideas how this task can be accomplished. First he tries one idea and for some reason it does not work. Without investigating why the first idea did not work, this person immediately proceeds to the next idea and the next, until one of the ideas does work.

What is wrong with this methodology? If one of the ideas works, then nothing is wrong. Eventually in a person's professional life, however, he will try all his ideas and none of them will work. What does he do now? This person is now in a situation in which all his ideas have failed and he does not have the vaguest idea why. Therefore, an individual should always make an attempt to determine why something does not work. Sometimes, of course, this will not be possible; however, knowing why something does not work leads to developing more contingency plans in the event of failure.

Suppose, for example, someone had two ideas for making process X work. Idea number one failed because of a firmware incompatibility. Idea two failed because of a system timing issue. Knowing this, a person is now in the position to develop a contingency plan to make the system work: he can redesign the firmware or alter the system timing. Which will cause the least amount of disruption to the other engineering groups involved in the project? Probably, redesigning the firmware will. Besides, when a supervisor asks why something does not work, it is always nice to be able to explain why rather than stand there with hands in your pockets and say "I dunno."

The second priceless act is what I have come to term the "design around a flaw" mentality. What happens here is that someone knows a flaw exists in a system and instead of correcting it, compensates for it. This primarily occurs in software design, but I have witnessed this practice in ASIC designs as well.

Here is an example. A person has written a large program with many subroutines and functions. When passed an integer, one function will multiply it by two and return the result. It has been determined by testing that when 2 is passed to this function, it returns 3 instead of 4. Rather than fix the function, the programmer (who knows that no integer times 2 will yield 3) simply adds a statement after the function call that, if the function returns 3, it should be changed to 4.

This is obviously a very simple example of a very poor design practice that, I am sorry to say, is practiced too much. The integrity of the overall system is in question if a flaw is allowed to persist within the system, unless there is some way to verify the extent to which a flaw exists. This is very difficult and often next to impossible to do.

In the preceding example, how does the programmer know that the only integer that will elicit an incorrect response from the function is 2? He probably does not. When a problem is found to exist on a system, fix it. On very large projects involving hundreds of lines of code or multiple signals entering from several different sources, finding a problem can take days or sometimes even weeks. In the end, however it is worth it because chances are that, eventually, the problem will be found to have greater scope than was originally observed, and this will necessitate its correction. Once the problem is corrected, an individual will end up having to find and remove "fix arounds" added to fix conditions that never would have occurred if the algorithm had worked correctly. (Incidentally, it will be very difficult to find all the "fix arounds" because they will not follow any logical reasoning — they were added solely to compensate for bugs in the system.)

Now that bad engineering design practices have been covered, it is time to mention one engineering design practice that is not practiced enough — modularization. Whenever an individual starts upon a new project of any great size, it is imperative to break the project down into functional blocks. On a large project, when an error is encountered, it helps greatly in the debugging process to be able to corral a mistake's origin to a specific area of the project.

In fact, for some projects, modularization is not optional. For example, no single person or design team could conceivably design an entire microprocessor. Although it might be possible for a single large group to design a microprocessor simultaneously, chances are that enormous duplication of effort would occur, and people would be tripping all over each other's "turf." The only rational way to tackle a big project is to break it down into blocks and develop specifications for each block that must be adhered to for each block to function within the system (or interface with its companion blocks).

Some would even argue that two measures of success exist for engineers. First, do they work on projects sufficiently complex to require modularization? Second, to which block are they assigned? If they are tasked to complete a module that is a "choke point" in the project (other modules cannot be tested and the project will be in a holding pattern until this module is successfully completed), chances are they are good engineers (or they have poor management).

ACKNOWLEDGMENTS

I would first like to thank Professor James Daly at the University of Rhode Island for sharing his publishing contacts with me, and providing some words of encouragement as well.

Thanks to my best friend, Mike Reed, who has been a source of inspiration. When I first proposed the idea of this book, his response was, "Oh yeah, I can see one on every coffee table in America." Mike is the distinguished winner of URI's Outstanding Undergraduate Engineering Design Award for an A/D nonlinearity tester. It is too bad he could not remember just what it was he designed to win the award when he went on his first job interview. (No, he did not get the job.)

A special thank-you to Roger McIntosh at McIntosh Analytical Systems, who always took time out to talk and share information on electronics, robotics, programming, and laboratory automation.

Thanks to all the fine people I work with in the molecular properties group at Pfizer Inc., especially Franco Lombardo and Chris Lipinski, whose support on many projects has been unwavering.

I would also like to thank two wizards of computational chemistry: Dr. Beryl Dominy (retired from Pfizer Inc.) and Dr. James F. Blake of Array Biopharma Inc. Beryl and Jim always made themselves available to share their expertise with a colleague in need, and in doing so helped to build the strength of the organizations that they served.

Thanks to Dr. Allen Chapman, one of my MBA instructors at Rensselaer Polytechnic, who taught me more than I can ever mention here.

My high school teachers at A. Crawford Mosley High School in Panama City, Florida contributed greatly to my becoming a productive person in society. With special thanks to Mr. Oltz, Dr. Deluzain, Ms. Brock, Ms. Stewart, Ms. Dolittle, and Ms. Colvin.

Words cannot express my appreciation to my family for their support. My aunt and uncle, Dave and Daisy Roberts, let me visit them in the Panhandle of Florida when the real world becomes too stressful. Summer visits to my father and stepmother on the beaches of Martha's Vineyard are equally relaxing. Finally, a special thanks to my mother and stepfather for unlimited visitation right to my pixie-bob kitten, Claudia, to whom I became deathly allergic. I named Claudia after the actress Claudia Black who stars in the sci fi series *Farscape*.

Introduction

A common mistake that people make when trying to design something completely foolproof is to underestimate the ingenuity of complete fools.

— Douglas Adams

SCOPE OF COVERAGE

The topic of pharmaceutical laboratory automation is a very all-encompassing one. It would be impossible for this text, or any text for that matter, to cover every possible facet of this broad discipline. Compromises, therefore, had to be made, the overriding goal of which was to make the text as broadly applicable and useful to as many laboratory scientists as possible. The areas in which this text deals can be subdivided into six categories: software, hardware, instrumentation, robotics, data acquisition, and data analysis. To a certain extent; some of the categories overlap, but each category merits consideration as a separate topic.

Software, the first category, is also the most controversial. The software examples in this text primarily focus upon Visual Basic (VB) because the audience will be primarily chemists and biologists at bachelor, masters, or Ph.D. levels, VB is the logical choice. Much of the target audience will never have taken a formal computer programming course, and their experience with programming will be rudimentary at best. VB is easy to learn for the novice but powerful enough to control a variety of complex systems. In recent years, Visual Basic has become ubiquitous over a variety of industries, primarily because any intelligent person can learn to use it well quite quickly.

Many readers will immediately be turned off because they feel VB is a substandard programming language. In the past this opinion had merit. However, as time has marched on and machines have grown faster, the rationale for this opinion is waning. "C" used to be much faster than any form of BASIC when processors were slow and interpreted languages like BASIC took far longer to execute. Now that processors are blazingly fast, it takes very little time for a modern computer to interpret code and then execute it. Native code compilers such as "C" do not enjoy the edge in performance they once did. For a desktop application, a noticeable difference barely exists between "C" and VB for the vast majority of tasks.

In the past VB was limited to the degree in which it could directly interface with the Windows operating system. As VB's popularity grew, clever programmers found ways to hack into the Windows API and Microsoft began building this functionality into VB. Now entire web sites and texts are dedicated to using Win API functions within VB. In "C" or "C++" language, little can be accomplished that could not be accomplished in VB. This is not to say that a performance "brick wall" does not exist within VB; however, very few users will ever hit it.

"C" and "C++" are, in fact, excellent programming languages, and are in many respects (mainly academic at this point) superior to VB. However, there is no point in spending months teaching laboratory scientists pointer arithmetic so they can program a task in one language when another language (albeit simpler) will suffice to get the task done.

Hardware is the second category on the list. In this text, hardware can be divided into two subcategories: electronic circuits designed by the user and electronic circuit cards purchased from a vendor to accomplish a specific task. Again, considering the audience, any electrical equipment designed and built must be simple and at the hobbyist level of complexity. In addition, any electrical equipment purchased must come with sufficient documentation to enable a user with limited experience to make use of the equipment. The hardware components for this text came from two sources, Keithley MetraByte and Computer Boards, Inc. Keithley has an outstanding reputation and excellent technical support —important considerations for the novice. Computer Boards makes some additional unique boards for Keithley's Metrabus system. Their support and documentation have proved to be top notch as well.

A reader with a formal or even hobbyist background in electrical circuits will have a great advantage in making use of the hardware development techniques provided within the text. Also, the programming examples used to control the cards from Keithley Metrabyte and Computer Boards, Inc. can be readily adapted to control a card from any given manufacturer.

Instrumentation used will vary in any given application based on a number of parameters. Since literally thousands of instruments are commercially available to perform a wide variety of tasks, the text centers on teaching the reader the ability to control any instrument by means of software (i.e., writing a device driver). When an instrument does not have an interface built in to control it, the text provides methods by which an interface can be constructed to achieve the level of control required. Most commercially available instruments can be controlled to some degree by means of a built in RS-232 or RS-485 interface. However, all of the functions may not be controllable by means of the built-in interface. This is where the methods provided by the text play an important role, giving the reader the ability to take control of his own destiny, so to speak, and automate any instrument required to perform his experiments.

Robotics, like instrumentation, is available from a wide variety of manufacturers to fulfill a diverse set of operational requirements. Unlike instrumentation, every function possible on a robotic system is accessible by software. (If a robot cannot manipulate all aspects of its functionality via software, it should not be considered for use in any automated system.)

One aspect of using robotics that is particularly irritating is that many manufacturers will not sell their drivers or source code to control their machines. Instead, they seek to sell the client a "canned" package capable of performing a set of tasks or some predetermined method. Although for some applications this may suffice, the purchaser is really being shortchanged because, should he seek to deviate from the preprogrammed application, he will have no means of doing so unless he contracts with the manufacturer to alter the software. The problem with allowing the manufacturer of a robotic system to program your applications is that soon the manufacturer will be selling your method to all of its clients. Even with a confidentiality agreement in place, it is probably unwise to have a contractor program any proprietary algorithms or methods.

Before purchasing any robotic system, make sure a set of drivers (usually a DLL, VBX, or OLE control) comes with the hardware to allow the user to program any custom application. If such drivers are not available, check to see if they can be purchased if a confidentiality agreement is put into place. Some laboratory robotic manufacturers, such as Zymark, will not sell source code under any circumstances.

Another difficulty involved in using robotics is that almost always the robot has to interface with other robots and instruments. A minimal application would typically consist of one syringe, one robotic arm, and one instrument. What becomes problematic is that each device cannot be programmed without regard to what the other devices are doing. For example, a robotic arm with a septum piercing needle attachment can be moved over a liquid solution contained in a bottle with a septum piercing cap. The robot can then be instructed to pierce the needle through the septum piercing cap. A syringe can then be told to aspirate a specified amount of the sample. The robotic arm can then be instructed to remove the needle from the bottle and move the sample over to an instrument for measurement.

If this sequence is programmed solely as a series of steps to be completed independently by each individual device, the following scenario would likely happen: the robotic arm would move over to the bottle and pierce it with the needle. The syringe would start to aspirate the liquid but before it could finish the robot would yank the needle out of the bottle and bring it over to the

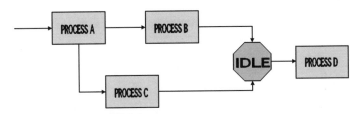

FIGURE I.1 Simultaneous tasks often require idling processes.

instrument. The syringe would have then aspirated a quantity of air because it had been removed from the sample container before the sample aspiration by the syringe was completed. The waiting instrument would be injected with air followed by a small amount of sample.

It is easy to see from this example that each individual device must have a means of keeping track of the status of the other devices. Simultaneous tasking is difficult because each and every subprocess must know the other subprocesses that must be completed before they are allowed to proceed. For example, consider the following specification. Start process A. After process A is started start process C. When process A is finished, start Process B. When process C and process B are completed, start process D. The flow chart in Figure I.1 illustrates what the controller software must do to make the process happen.

In most contemporary programming classes few students are taught the discipline of queuing and idling processes, yet the mastery of these processes is essential for the construction of any automated system. Idling processes fall into two categories. The first halts the program until a specified period of time has elapsed. The second, which is by far more useful, halts the system and allows processes that have been previously spawned to continue to run during the idling process. In fact, the termination of the idling process may depend upon the completion of one of the previously started processes.

Data acquisition is a topic that actually spans two other categories — instrumentation and hardware. Data are typically captured from an instrument or logged by means of some hardware interface. When the term data acquisition is mentioned, an individual tends to think in terms of copious amounts of numerical data filling up spreadsheets. However, a sensor telling a system that a particular vessel has been filled with a test fluid is data acquisition as well.

Data analysis is the last category in the list and probably one of the most important. In this text, data analysis is performed by means of two software packages, Excel and Origin. Excel is, of course, the most popular spreadsheet package. Excel's real strength is the fact that its macro (or scripting) language is a derivative of Visual Basic termed "Visual Basic for Applications" or VBA. It is VBA that gives Excel an enormous amount of power. A note of caution is in order in this area, however. A custom control (VBX/OCX) meant to run under stand alone Visual Basic (VB) may not be compatible with VBA. (Typically, the control will install and be accessible from the palette, but some of the properties and methods will not be accessible or will not function. It is best to check with the software manufacturer if an individual wishes to use a VB control under VBA.)

Although Excel coupled with VBA is no doubt a very formidable tool, it is lacking in some areas, notably graphics and sophisticated data analysis. To accomplish tasks in which Excel does not excel, Origin has been used. An entire section on data analysis using Origin has been included in this text. Origin was selected for three primary reasons. First, Origin has a scripting (macro) language called Labtalk, which is based on "C", just as VBA is based on VB. Second, Origin, like Excel, works with data in a spreadsheet-like format, with which many laboratory scientists are familiar and comfortable. While Matlab is a fine program with similar functionality as Origin, programming in Matlab requires all algorithms be put in matrix-like format. In addition, an individual must specify scalar vs. matrix operations. This is not an easy task for a novice programmer,

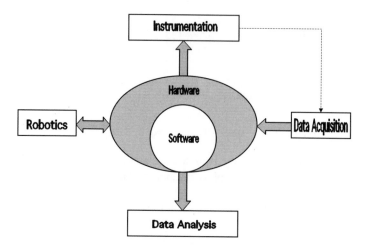

FIGURE I.2 The relationship of automation categories, from the perspective of this text.

nor do the concepts of such programming come easily. Third, Origin is a very well thought-out program with fine technical support. In essence, Origin can be thought of as Excel on steroids.

While it is true that many of the functionalities not included in Excel can be purchased as "add-ins" from third party vendors, the use of such packages leaves some degree of question as to the accuracy of the tools, and the compatibility of such packages to VBA programming techniques. People should be concerned whenever working with "nonnative" functions in any application.

Figure I.2 displays the interrelationship between the various components of automated systems. At the core of any automated project is the software. It is from the software that all aspects of an automated assay are controlled. (In large-scale industrial automation, hardware [such as PLCs, ASICs, PALs, etc.] may in fact control all systems, but the cost and complexities of hardware control put it beyond the reach of small scale laboratory automation.) Surrounding the software is a layer of hardware that enables the software to communicate with the peripheral devices such as robots and instrumentation. In some instances the hardware layer is barely discernible, such as when an instrument has a built-in RS-232 serial interface. The hardware is still present, performing translations of the lower level commands. Notice the software–data analysis connection does not pass through a hardware interface because the software can directly analyze the data once they have been acquired.

The other automation categories surround the software/hardware controlling core in the preceding diagram. Notice the direction of the arrows that indicate communication. The robotics will always communicate bidirectionally with the controlling core. The controlling core will submit results to an algorithm for data analysis. The communication may be unidirectional for simple applications as shown. It may also be bidirectional if the controlling core is sophisticated enough to take corrective action based on analyzed results. The controlling core will send commands to the instruments to set them up and take measurements, so if the acknowledgements of commands are ignored, communications to the instrumentation are unidirectional. However, instrumentation will also provide data to the controlling core.

The data acquisition rectangle may seem unnecessary to some because, in many instances, all the data will come from the instrumentation. This is not always the case, however. Sometimes data will be acquired by means of custom designed hardware to check a specific task. In other instances commercially available data acquisition boards will be used to acquire data. When data come directly from an instrument, the instrument may continually send the data all the time (in which case the data need to be polled at an appropriate interval), or the instrument may need to be prompted by the controlling core for the information. The dashed line from the instrumentation to the data

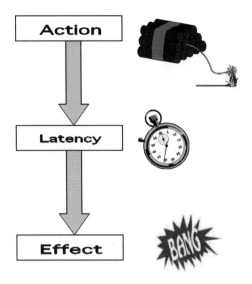

FIGURE I.3 The action latency effect (ALE) principle.

acquisition represents data that have been acquired from instrumentation by any of the preceding means.

From Figure I.2 it should be clear that a number of things must happen before a command issued at the software level translates into an action on the part of a device. There are delays in software (on the millisecond level), delays in hardware (on the micro/nanosecond level), and delays in communication to and from the controlling core. Sometimes these delays are inconsequential, but more often they must be accounted for in the design of the system.

An example of where delays are unimportant would be data analysis. An equation can be calculated (the act of calculation will produce some finite delay), after which the next equation will be calculated and so forth. Each successive action (in this case calculation) is performed without regard of any delay incurred by the previous action (calculation).

This is not the case in hardware/software interaction, however. Suppose a command is sent to a piece of hardware that is instructed to count the frequency of a certain signal. It will probably take some time for the frequency counter to sample the frequency and return a result. Suppose the very next software command requires the value of the frequency counter to perform an action. If the interval that it takes to count the frequency exceeds the interval it takes the software to process the next command, the process is in trouble if it does not take into account the latency of the hardware.

The ironic thing about automation is that on an internal level much of the control logic must be designed to handle delays. These delays are, of course, in the fractions of seconds, and not nearly comparable to the delays an individual would incur doing the process by hand. The Action->Latency->Effect process (Figure I.3) cannot be ignored for all but the most basic automated systems.

WHAT CONSTITUTES AN AUTOMATED SYSTEM?

So what is an automated process? In the basic sense, an automated system is a process that acquires input and, based, on that input makes decisions. The decisions the process renders lead to actions taken on the part of the process to produce some intended result. Taken collectively with input and decisions rendered, the process being executed is an algorithm. An automated system in the broadest sense has one algorithm to complete the task desired (see Figure I.4). However, it is most common

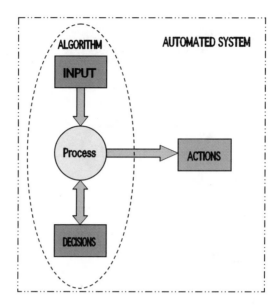

FIGURE I.4 Basic automated system.

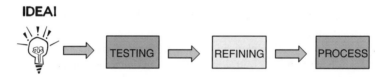

FIGURE I.5 Birth of a process.

to refer to an automated system as having many algorithms, each of which performs a specific task in support of completing the overall process.

From where do these processes that people seek to automate come? Usually it all starts with an idea that someone has to accomplish a task in a novel way. The idea then undergoes a period of testing in which the theories that support the process are tested and found to be valid or invalid. If the idea can be made to work in a *reliable* and *reproducible* manner, the process will then be subject to a period of refinement. Here it will be learned how to make the process work in the most cost-effective and reliable manner. When developing a system, it is crucial to know what may or may not occur. Knowing that X may occur and, if X does occur, then that corrective action Y should be taken, is critical for the success of any automated system. Once all the possibilities are accounted for, the process is ready to be automated. (See Figure I.5.)

Many times in an organization it is not the large tasks that lead to lapses in productivity. Usually in a large organization the large "bread and butter" tasks are well funded and carried out in an efficient manner. Behind the scenes, however, there are always ancillary tasks that support the overall goals of the organization. Many of these tasks can lead to great strides in the achievement of the goals of the organization — if they can be carried out at a level where they can be utilized by everyone in the organization who needs them.

For example, pharmaceutical companies have no trouble manufacturing and packaging the drugs they have discovered. However, many tasks crucial to drug discovery are carried out in small laboratories manned by as few as three to six people. Laboratory results on measurements such as solubility, pKa, ElogD, permeability, potency, toxicity, and stability are crucial to identifying early stage compounds with the potential to become drug candidates. Why devote resources to a compound with poor solubility and no chance of getting into a person's blood stream, or a compound so unstable its chemical composition changes in a month?

Small tasks such as these can make a tremendous difference in the success of a research and development department in a drug discovery organization. But to make a difference, the laboratories doing the measurements must be able to do them on a high throughput basis. Ideally, every compound in an organization's repository should have data available on the measurements listed previously. In addition, a compound should be able to be tested for any of these conditions with a 48-hour or less turnaround time.

To have such large scale efficiencies is impossible without the use of automation. Not only must the assays that conduct the measurements be automated, but the supporting infrastructure around them must be automated as well. The sample repository must be capable of delivering the samples to be tested in a timely manner. The information resources department must be able to post the results to potential clients quickly, and in such a manner that a Ph.D. in computer science is not needed to access the results. A large variety of tasks must be integrated for true performance gains to be realized.

This text was meant to aid those with innovative ideas for automating new drug development assays to accomplish their goal with a minimum of outside assistance. A number of automated systems were developed that led to the formation of this book. It is the author's hope that many involved in the field of laboratory automation could benefit from an all inclusive reference of some of the more essential information.

EXAMPLES OF AUTOMATED SYSTEMS THAT LED TO THIS TEXT

Descriptions of four assays that were automated by the author are presented in this section. A system level view is given to the reader to provide him with some insight as to what applications the methods in this text might be used to automate. The description is sufficient to provide the reader with an idea of the overall method but sufficiently vague to protect intellectual property and ready duplication of the methods described. The shading of the assay diagrams is as follows: dark denotes instrumentation, grey denotes data analysis, and light denotes physical systems such as robots and syringes.

The block diagram in Figure I.6 is that of an automated turbidimetric (or kinetic) solubility assay. The principal components in this assay are (1) a Hach turbidimeter, (2) three Kloehn syringe pumps, and (3) a Tecan RSP 9000 robotic sample processor. The basis for this assay is to detect light scattering by particulate matter in a given test solution. Here a baseline reading of turbidity units is taken of a given liquid. The RSP9000 and syringe pump then incrementally add compound in solvent to the given test solution. After each successive addition another turbidity reading is taken. If the turbidity rises to some predetermined level above the baseline of the test solution, the compound is declared insoluble in the test solution at that concentration level. The assay is terminated for each compound when a maximum of 13 additions has been made or the compound has been found to be insoluble at a given concentration. Up to 384 compounds may be analyzed, one after another, in a single assay run.

An automated pKa assay is illustrated in Figure I.7. The principal components in this assay are (1) a SpectraMax 250 scanning spectrophotometer, (2) a Cavro syringe pump, (3) a Gilson robotic arm, and (4) a Labsystems multidrop microplate dispensing unit. The basis for this assay is to detect a perturbed chromophore over a range of pH buffer values and then fit a curve to absorbance vs. pH values to determine the pKa value. A Labsystems multidrop unit fills a UV clear microplate with an incremental range of UV transparent pH Buffers. A Gilson robot and Cavro syringe pump then add a fixed amount of compound in solvent to each pH buffer well on the microplate. A 96-well microplate can be utilized row-wise (12 pH buffers, 8 compounds per plate) or column wise (8 pH buffers, 12 compounds per plate), depending upon the precision or speed of

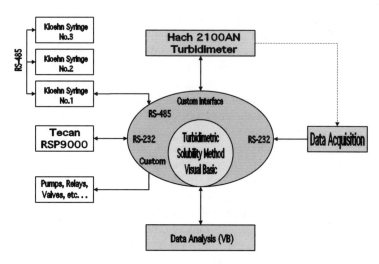

FIGURE I.6 Turbidimetric solubility assay.

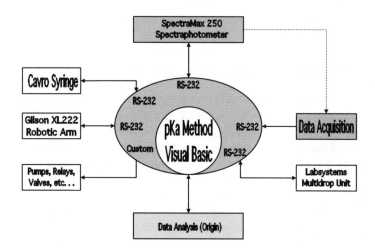

FIGURE I.7 Spectroscopic pKa assay.

throughput desired. Notice that the data analysis here is done within Origin while the robotic and instrument control is accomplished with Visual Basic.

An automated platereader solubility assay is illustrated in Figure I.8. The principal components in this assay are (1) a Tecan SLT platereader, (2) a Kloehn syringe pump, (3) a Tecan RSP9000 robotic sample processor, and (4) a Labsystems multidrop microplate dispensing unit. As with its cousin, the turbidimetric solubility assay, the basis for this assay is to detect a light scattering of particulate matter that comes out of solution. Instead of a turbidimeter, a platereader capable of detecting light scattering is used. Unlike the turbidimetric solubility assay that would test solubility over a range of values, each well in the platereader solubility assay represents a single concentration of drug in solution in which it is in solution or out of solution. Notice that the Kloehn syringe is "daisy chained" off the RSP 9000 unit. The Tecan RSP 9000 allows several devices to be controlled through the robotic sample processor. However, a firmware problem prevents use of more than one Kloehn syringe off the RSP9000 unit, which is why a separate RS-485 driver was used to control three Kloehn syringes in the turbidimetric solubility assay.

Figure I.9 is that of an automated ELogD assay. The principal component of this assay is an Agilent HPLC "Chemstation" (which has several components, syringes, pumps, valves, etc.). This

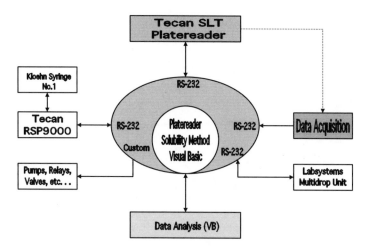

FIGURE I.8 A platereader solubility assay.

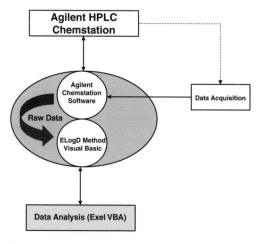

FIGURE I.9 An ElogD assay.

is a rather unique assay in that the controlling core is shared by two software applications. Here Visual Basic and the Agilent Chemstation software share control of this process. This is possible by using methods such as OLE, DDE, and SendKeys to transmit information back and forth between the two controlling applications. The Agilent Chemstation software has a macro language that allows automation of many of the functions of the HPLC. The data obtained from the Chemstation is passed to Excel where it is analyzed by an Excel VBA program. It is then well formatted into reports for laboratory notebooks, and saved as comma-delimited files into network directories that automatically upload the data to the corporate database (Oracle).

The construction of such assays is not an easy task. In addition to the programming difficulties involved, an individual must deal with hardware/software integration issues and the plethora of problems that go with them. Of course people must also contend with getting the process to work "just right" completely unattended because no human being will be there to intervene when anything goes wrong. Although developing such processes by nature is very difficult, some techniques can ease the pain of development along the way. This book is a collection of the techniques proven to be most useful in the construction of such assays.

Table of Contents

1 Microsoft Excel Visual Basic Macros

1.1 THE IMPORTANCE OF CONSISTENT LABORATORY NOTEBOOKS AND REPORTS

Not so long ago the only purpose of keeping a laboratory notebook was to keep a record of the procedures followed when performing an experiment. Should the experimenter make a valuable discovery, a record had been kept so that the compound or method could be reproduced (hopefully by anyone). Laboratory notebooks, however, are now so much more important than they were in the past. Synthesis procedures for many compounds are often much more complicated than they were in the past and more instrumentation is utilized to produce a chemical entity. Still more important is the ability to protect legally anything that has been discovered. In a patent dispute the creator must be able to prove not only that he has discovered the chemical entity, but also that he has recognized its potential and established a pattern of testing within one or more therapeutic areas. Having good records available for review by government agencies such as the FDA is equally important.

Reports are of limited value unless they are consistent. If a person has a stack of 100 reports and wishes to find parameter X in all of them, it is nice to be able to look in the same place and find the parameter in all the reports. Even better still, a computer program could be used to extract all the information for the user and present it in any form he wanted.

When reports are prepared by hand it is only natural that their content will vary slightly from report to report. Even though people are creatures of habit, it is easy to forget to include a parameter here or there, have the scale of a graph on report A differ from that of report B, etc. The problem is aggravated exponentially when two different people are running the same method and each is preparing their own reports. The problem is further exacerbated by the fact that many laboratory scientists do not keep up with their notebooks. Then, at the last minute, they must catch up and prepare a whole series of legal notebook entries while feeling the pressure of having too little time to give each notebook entry the attention it deserves.

This is dangerous when a problem is encountered and a third party begins looking at the data. Did a scientist create a plot with different scale axes to minimize a flaw in a compound, or was it a typical report variation? Did a scientist leave out a parameter from a report by mistake, or was this done because this parameter deviated from the results the scientist was hoping to see? A lawyer could have a field day with any number of honest mistakes, potentially costing millions.

The solution to this problem is to automate the preparation of laboratory reports. The computer will never get tired and forget to include a parameter and does not possess the capacity to show bias. All reports will then be uniform, which is good for everyone involved. It is good for the company if it comes under legal scrutiny. It is good for the scientist if he needs to look through the data. It is good for the database people, who require consistently formatted files for upload into the corporate databases. It is good for the corporate legal people because legal notebooks are prepared in a timely, consistent, and efficient manner. In short, it is good for everyone except those who may seek to bring litigation against the organization.

Ideally, this process should start right after the data have been collected, at which point they would undergo some form of analysis and be put into a standardized report template. The purpose of Section 1.1 and Section 1.2 is to empower readers to be able to automate the creation of their own laboratory reports using Microsoft Excel. Although other spreadsheet products exist, some of which are superior to Excel, the acceptance of Excel is so widespread and it has such an easily used scripting language (Visual Basic for Applications [VBA]) that it makes sense to create most reports in this application.

If readers can master the techniques in Section 1.1 and Section 1.2 they will be able to (1) automate data analysis tasks, (2) automate the creation of legal notebook pages and reports, and (3) create files that can readily be uploaded to a corporate database.

1.2 RUNNING VBA MACROS IN EXCEL

First, make sure the VB toolbar (Figure 1.1) is visible within Excel. It is from here that the VBA programming environment is entered. Look under the menu View->Toolbars and make sure Visual Basic is checked in Excel; otherwise, this toolbar will not be visible from within Excel.

FIGURE 1.1 The Visual Basic toolbar.

The user can directly enter the Visual Basic programming environment by pressing the VB editor button (fourth from left). This will allow the user to look at all the VBA code modules present within every workbook currently loaded into Excel (even hidden workbooks). In addition, the user can design forms and directly manipulate the properties of various objects contained within the workbooks.

The project explorer is probably the most important window to have open when operating within the VBA environment. It is from here that all of the workbooks that are loaded in Excel can be seen —even those hidden from sight. In addition to being able to access the workbooks, the user may create or access new modules and forms that reside or are to be placed in the workbook. The project explorer can be opened by selecting View->Project Explorer, or from the keyboard via CTRL+R or Alt+V+P.

In Figure 1.2, the workbook "HPLC_Analysis.xls" has a module and a form that reside within the workbook. Additional forms can be added to any project from the menu by selecting

FIGURE 1.2 The project explorer window.

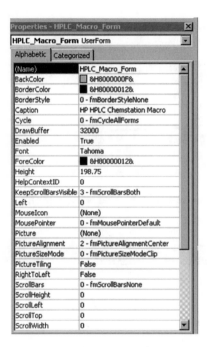

FIGURE 1.3 The properties window allows static parameters to be set at design time.

FIGURE 1.4 The VBA toolbox.

Insert->User Form or from the keyboard via Alt+I+U. Similarly, additional modules can be added to any project from the menu by selecting Insert->Module or from the keyboard via Alt+I+M.

The properties window (Figure 1.3) is very useful in that it allows the user to set most properties of various objects available in the VBA design environment at the time of design, as opposed to during program execution. This saves the programmer from writing a lot of unnecessary code. The properties window can be brought up most easily by pressing the F4 button on the keyboard. It can also be brought up from the menu by selecting View->Properties Window Form or from the keyboard via Alt+V+W.

The toolbox is only available when a form is being designed within the workbook. If the active window is a code module, the toolbox will not appear. The toolbox can be brought up from the menu by selecting View->Toolbox, from the keyboard via Alt+V+X, or by pressing the toolbox icon in the VBA toolbar (Figure 1.4).

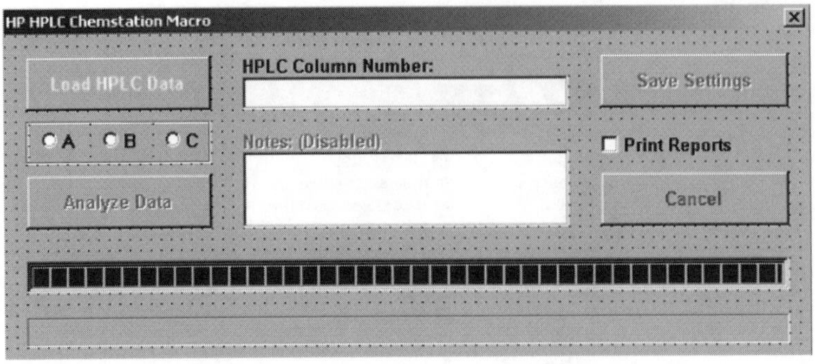

FIGURE 1.5 A sample form that could be built in VBA.

FIGURE 1.6 The immediate window is useful in debugging macros.

The toolbox shows all the components available for use in forms such as OCX, VBX, and OLE controls. The user can add packages of aftermarket controls to this palette by right clicking and selecting additional controls and checking off the desired controls.

Figure 1.5 shows an example form that could potentially be built using the objects from the toolbox. Such forms can be invaluable for setting starting parameters for a macro, and default values can be stored to the Windows Registry as will be shown later in the text.

The Immediate Window is of great value in debugging macros. From this window, query the value of any variable when the macro has been stopped with the STOP statement by typing Debug.Print VariableName. The user can also have different parameters printed at different times during program execution by adding the Debug.Print command at any point in the program from which it would be desirable to report out a particular parameter. The immediate window can be brought up from the menu by selecting View->Immediate Window from the keyboard via Alt+V+I or the shortcut CTRL+G (see Figure 1.6).

1.3 SETTING UP A MICROSOFT EXCEL VISUAL BASIC MACRO

Wherever the starting point in any macro may be, it is advantageous to have the following as the first line of code executed:

```
'Disable Dialog Box Type Alerts in Excel

Application.DisplayAlerts = False
```

This simple piece of code will prevent Excel from displaying any dialog boxes for the user such as "Do you want to replace the existing file?" followed by a yes and no button. The requirement of user intervention that was mentioned previously is now eliminated. Of course, if a mistake is made in the code, files may inadvertently be overwritten, or some unanticipated damage may occur. However, to develop a truly automated macro this statement is a must.

1.4 GETTING THE DATA INTO EXCEL

Chances are the application being used will export a data file from which an analysis is required. The two most common file formats are comma and text delimited, although many applications now have the option to export a native Excel file. The following subroutine will allow a user to bring a text or comma-delimited file into Excel:

```
Sub Load_Data()

'Subroutine to Load text or comma delimited file into a
worksheet

Dim vFileName As String

Dim nochoice As Integer

Dim WindowCaption$

Dim ErrorBoxCaption$, ErrorBoxMsg$

On Error GoTo FileOpenErr

'Set Up Captions and Messages

WindowCaption$ = "Caption Desired for Load Dialog Box"

ErrorBoxCaption$ = "Caption Desired for Error Msg Box"

ErrorBoxMsg$ = "Message Desired for Error Msg Box"

vFileName = Application.GetOpenFilename("Comma Delimited Files
(*.csv),.csv, Text Files (.txt), *.txt",, WindowCaption$)

If vFileName <> "" Then

Workbooks.Open FileName: = vFileName

End If

'File was chosen and loaded into worksheet - now exit

Exit Sub

FileOpenErr:

'If no file is chosen you end up here

nochoice = MsgBox(ErrorBoxMsg$, vbOKOnly, ErrorBoxCaption$)

End Sub
```

The preceding subroutine can be executed by selecting the item shown in Figure 1.7, which will result in a dialog box being displayed (see Figure 1.8). If no file is selected, the message box shown in Figure 1.9 will appear, indicating an error.

This subroutine can be modified in the following ways. The user can set up custom captions and error messages by modifying the string variables WindowCaption$, ErrorBoxCaption$, and ErrorBoxMsg$. In addition, the importation of additional file types is possible by modification of the Application.GetOpenFilename statement. As in the previous example, additional file types need to be separated by commas. Finally, if the filename is already known and it is unnecessary for the

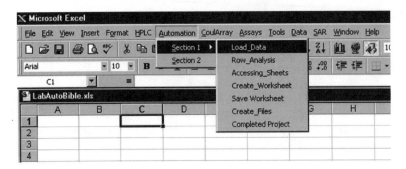

FIGURE 1.7 Running the load data macro.

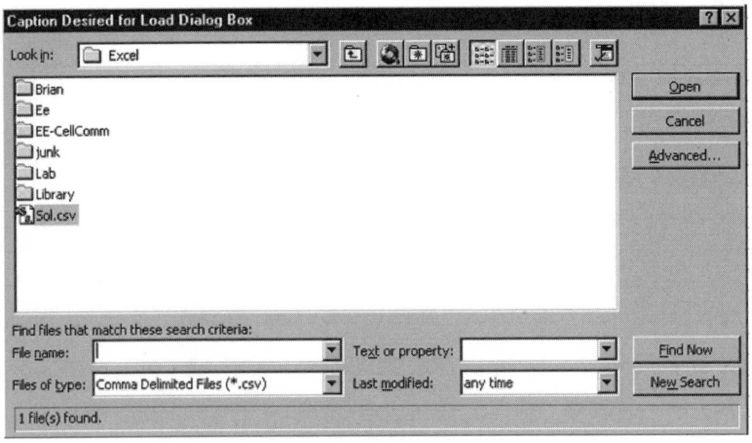

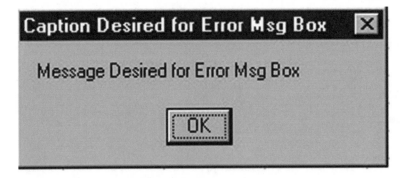

FIGURE 1.8 File loading dialog box called from Excel macro.

Caption Desired for Error Msg Box ☒

Message Desired for Error Msg Box

[OK]

FIGURE 1.9 Standard message dialog box.

user to choose it, the filename can be hardcoded into the variable vFileName and eliminate the Application.GetOpenFilename command as follows:

```
vFileName = "C:\TEMP\JUNKDATA.CSV"
```

This subroutine works fine for bringing data into Excel but a word of caution is in order here. If, after bringing the data into Excel, the program modifies the data and Excel's AutoSave feature is enabled, the original data file brought into Excel will be overwritten and lost. To preserve the original file to be analyzed (always a good idea, especially if sample files are needed to track down a bug), a modified version of the subroutine can be used:

```
Sub Load_Data_Preserve()

'Subroutine comma delimited file into a worksheet and save
the worksheet

'immediately to a new filename preserving the integrity of
the original data

Dim vFileName As String

Dim nochoice As Integer

Dim WindowCaption$

Dim ErrorBoxCaption$, ErrorBoxMsg$

On Error GoTo FileOpenErr

'Set Up Captions and Messages

WindowCaption$ = "Caption Desired for Load Dialog Box"

ErrorBoxCaption$ = "Caption Desired for Error Msg Box"

ErrorBoxMsg$ = "Message Desired for Error Msg Box"

vFileName = Application.GetOpenFilename("Comma Delimited Files
(*.csv),.csv, Text Files (.txt), *.txt",, WindowCaption$)

If vFileName <> "" Then

Workbooks.Open FileName: = vFileName

'Now Save the Original Data As a New File

ActiveSheet.Name = "JUNKDATA"

ActiveWorkbook.SaveAs FileName: = "C:\TEMP\JUNKDATA.CSV,"
FileFormat: = xlCSV, _

Password: = "", WriteResPassword: = "", ReadOnlyRecommended:
= False, CreateBackup: = False

End If

'File was chosen and loaded into worksheet - now exit

Exit Sub

FileOpenErr:

'If no file is chosen you end up here

nochoice = MsgBox(ErrorBoxMsg$, vbOKOnly, ErrorBoxCaption$)

End Sub
```

Here the file is imported into Excel, and the worksheet is immediately renamed to "JUNK-DATA" and saved as a comma-delimited file to the path "C:\TEMP\JUNKDATA.CSV." In order for this subroutine to work on a machine, the path "C:\TEMP" must exist. Now the integrity of the original data file is preserved, the modifications being made are stored under "C:\TEMP\JUNK-DATA.CSV," and when the macro is complete, the file can be renamed appropriately and resaved.

One last point worth mentioning is how to control the default directory that the FileOpen (or FileSave for that matter) dialog box shows upon activation. This can be done by means of the ChDir and ChDrive commands. ChDir will change the default system directory that will appear when the FileOpen/Save dialog box appears. It will not change the default drive, even if the drive is specified in the directory path. Only the ChDrive command can do that. For example:

```
ChDrive "C" ' Make "C" the default drive

ChDir "C:\TEMP\OLDDATA" 'Change the default directory
```

Note: if the drive is to be changed, the ChDrive command must always precede the ChDir command.

1.5 FINDING AND EXTRACTING DATA

From the website, http://www.pharmalabauto.com, load the Assay Results.xls file. The worksheet depicted in Figure 1.10 should appear. Here is a typical file of results for samples run for drug screening properties. The first two columns contain the properties' sample number and lot number.

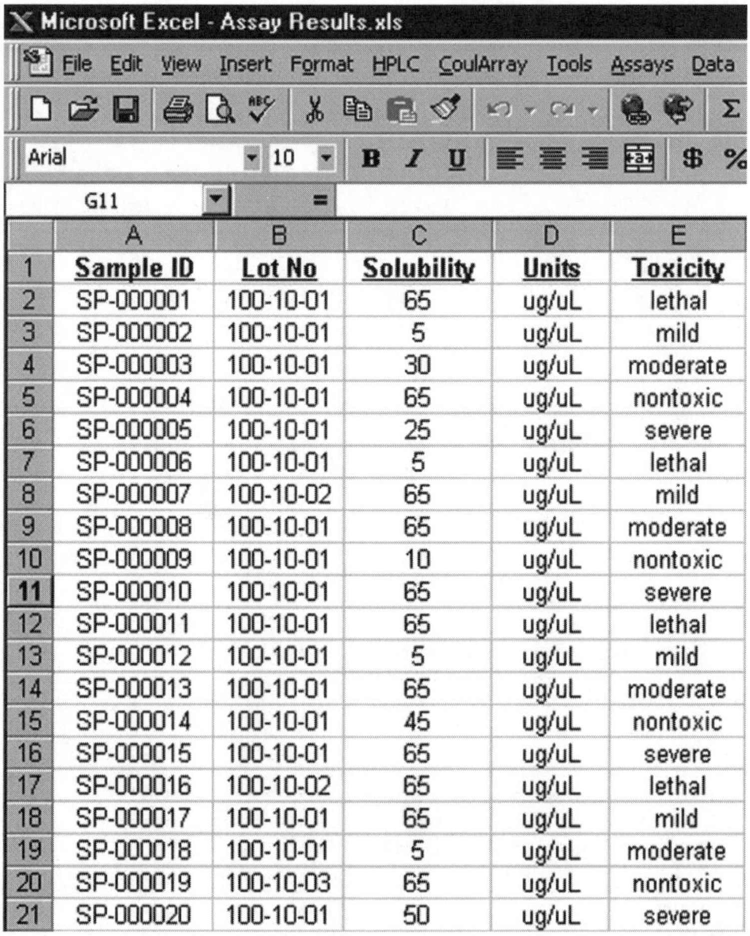

	A	B	C	D	E
1	**Sample ID**	**Lot No**	**Solubility**	**Units**	**Toxicity**
2	SP-000001	100-10-01	65	ug/uL	lethal
3	SP-000002	100-10-01	5	ug/uL	mild
4	SP-000003	100-10-01	30	ug/uL	moderate
5	SP-000004	100-10-01	65	ug/uL	nontoxic
6	SP-000005	100-10-01	25	ug/uL	severe
7	SP-000006	100-10-01	5	ug/uL	lethal
8	SP-000007	100-10-02	65	ug/uL	mild
9	SP-000008	100-10-01	65	ug/uL	moderate
10	SP-000009	100-10-01	10	ug/uL	nontoxic
11	SP-000010	100-10-01	65	ug/uL	severe
12	SP-000011	100-10-01	65	ug/uL	lethal
13	SP-000012	100-10-01	5	ug/uL	mild
14	SP-000013	100-10-01	65	ug/uL	moderate
15	SP-000014	100-10-01	45	ug/uL	nontoxic
16	SP-000015	100-10-01	65	ug/uL	severe
17	SP-000016	100-10-02	65	ug/uL	lethal
18	SP-000017	100-10-01	65	ug/uL	mild
19	SP-000018	100-10-01	5	ug/uL	moderate
20	SP-000019	100-10-03	65	ug/uL	nontoxic
21	SP-000020	100-10-01	50	ug/uL	severe

FIGURE 1.10 Worksheet with typical data.

The last three columns contain screening information on solubility and toxicity of the compounds. A typical application would involve extracting this information and writing it into arrays where it could be analyzed and manipulated into a desired format. The big question here is how an individual extracts this information using a VBA macro that can be universally applied to any type file.

The first step in this process is to call attention to the worksheet from which the data are to be extracted. This is typically done using the command Workbooks("WorkbookName.xls").Activate (for the sample file it would be Workbooks("Assay Results.xls").Activate). A bug in Microsoft Excel can cause headaches to someone trying to accomplish this simple task. This is termed henceforth as the "hide file extensions for known file types" bug.

Here is how the bug occurs. Double click on "My Computer"; then, under the view menu item, select options and choose the view tab. There will be a check box that says "hide file extensions for known file types." If this box is checked, then workbooks in Microsoft Excel will have a "filename" caption at the top. If the box is unchecked, then the workbooks will have a "filename.xls" caption.

Now, if the developer always uses the syntax Workbooks("Assay Results.xls").Activate, the macro will always function correctly. The problem lies in the developer using the syntax Workbooks("Assay Results").Activate (omitting the ".xls" in the workbook name). This syntax will work on a machine where the "hide file extensions for known file types" is checked. However, if the same macro is copied or used on a different machine where the "hide file extensions for known file types" option is not checked, an error will result because Excel will look for a worksheet with the caption "Assay Results" and will only find a worksheet with the name "Assay Results.xls." A way around this problem is given by using the following subroutine:

```
Sub ActivateWorkbook(wbname$)

'Activate the desired workbook

'Compensates for Hide Extensions of Known File Type Error 'in
Excel

On Error GoTo ChangeStatts

Workbooks(wbname$).Activate

Exit Sub

ChangeStatts:

Select Case Err

    Case 9

        wbname$ = wbname$ & ".xls"

        Workbooks(wbname$).Activate

        Debug.Print "New Name: "; wbname$

    Case Else

        dummy = MsgBox("An Unanticipated Error No." & Str(Err)
        & " Occurred !", vbOKOnly, "Error")

End Select

End Sub
```

Use this subroutine whenever a different workbook must be activated and pass the workbook name using wbname$ without the ".xls" suffix. If the ".xls" suffix is required, an error will be triggered and the error handling mechanism within the subroutine will automatically add the suffix ".xls" to the workbook name. This will allow worksheets to be activated trouble free from any computer regardless of the "Hide File Extensions for known file types" setting on a particular machine.

Once the proper workbook has been activated it is time to go about the process of extracting the required information from its worksheets. The first step in this process is to determine how many rows of data are in the worksheet to analyze. Chances are that not every file for analysis will have the same number of rows. The next three functions will accomplish this. The first two functions will return the row and column number, respectively, when a worksheet address is passed to them in "R1C1" type format.

```
Function FindRow(addr$) As Integer

Dim i As Integer

Dim strt As Integer, noc As Integer

'Pass Current Address to this Function and it will Return the
Row Number

For i = 1 To Len(addr$)

   If Mid$(addr$, i, 1) = "R" Then

      strt = i + 1

   End If

   If Mid$(addr$, i, 1) = "C" Then

      noc = i - strt

   End If

Next i

FindRow = Val(Mid$(addr$, strt, noc))

End Function

Function FindCol(addr$) As Integer

Dim i As Integer

Dim strt As Integer, noc As Integer

'Pass Current Address to this Function and it will Return the
Column Number

For i = 1 To Len(addr$)

   If Mid$(addr$, i, 1) = "C" Then

      strt = i + 1

      noc = Len(addr$) - i

   End If

Next i

FindCol = Val(Mid$(addr$, strt, noc))

End Function
```

The next function requires the use of the preceding "FindRow" function to determine the total number of rows in a worksheet.

```
Function TotalRows(wkbook$, sheet$) As Integer
'Determines the total number of rows of data in a worksheet
Dim i As Integer 'Rows
Dim j As Integer 'Cols
Dim lastcell$
'Ignore the error message that appears on blank cells
On Error Resume Next
'Workbooks(wkbook$).Activate
Call ActivateWorkbook(wkbook$)
Worksheets(sheet$).Select
LastRow = 0
   For i = 1 To 18000
   If i > (LastRow + 50) Then
      TotalRows = LastRow
      'Debug.Print sheet$, "Last Row = "; TotalRows
      Exit Function
   End If
   For j = 1 To 20
   'Set variable v equal to value contained in cell
   If Worksheets(sheet$).Cells(i,j).Value <> "" Then
      'If a cell has text then log position
      Worksheets(sheet$).Cells(i,j).Select
      lastcell$ = Selection.Address(ReferenceStyle: = xlR1C1)
      LastRow = FindRow(lastcell$)
      Exit For
   End If
   Next j
 Next i
End Function
```

The last function works by passing the workbook name (wkbook$) and the sheet name (sheet$) to the TotalRows function. The function then looks down the worksheet row by row across the first 20 columns. If any data are found in a row, their position is logged. If the function searches through 50 rows and does not find any additional data, then it assumes the last row in which it logged data is the last row of data in the worksheet.

Notice that the contents of each cell are referenced by means of the (.Value property) (**Worksheets(sheet$).Cells(i, j).Value**). The intuitive property name would be (.Text) because it would be used to extract data from a textbox or another object that holds data. Unfortunately, no such property exists for Excel worksheet cells. The (.FormulaR1C1) property can also be used to extract the value of a worksheet cell in many circumstances.

With reference to the topic of data in cells, it is important to mention that the value of data extracted from cells can be inadvertently rounded to a particular decimal place if the cell size is too small to hold the precision (or decimal places) of the number it contains. To ensure the contents of a cell will not be unintentionally rounded when its contents are extracted to a variable, it is important to (.AutoFit) the contents of the cell if this has not already been done. The following code snippet gives an example of one way to accomplish this:

```
'Prior to Searching for Data Expand all Cells to Full size
'Rounding from cells too small will create search errors
    ActiveWindow.Panes(1).Activate
    Cells.Select
    range("A8").Activate
    Cells.EntireColumn.AutoFit
```

Using these subroutines and functions it is easy to construct a subroutine that will find the number of rows in the name "Assay Results.xls" workbook.

```
Sub Row_Analysis()
'Worksheet will ask user to load a worksheet
'Subroutine will then report the number of rows in the worksheet
Dim dummy As Integer
Dim MsgBoxCaption$, MsgBoxMsg$
Call Load_ExcelData
Call ActivateWorkbook("Assay Results")
Total_Samples = TotalRows(ActiveWorkbook.Name,
ActiveSheet.Name) - 1
MsgBoxCaption$ = "Analysis of Worksheet"
MsgBoxMsg$ = "There are " & Str(Total_Samples + 1) & " rows
in this worksheet!"
dummy = MsgBox(MsgBoxMsg$, vbOKOnly, MsgBoxCaption$)
'Find_Heading ("Lot No")
'Call Data_Extraction
'Call Data_Analysis
'Call Create_Worksheet
'Call SaveWorksheet
End Sub
```

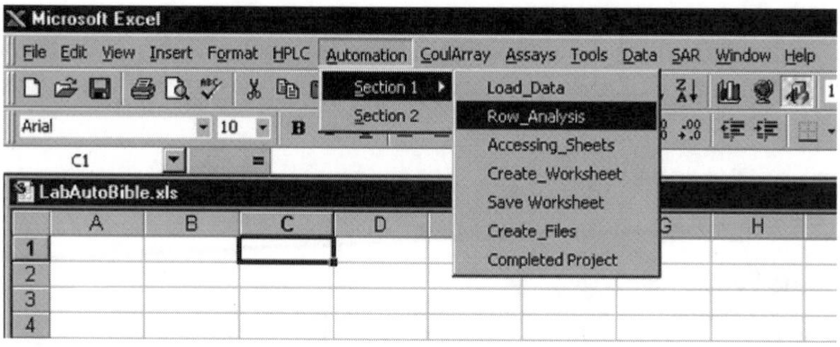

FIGURE 1.11 Running the row analysis macro.

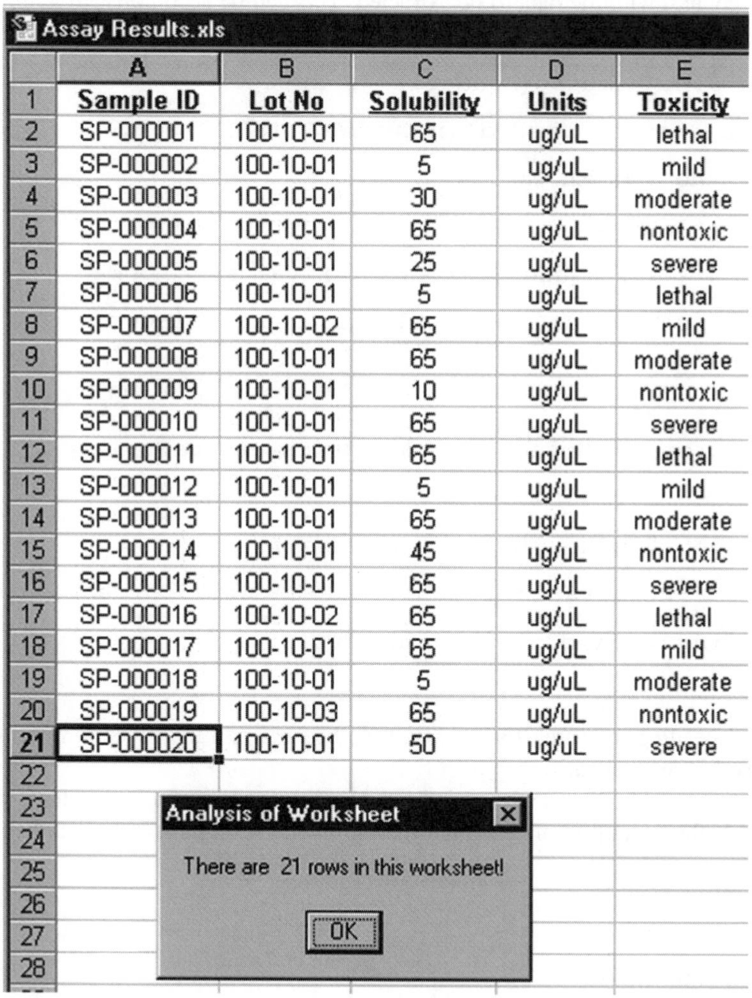

FIGURE 1.12 The row analysis macro run on the sample worksheet.

The selection process to run the "Row_Analysis" macro is illustrated in Figure 1.11. Running the macro "Row_Analysis" on the file "Assay Results.xls" will yield the result shown in Figure 1.12. Keep in mind that the number of rows is not the same as the number of lines of data. Here there are 21 rows, but one row comprises column headings. The total number of rows of data (or in this case samples analyzed) is 20. It is often convenient to store this number in a global variable, which is done in the Row_Analysis subroutine. This way, when writing loops to manipulate the data, the ending parameter for the loop is available without worrying about passing it from subroutine to subroutine.

```
'At top of Module

Global Total_Samples As Integer

'In Row_Analysis Sub

Total_Samples = TotalRows(ActiveWorkbook.Name,
ActiveSheet.Name) - 1
```

Now that the number of rows of data is known, data extraction is rather easy. The first step is to identify the column with the data to be extracted. This can be accomplished with the following function:

```
Function Find_Heading(header$) As Integer

'Finds the header passed in header$

'Returns the column Number

Dim CurrentAddress As String

Cells.Find(What:=header$, LookIn:=xlValues, LookAt_

:=xlPart, SearchOrder:=xlByColumns, SearchDirection:=xlNext,
MatchCase:=_

False).Activate

CurrentAddress = Selection.Address(ReferenceStyle:=xlR1C1)

'Debug.Print CurrentAddress

'Debug.Print FindRow((CurrentAddress))

Debug.Print header$; " Column: ", FindCol((CurrentAddress))

Find_Heading = FindCol((CurrentAddress))

End Function
```

The column heading to search for is passed via header$ and the function returns the column number that the header is located in.

It is important to note that the function Find_Heading will not find a header if the column it is in is hidden. If the capability of searching through hidden columns is desired then

```
LookIn:=xl Values

Must be changed to

LookIn:=xlFormulas
```

For the function Find_Heading this code would have to be changed from:

```
Cells.Find(What:=header$, LookIn:=xlValues, LookAt_

:=xlPart, SearchOrder:=xlByColumns, SearchDirection:=xlNext,
MatchCase:= _

False).Activate
```

To

```
Cells.Find(What:=header$, LookIn:= xlFormulas, LookAt _
:=xlPart, SearchOrder:=xlByColumns, SearchDirection:=xlNext,
MatchCase:= _
False).Activate
```

Now to extract the data into arrays a subroutine can be constructed as follows:

```
Sub Data_Extraction()
'Extracts data from "Assay Results" Workbook into arrays
Dim ii As Integer, numbrows As Integer
'*** Variable to contain data ***
Dim Samp(100) As String, Lot(100) As String
Dim Sol(100) As String, Unit(100) As String
Dim Tox(100) As String
'*** Column Headers ***
Dim col_samp As Integer, col_lot As Integer
Dim col_sol As Integer, col_unit As Integer
Dim col_tox As Integer
'Find column headers for data on worksheet
col_samp = Find_Heading("Sample ID")
col_lot = Find_Heading("Lot No")
col_sol = Find_Heading("Solubility")
col_unit = Find_Heading("Unit")
col_tox = Find_Heading("Toxicity")
'Find number of rows
numbrows = TotalRows(ActiveWorkbook.Name, ActiveSheet.Name)
'Use loop to extract data
For ii = 2 To numbrows
    Samp(ii - 1) = Worksheets(1).Cells(ii, col_samp).Value
    Lot(ii - 1) = Worksheets(1).Cells(ii, col_lot).Value
    Sol(ii - 1) = Worksheets(1).Cells(ii, col_sol).Value
    Unit(ii - 1) = Worksheets(1).Cells(ii, col_unit).Value
    Tox(ii - 1) = Worksheets(1).Cells(ii, col_tox).Value
    Debug.Print (ii - 1), Samp(ii - 1), Lot(ii - 1), Sol(ii -
    1), Unit(ii - 1), Tox(ii - 1)
```

```
Next ii

End Sub
```

Once the total number of rows of data has been determined and the column position of each type of data is found using the "Find_Heading" function, it is a simple matter to use a loop to extract the data into an array.

Notice that the loop starts with "2," because, in the assay results worksheet, the data begin on row 2 for all types of data. This may not always be the case. It may be necessary to search for the starting row for each data type; a derivative of the Find_Heading function could be created to return the starting row.

Another point worth mentioning is that in this subroutine all the arrays to hold the data are declared locally with the Dim command within the subroutine. Oftentimes such data will need to be accessed by multiple subroutines within the macro and should be declared globally at the module level. The user could do this by putting statements such as the following at the beginning of the module:

```
Global Samp(100) As String, Lot(100) As String

Global Sol(100) As String, Unit(100) As String

Global Tox(100) As String
```

These variables would then retain their value and be accessible to any subroutine or function within the module or worksheet. Also examine the following statement:

```
Worksheets(1).Cells(ii, col_samp).Value
```

ii is the row number, which is incremented with each loop, and col_samp is the column number found using Find_Heading.

Note: it is possible to call Find_Heading in this statement each time the loop is run but it is faster and more readable to run the function once at the start of the subroutine and determine the column locations.

Worksheets(1) is the first worksheet whose default name is "Sheet1." Other methods of referencing sheets will be covered in the final sample application. It is very important to specify which worksheet and workbook data are being extracted from so that incorrect data will be extracted.

To test the preceding subroutine, uncomment (remove the ') the following lines to the end of the Row_Analysis subroutine and execute Row_Analysis as a macro:

```
Find_Heading ("Lot No")

Call Data_Extraction
```

1.6 ANALYSIS AND MANIPULATION OF DATA

Once the data are safely stored into arrays, more than likely some kind of analysis is desired. For the sake of discussion suppose it is desired to extract the samples from the assay results worksheet that have a solubility = >50 and have a toxicity that is not severe or lethal. Under these guidelines, only the following samples should be extracted: SP-000004, SP-000007, SP-000008, SP-0000013, SP-000017, SP-000019.

To accomplish this task, first define the following variables as global at the beginning of the module:

```
Global ExtractSamp(100) As Boolean

Global Samp(100) As String, Lot(100) As String
```

```
Global Sol(100) As String, Unit(100) As String

Global Tox(100) As String
```

The ExtractSamp Boolean value will be set "true" if the sample should be extracted and "false" if it should not be extracted. A subroutine that will accomplish this is:

```
Sub Data_Analysis()

'Finds samples whose solubility is = > 50 ug/uL AND

'whose toxicity is not lethal or severe

Dim ii As Integer, numbrows As Integer

numbrows = TotalRows(ActiveWorkbook.Name, ActiveSheet.Name) - 1

For ii = 1 To numbrows

    If Val(Sol(ii)) > = 50 And Tox(ii) <> "lethal" And Tox(ii)
    <> "severe" Then

    ExtractSamp(ii) = True

    Else

    ExtractSamp(ii) = False

End If

Debug.Print ii, ExtractSamp(ii)

Next ii

End Sub
```

To test the preceding subroutine, uncomment (remove the ') the following line to the end of the Row_Analysis subroutine and execute Row_Analysis as a macro:

```
Call Data_Analysis
```

1.7   ACTIVATING SHEETS WITHIN DIFFERENT WORKBOOKS

Up until now, the focus has resided on one workbook using the default worksheet "Sheet1". Such will be the case in only the simplest macros. When creating a custom report, it is nice to do so with a fresh worksheet in a new workbook. Because this will involve referencing multiple worksheets and workbooks, an explanation on the various ways to access worksheets and workbooks is in order here.

As mentioned previously, to make a workbook active (that is receiving focus), call the ActivateWorkbook subroutine passing the workbook name as a string. It is a good idea to call the ActivateWorkbook subroutine at the start of any function or subroutine because it will make clear which workbook the object data are being extracted and/or manipulated in. This is true even if the workbook is already active from a call to ActivateWorkbook in a previous function or subroutine. To someone unfamiliar with the code, seeing which workbook is the attention of focus at the start of a subroutine makes debugging code much easier.

It is also a good idea to keep track of workbook names by using a global variable (which is appropriately named to reflect the worksheets function) at the beginning of a module. The workbook name can then be stored into the global variable using the (.Name) property. For example:

```
'Place at top of Module

Global Testbook As String

'Add this code after loading, creating, or renaming 'Testbook

Testbook = ActiveWorkbook.Name
```

One more word about workbooks: the statement ActiveWorkbook.Name = "NewWork-bookName" will not work because the (.Name) property for a workbook is read only. The only way to change the name of a workbook is to save the workbook with a new name. A method of accomplishing this will be shown in the sample subroutine Accessing_Sheets.

Once attention has been called to the appropriate workbook, the user must specify with which worksheet he wishes to work. Two methods are available to do this: Worksheets and Sheets.

One way to specify a particular sheet is by its index number. Default settings for a new sheet could be Sheet1 = 1, Sheet2 = 2, ... etc., for example:

```
Worksheets(1).Activate

Sheets(1).Activate
```

Another method is to use the actual name of the sheet, shown in the tab attached to the sheet at the bottom of the Excel screen.

```
Worksheets("Sheet1").Activate

Sheets("Sheet1").Activate
```

Sheets can also be renamed to whatever is appropriate:

Method 1: (two steps)

```
Sheets(1).Activate

ActiveSheet.Name = "Meaningless Data"
```

Method 2: (one step)

```
Worksheets(1).Name = "Meaningless Data"
```

At times it may be convenient to populate a sheet with values that will be used in interim calculations but will never be present in the final report. Such sheets can be deleted easily by using the delete method:

```
Worksheets(1).Delete

Worksheets("Meaningless Data").Delete
```

or

```
Sheets(1).Delete

Sheets("Meaningless Data").Delete
```

The following subroutine illustrates the points made earlier; it can be run as a standalone (without being called from another subroutine). Also see Figure 1.13.

```
Sub Accessing_Sheets()

'A demonstration on accessing worksheets

Dim sheetref As Variant
```

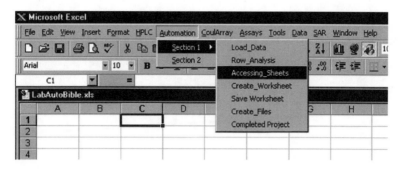

FIGURE 1.13 Running the macro that shows how to access different sheets.

```
Dim ii As Integer
'Open a new workbook
Workbooks.Add
'This will not work - read only property
'ActiveWorkbook.Name = "SheetTester"
'The only way to change a workbook name is to save it as a
new filename
ActiveWorkbook.SaveAs FileName:="C:\TEMP\Bookxls.xls",
FileFormat:=xlNormal _
, Password:="", WriteResPassword:="", ReadOnlyRecommended:
=False,_
CreateBackup:=False
'Save Worksheet Name to Global Variable at top of Module
Testbook = ActiveWorkbook.Name
'Methods for activating Sheets
Debug.Print "Workbook Name: ", Testbook
For ii = 1 To 6
   Select Case ii
      Case 1 To 3
         sheetref = ii
      Case 4
         sheetref = "Sheet3"
      Case 5
         sheetref = "Sheet2"
      Case 6
         sheetref = "Sheet1"
```

```
        End Select
        Worksheets(sheetref).Activate
        'Stop
    Next ii
    For ii = 1 To 6
        Select Case ii
            Case 1 To 3
                sheetref = ii
            Case 4
                sheetref = "Sheet3"
            Case 5
                sheetref = "Sheet2"
            Case 6
                sheetref = "Sheet1"
        End Select
        Sheets(sheetref).Activate
        'Stop
    Next ii
    'To rename sheets method 1
    For ii = 1 To 3
        Sheets(ii).Activate
        ActiveSheet.Name = "NewName" & Str(ii)
    Next ii
    'Stop
    'To rename sheets method 2
    For ii = 1 To 3
        Worksheets(ii).Name = "NewName2" & Str(ii)
    Next ii
    Stop
    'To delete unneeded sheets
    Sheets(1).Delete
    Worksheets("NewName2 2").Delete
    End Sub
```

This subroutine shows how to use the methods outlined previously to activate, rename, and delete worksheets. To pause the macro at any point, the user can "uncomment" the stop commands or add additional stop commands at any point desired.

1.8 CREATING A CUSTOM REPORT WORKSHEET

With a subroutine written to select only those samples of interest and the ability to create and manipulate data between multiple sheets, it is now possible to create a custom report of the information that is desired. The following subroutine does that; an explanation of the code follows:

```
Sub Create_Worksheet()

'Creates a Custom Worksheet with only desired data

Dim ii As Integer 'Loop index

Dim rowptr As Integer 'Row Pointer

Dim header$ 'Heading above each column

'Open a new workbook

Workbooks.Add

'Save the new workbook and rename it with a pertinent name

ActiveWorkbook.SaveAs FileName:="C:\TEMP\Custom Report.xls",
FileFormat:=xlNormal_

, Password:="", WriteResPassword:="", ReadOnlyRecommended:
=False,_

CreateBackup:=False

'Save Worksheet Name to Global Variable

DataReport = ActiveWorkbook.Name

'Rename Report Worksheet

ActiveSheet.Name = "Report"

'Create Report Heading

Range(Cells(1, 1), Cells(1, 1)).Select

'Write Appropriate Label

ActiveCell.Value = "XYZ Research Laboratories"

'Make Label Bold

Selection.Font.Bold = True

Range(Cells(2, 1), Cells(2, 1)).Select

ActiveCell.Value = "Group ABC 123"

Selection.Font.Bold = True

Range(Cells(3, 1), Cells(3, 1)).Select

ActiveCell.Value = Format(Date$, "dddd, mmm d yyyy")
```

```
Selection.Font.Bold = True

Range(Cells(4, 1), Cells(4, 1)).Select

ActiveCell.Value = Format(Time, "hh:mm:ss AMPM")

Selection.Font.Bold = True

'Create Worksheet Headers

'There are 5 columns to Write Labels For

rowptr = 5 ' Labels on row No. 5

For ii = 1 To 5

    Select Case ii

    Case 1

        header$ = "Sample ID"

    Case 2

        header$ = "Lot No."

    Case 3

        header$ = "Solubility"

    Case 4

        header$ = "Units"

    Case 5

        header$ = "Toxicity"

    End Select

    'Select Proper Range (Single Cell)

    Range(Cells(rowptr, ii), Cells(rowptr, ii)).Select

    'Write Appropriate Label

    ActiveCell.Value = header$

    'Make Label Bold

    Selection.Font.Bold = True

Next ii

'Now write in Selected Data

'Loop For Rows

For ii = 1 To Total_Samples

    If ExtractSamp(ii) = True Then

        'increment row pointer

        rowptr = rowptr + 1

        'Write in Data
```

```
        Worksheets("Report").Cells(rowptr, 1).Value = Samp(ii)

        Worksheets("Report").Cells(rowptr, 2).Value = Lot(ii)

        Worksheets("Report").Cells(rowptr, 3).Value = Sol(ii)

        Worksheets("Report").Cells(rowptr, 4).Value = Unit(ii)

        Worksheets("Report").Cells(rowptr, 5).Value = Tox(ii)

      End If

    Next ii

  End Sub
```

To test the preceding subroutine, select from the menu bar illustrated in Figure 1.14. The workbook and worksheet shown in Figure 1.15 will be produced. This is a report that would be typical of something put into a legal notebook at a pharmaceutical company. The report heading contains, line by line: company name, group name, today's date, and time processed. In practice, if these data came from a particular screen, the time and date should be extracted from a file showing when the sample was analyzed, rather than the date it was compiled. The reason for this is that when defending a patent it is important to show not only that a compound was created, but also that analyses were being done on the compound for a particular therapeutic area. Exact dates and times of these analyses are crucial.

The command Range(Cells(row, col), Cells(row, col)).Select selects a range of cells upon which to work. In this case only one cell is selected to add text and boldface type. Also, the Date$ and Time functions come in handy when creating such reports. They can be formatted in a variety of ways with the format command.

The column headers are created starting with row 5. The current row position is maintained in the macro by the rowptr (row pointer) variable. The loop with ii is used to index each column and a select case statement assigns header$ to equal the proper heading for that column. Finally, at the end of the loop the proper cell is selected and formatted with the font made bold.

The last loop writes in the data previously selected in the Data_Analysis subroutine. The rowptr variable is again used to keep track of the current row. If the global Boolean variable ExtractSamp(ii) = True, then the sample should be extracted. The data are then written to the worksheet, and the row pointer (rowptr) is incremented. This action is performed through the loop on all the samples, the total number of which was stored in the global variable Total_Samples declared at the module level.

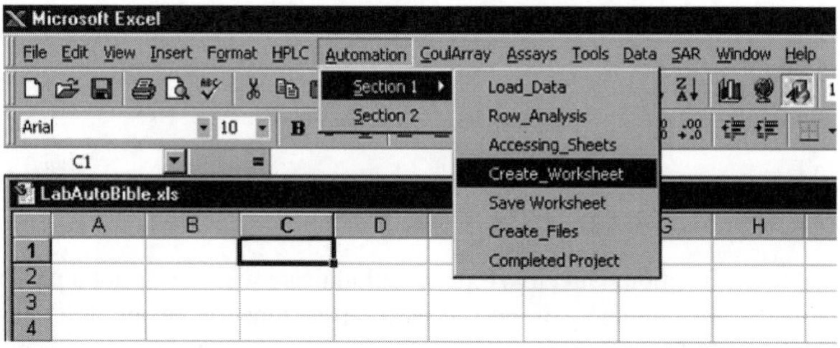

FIGURE 1.14 Running a macro to create a sample laboratory report.

	A	B	C	D	E
	Custom Report.xls				
1	**XYZ Research Laboratories**				
2	**Group ABC 123**				
3	**Wednesday, Jan 17 2001**				
4	**11:30:04 AM**				
5	**Sample ID**	**Lot No.**	**Solubility**	**Units**	**Toxicity**
6	SP-000004	100-10-01	65	ug/uL	nontoxic
7	SP-000007	100-10-02	65	ug/uL	mild
8	SP-000008	100-10-01	65	ug/uL	moderate
9	SP-000013	100-10-01	65	ug/uL	moderate
10	SP-000017	100-10-01	65	ug/uL	mild
11	SP-000019	100-10-03	65	ug/uL	nontoxic
12					

FIGURE 1.15 Example report created by means of an Excel VBA macro.

1.9 SAVING THE CUSTOM REPORT WORKSHEET

It is important to give some thought on how to save the reports generated by the macros. At some point they may or may not need to be looked at again. How will they be sorted and stored? What will they be named? In what format will they be saved? The following functions prove very useful in creating and managing an efficient storage method from within a macro without user intervention.

The function FileExists will indicate whether or not a particular filename exists on the system in a particular location. This is essential information to know to prevent the inadvertent overwriting of existing files when running a macro with the DisplayAlerts function turned off. The full path of the file (example "c:\temp\junk.txt") is passed to the function as a string and the function will return true if the file exists on the system in that location.

```
Function FileExists(full_path As String) As Boolean

    If Dir(full_path) <> "" Then

        FileExists = True

        'Debug.Print "File Exists"

    Else

        FileExists = False

        'Debug.Print "File Doesn't Exist"

    End If

End Function
```

Its first cousin, the function, DirectoryExists, will indicate whether or not a particular directory exists on the system. The directory is passed as a string function (example "c:\temp") and the function will return true if the directory is already present on the system.

To utilize Directory Exists Function the following Library MUST be installed

```
'Library Scripting

'"Microsoft Scripting Runtime" must be checked in References
```

```
'C:\WINNT\system32\scrrun.dll

'Microsoft Scripting Runtime

'Use F2 to view Objects — Should be checked in References

Function DirectoryExists(full_path As String) As Boolean

Dim m_FileSys As FileSystemObject

Set m_FileSys = New FileSystemObject

    DirectoryExists = m_FileSys.FolderExists(full_path)

End Function
```

The next subroutine is very handy. It will create a directory to store data in if the directory does not already exist. The path with the directory name to be created is passed as a string in path$ (example: "c:\temp\new_dir") and the directory will be created on the system if it does not already exist. (For the example path$, the directory "new_dir" would be created if it did not already exist on the system.)

One shortcoming of the MkDir statement (make directory) is that, to make a directory, the directory immediately beneath it must exist. In other words, when calling the command to create the directory, the program's active path must be the directory in which the new directory is to be created.

To get around this shortcoming, the Sub CreateDataDir breaks the Path$ passed to the subroutine up into successive paths leading to the last directory to be created. The successive paths are stored in the string array pathstep (up to 24 directories may precede the directory being created). To create the Path$ passed to the subroutine, the number of paths preceding the final path are determined, and then each successive path is checked to see if it exists. If a path does not exist, then it is created, and every successive path to the final path is also created until the final Path$ passed to the subroutine is complete.

```
Public Sub CreateDataDir(Path$)

Dim Result As Boolean

Dim pathstep(25) As String, ii As Integer, msp As Integer

'Debug.Print "Full Path: "; Path$

'Determine all Subpaths

For ii = 1 To Len(Path$)

   If Mid(Path$, ii, 1) = "\" Then

      pathstep(idx) = Mid(Path$, 1, (ii - 1))

      'Debug.Print idx; ":"; pathstep(idx)

      idx = idx + 1

   End If

   msp = idx - 1

   'If ii = Len(Path$) Then

      'Debug.Print "Last Char:"; Mid(Path$, ii, 1)

   'End If
```

```
Next ii
'Change To Proper Drive
ChDrive pathstep(0)
'Change to Proper Directory
ChDir pathstep(0)
'Check Each Step in the full path, if a directory doesn't exist
'create it along the way to the final path
For ii = 1 To msp
    Result = DirectoryExists(pathstep(ii))
    Select Case Result
        Case True
        'Debug.Print "Path Found !!! Do Not Recreate!"
        'MsgBox "Path Found !!", vbExclamation, "System Error"
        'Log Error Here
        'Debug.Print ii; "of"; msp; ""; pathstep(ii); " Exists!!"
        Case False
        'Debug.Print "Path Not Found !!! Creating Dir!"
        MkDir pathstep(ii)
        ChDir pathstep(ii)
        'Debug.Print ii; "of"; msp; ""; pathstep(ii); "
        Created!!"
    End Select
Next ii
End Sub
```

Now that how to create directories and determine if files and/or directories exist on a system is clear, the question arises as to a good method for naming filenames. It is preferential to have a filename that makes the following things immediately apparent to anyone looking at the file: (1) the instrument name, (2) the instrument number (if more than one instrument of the same type exists in the organization), (3) the date the file was created, and (4) the type of experiment data the file contains.

A great way to create the date for the filename is by using the Date$ and Time$ system variables with the Format command.

```
Fndate$ = Format(Date$, "yyyymmdd") & Format(Time$, "hhmmss")
```

Here a date is stored in the string variable fndate$, which is a four-digit year, two-digit month, two-digit day followed by a two-digit hour, two-digit minute, and two-digit second. It is immediately apparent to anyone looking at the file not only what day it was created but also what time it was created. This information can be crucial for patent defense purposes. Another advantage of using such a format to represent the date in the filename is guaranteed uniqueness. Since seconds are included in the Time$ function, no two filenames will be alike because even the fastest computers will be unable to store two files within a second.

The following subroutine will create a filename when all the elements are passed to it. The filename is created with "_" between each piece of information that follows the format instrumentname_instrumentnumber_date_datatype.

```
Function CreateFileName(iname$, inumb%, thedate$, dtype$) As
String

'Function will create a filename based on parameters passed

'iname$ = instrument name

'inumb% = instrument number

'thedate$ = date of analysis

'dtype$ = type of data

CreateFileName = iname$ & "_" & Trim$(Str(inumb%)) & "_" &
thedate$ & "_" & dtype$

End Function
```

A sample call to this function could be:

```
Filename$ = CreateFileName("Agilent",17,Fndate$,"HPLC")
```

With the appropriate filename stored in the string variable Filename$, it is time to save the worksheet in the desired format. Some thought should be given to where the data are stored on the machine. One method would be:

```
C:\datafiles\assay\2000\Jan\todaysdate\Filename$
```

Here files are stored in a section of the hard drive called "datafiles" followed by the assay they were acquired on. The date of creation is specified in subsequent folders. A folder specifying today's date can be created using the code:

```
Dir_For_Files = Format(Date$, "mmm d yyyy") & "\"
```

The workbook can then be saved using the commands:

```
ActiveWorkbook.SaveAs FileName:="C:\TEMP\JUNK.XLS",
FileFormat:=xlWorkbookNormal,_

    Password:="", WriteResPassword:="", ReadOnlyRecommended:
    =False, CreateBackup:=False
```

A copy of the original workbook can also be made and saved under a different filename using the (.SaveCopyAs) method demonstrated in the following subroutine:

```
Sub Copy_n_Save()

'Copy Data Sheets to a New Workbook and Save Them

'Create a New Workbook

Dim Archive_Name$

Archive_Name$ = "c:\temp\junk" & ".xls"

ActiveWorkbook.SaveCopyAs (Archive_Name$)
```

```
End Sub
```

Employing all the methods discussed previously, the Sub SaveWorksheet will save the custom report created in the macro in an "intelligent manner" (see Figure 1.16).

```
Sub SaveWorksheet()
```

'Subroutine saves the Custom Report Workbook in an "intelligent manner"

```
Dim FileName$, Fndate$
```

```
Dim Path$
```

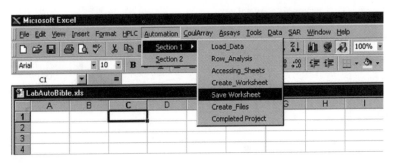

FIGURE 1.16 Running the save worksheet macro.

```
Dim FullPath$, FullPathE$
```

'Create an Intelligent File Name

```
Fndate$ = Format(Date$, "yyyymmdd") & Format(Time$, "hhmmss")
```

```
FileName$ = CreateFileName("Agilent", 17, Fndate$, "HPLC")
```

'Create an Intelligent File Location

```
Dir_For_Files = Format(Date$, "mmm d yyyy")
```

```
Path$ = "c:\datafiles\assay\" & Format(Date$, "yyyy") & "\" & _
```

```
Format(Date$, "mmm") & "\" & Dir_For_Files & "\"
```

'If the Path doesn't exist -> make the directories!

```
Call CreateDataDir((Path$))
```

'The full path consists of the path AND the filename

```
FullPath$ = Path$ & FileName$
```

```
FullPathE$ = Path$ & FileName$ & ".xls"
```

'Debug.Print Path$, FileName$, FullPath$

'If the FileName does not exist then write the file

```
Select Case FileExists(FullPathE$)
```

```
   Case True
```

   'The File already Exists - Error

   'Debug.Print "Error: File Already Exists"

```
      Case False

      'The File Does not Exist - Write it out

      'Debug.Print FullPath$; " Written"

      ActiveWorkbook.SaveAs FileName:=FullPath$, FileFormat:
      =xlNormal_

      , Password:="", WriteResPassword:="",
      ReadOnlyRecommended:=False,_

      CreateBackup:=False

   End Select

   'Save Worksheet Name to Global Variable

   DataReport = ActiveWorkbook.Name

   'Public Sub CreateDataDir(Path$)

   'Function FileExists(full_path As String) As Boolean

   End Sub
```

Running the preceding subroutine will save the ActiveWorkbook to a directory with a format like:

```
C:\datafiles\assay\2001\Jan\Jan 31 2001
```

(month, day, and year in path will depend upon the day it is run). The filename will be something like (depending upon date and time):

```
Agilent_17_20010131123022_HPLC.xls
```

1.10 CREATING AN UPLOADABLE FILE FOR A DATABASE

Storing the data locally in custom reports on the user's hard disk may be sufficient for some applications. However, data are useless if key people and groups who can benefit from their use cannot access them. At some point, it would probably be beneficial to have all the data collected by an assay accessible by everyone in the organization through a relational database. Detailing setting up a database on the various platforms available (Oracle, Sybase, Access, etc.) is beyond the scope of this book; however, all databases are usually set up to accept data from files set up in certain ways.

The most common way data are loaded into a database is using a delimited file (usually comma) that has its first line reserved for the column names that the data will be going to within the database. A common way to name columns is in capital letters with words separated by underscores, for example:

```
SAMPLE_NO
```

```
SOLUBILITY
```

```
SOLUBILITY_UNITS
```

```
LOT_NO
```

Having the column names in all capital letters is a good practice because if someone is querying for the data in a database that provides case-sensitive queries, he may not know if any of the letters are capitalized, only the first letter is capitalized, or every distinct word is capitalized. This avoids confusion for everyone and helps minimize frustrated users.

The easiest way to create an uploadable file is by creating a worksheet, populating the first row with the appropriate column names, and then writing the data down each row of the worksheet for each sample run. The worksheet is then saved as a comma-delimited file. The following subroutine accomplishes this:

```
Sub Create_File()
'Subroutine Creates Uploadable Database File via Worksheet
Dim ii As Integer 'Loop index
Dim rowptr As Integer 'Row Pointer
Dim header$ 'Heading above each column
'Open a new workbook
Workbooks.Add
'Save the new workbook and rename it with a pertinent name
ActiveWorkbook.SaveAs FileName:="C:\TEMP\UPLOADABLE.CSV",
FileFormat:=xlCSV,_
Password:="", WriteResPassword:="", ReadOnlyRecommended:
=False, CreateBackup:=False
'Save Worksheet Name to Global Variable
UpFile = ActiveWorkbook.Name
rowptr = 1 'DataBase Column Names Start on Row No. 1
For ii = 1 To 5
   Select Case ii
   Case 1
      header$ = "SAMPLE_NO"
   Case 2
      header$ = "LOT_NO"
   Case 3
      header$ = "SOLUBILITY"
   Case 4
      header$ = "SOL_UNITS"
   Case 5
      header$ = "TOX"
   End Select
   'Select Proper Range (Single Cell)
   Range(Cells(rowptr, ii), Cells(rowptr, ii)).Select
   'Write Appropriate Label
   ActiveCell.Value = header$
```

```
Next ii

'Now write in Selected Data

'Loop For Rows

For ii = 1 To Total_Samples

    If ExtractSamp(ii) = True Then

        'increment row pointer

        rowptr = rowptr + 1

        'Write in Data

        '*.csv files only have 1 worksheet !!!

        Worksheets(1).Cells(rowptr, 1).Value = Samp(ii)

        Worksheets(1).Cells(rowptr, 2).Value = Lot(ii)

        Worksheets(1).Cells(rowptr, 3).Value = Sol(ii)

        Worksheets(1).Cells(rowptr, 4).Value = Unit(ii)

        Worksheets(1).Cells(rowptr, 5).Value = Tox(ii)

    End If

Next ii

'ReSave the workbook after it is populated with data

ActiveWorkbook.SaveAs FileName:="C:\TEMP\UPLOADABLE.CSV",
FileFormat:=xlCSV,_

Password:="", WriteResPassword:="", ReadOnlyRecommended:
=False, CreateBackup:=False

End Sub
```

Figure 1.17 illustrates running the preceding subroutine. This subroutine produces a workbook shown in Figure 1.18. This workbook is saved as a comma-delimited file under "c:\temp\upload-able.csv". The file could just as easily have been stored in an "intelligent" manner using date/time methodology used when saving the report worksheet. Usually, uploading files are written to the same directory over and over again. When a server detects a new file it immediately uploads it. Most servers check for new files every second or so, so accidentally overwriting another file usually is not a problem.

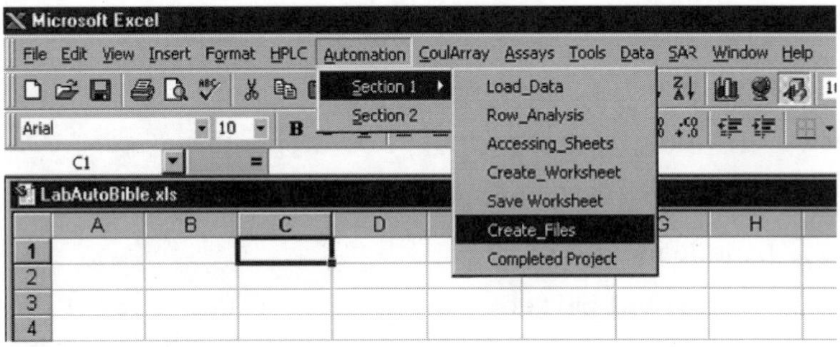

FIGURE 1.17 Running a macro that creates comma-delimited files from worksheets.

UPLOADABLE.CSV

	A	B	C	D	E
1	SAMPLE_NO	LOT_NO	SOLUBILITY	SOL_UNITS	TOX
2	SP-000004	100-10-01	65	ug/uL	nontoxic
3	SP-000007	100-10-02	65	ug/uL	mild
4	SP-000008	100-10-01	65	ug/uL	moderate
5	SP-000013	100-10-01	65	ug/uL	moderate
6	SP-000017	100-10-01	65	ug/uL	mild
7	SP-000019	100-10-03	65	ug/uL	nontoxic
8					
9					
10					

FIGURE 1.18 A comma-delimited file suitable for upload to a database created from a worksheet.

Another way of creating an uploadable file is by manually constructing each row with delimiters and then writing the file out using the Print# command. This method is typically used when creating uploadable database files from Visual Basic in an environment outside Excel where worksheets are not available. In addition, the Append method can be utilized to add more rows to an existing file when this method is employed. An example subroutine follows:

```
Sub Manual_Write()
'Subroutine Manually writes a Comma Delimited File
Dim delim$, CR$, ii As Integer
Dim LineHeader$
Dim rowdata(100) As String
'Set up Delimiter Character
delim$ = "," 'This is easiest to understand
'Or
delim$ = Chr(44) 'Chr(44) = ASCII code for ","
'Either will work
'Set Up carriage return string
CR$ = Chr$(13)
'Construct Line Header
LineHeader$ = "SAMPLE_NO" & delim$ & "LOT_NO" & delim$ &_
"SOLUBILITY" & delim$ & "SOL_UNITS" & delim$ & "TOX"
'Construct Row Data & Write CSV file
Open "C:\TEMP\MUPLOADABLE.CSV" For Output As #1
'If this file already existed and one wished to append
'more data to it use this command instead:
'Open "C:\TEMP\MUPLOADABLE.CSV" For Append As #1
'Print Out Header with Database Column Names
Print #1, LineHeader$
'Construct Comma Delimited Row Data and Print to File
```

```
For ii = 1 To Total_Samples

    If ExtractSamp(ii) = True Then

        'increment row pointer

        rowptr = rowptr + 1

        rowdata(rowptr) = Samp(ii) & delim$ & Lot(ii) & delim$ &_

        Sol(ii) & delim$ & Unit(ii) & delim$ & Tox(ii)

        Print #1, rowdata(rowptr)

    End If

Next ii

Close #1

End Sub
```

Although this method is much more complicated to implement, it provides the ultimate in flexibility in terms of being able to structure the file exactly as desired.

While on the subject of writing files, one more subroutine comes in very handy. It is used to log errors to a file on disk, which can be crucial if a process is running unattended and an error occurs:

```
Public Sub Log_Error(ErrorTxt$)

Open "c:\temp\error.txt" For Append As #1

Print #1, ErrorTxt$

Close #1

End Sub
```

Pass the text to be stored in the file specified in the Open statement in the string variable ErrorTxt$. If the file does not exist it will be created. If the file exists, then the information in ErrorTxt$ will be appended to the end of the file.

A completed project has been created using the preceding methods and subroutines. It can be run from the menu bar as shown in Figure 1.19.

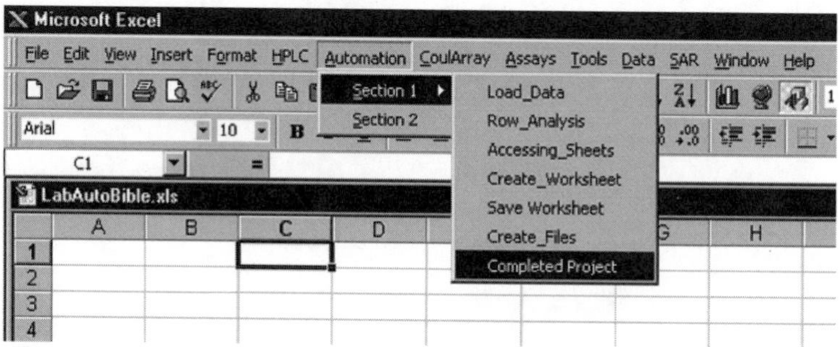

FIGURE 1.19 Running the complete project.

2 Advanced Microsoft Excel VB Macro Techniques

2.1 USING HIDDEN WORKBOOKS TO STORE AND RUN MACROS

When constructing a macro for any purpose, it would be most intuitive to place the macro code within the workbook where the data will be manipulated. In fact, that is exactly what is done when a macro is recorded using the macro recorder. While this is the easiest thing to do, it is not the best thing to do.

The problem with storing a macro within a workbook is that when data are brought into the workbook and that workbook is saved as a workbook, these data will be present the next time the macro is run. For example, suppose a macro from a workbook is run and it creates a row of data from every piece of data from a (*.csv) file. The first time, the macro is run using a *.csv file that contains 50 pieces of data. Now 50 rows of data are present in the workbook and the workbook is saved. Suppose the macro is run a second time on a dataset that has 25 rows of data. Now the workbook will still have 50 rows of data: 25 from the most recent file run and 25 that were not overwritten from the previous run.

It is best to store macros that will be run over and over again in their own workbooks, which will be hidden from view and never have data written to them. Even if the user is told specifically never to resave the workbook containing the macro, quite often this will happen. If the macro is stored in a hidden worksheet, however, the user does not "see" it and thus will not corrupt it by changing and then saving it.

To create a hidden workbook in which to store a macro, simply create a new workbook named "HideTest.xls" and save it to the XLStart directory (see Figure 2.1). By saving the workbook to the XLStart directory, the workbook will be loaded every time Excel is started. Thus the macro will be available for use in every Excel session. Now choose Window->Hide (or Alt+W+H) to hide the newly created workbook. Close Excel and when prompted choose "Yes" to save changes made to "HideTest.xls" (the change made was setting its hidden property to true).

Conversely, a workbook that is hidden and has been loaded into Excel can be unhidden by choosing Window->Unhide. Doing so will produce the dialog box shown in Figure 2.2 in which the user can select the workbook he wishes to "unhide." Once a hidden workbook has been created, the programmer can access it by entering the Visual Basic environment accessed directly from Excel by the VB toolbar.

The controls on the toolbar shown in Figure 2.3 from left to right are: (1) run macro, (2) record macro, (3) pause/resume macro, (4) VB editor, (5) control toolbar, and (6) design mode. If this toolbar is not present in Excel it can be brought forth by selecting View->Toolbars->Visual Basic.

Pressing the button for the VB editor will place the user in the Visual Basic for Applications (VBA) design environment; the hidden workbook is clearly visible in the Project Explorer window on the left. Here the workbook "AdvMacTech.xls" is shown as the VBAProject at the very top of the project explorer window (see Figure 2.4).

A module that can run VBA code can easily be added to this project through Insert->Module (see Figure 2.5). (Note that the VBAProject (AdvMacTech.xls) must be selected in the project explorer window when this menu item is selected or the module will be added to some other VBAProject.)

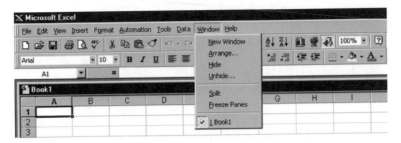

FIGURE 2.1 Process to hide and unhide workbooks.

FIGURE 2.2 The unhide dialog box.

FIGURE 2.3 The Visual Basic toolbar.

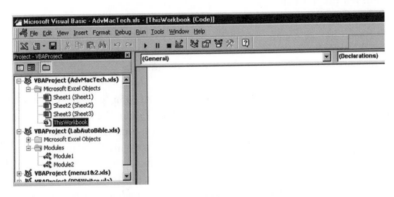

FIGURE 2.4 Accessing a hidden workbook.

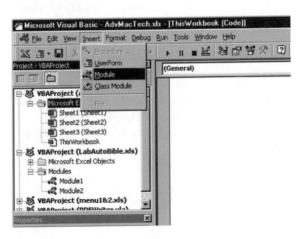

FIGURE 2.5 Adding a module to a hidden workbook.

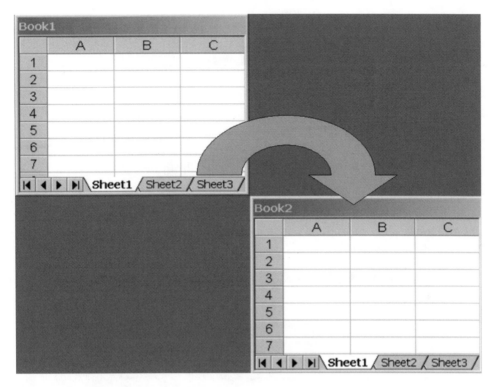

FIGURE 2.6 Workbook1 (hidden) accesses values only in Workbook2 (visible).

It is from the module in the hidden workbook that macro code should be written. In this manner the hidden workbook addresses the values in the visible workbooks and manipulates them. The hidden workbook module (Figure 2.6) does not access or import any values in its own worksheets and is therefore incorruptible. The module in the hidden workbook directly references the *values* in the visible workbooks, but the visible workbooks do not share any global variables with the code in the hidden workbook. This is an important distinction. Running scripts in different workbooks and sharing global variables *between workbooks* will be discussed later.

2.2 ADDING CUSTOM MENU ITEMS TO RUN MACROS

While macros can be run using Tools->Macro->Macros from the menu, this method of macro execution poses some difficulties for the user. First, when the user accesses macros in this manner, every subroutine and function in the workbook of interest (or all workbooks for that matter) will be listed as an option to be run. This can be confusing and may overwhelm the end user. A macro of any degree of complexity will have 50 or more functions and subroutines. It is unwise to require the user to sort through them all to find the one that "begins" the macro (see Figure 2.7).

Clearly, a better option is to have the ability to run each macro contained in a hidden workbook from a menu item! There are many ways to go about doing this, but one of the best is available free as a shareware program available from John Walkenbach & Associates, which can be downloaded from http://www.j-walk.com/ss/excel/tips/tip53.htm. The file menumakr.xls is also contained on the website, http://www.pharmalabauto.com.

Once this file is downloaded it contains a workbook that looks like the one shown in Figure 2.8. This figure shows how the menumakr program was used to construct menus for the programs run in Chapter 1 of this book. The five columns in this worksheet and their functions are as follows:

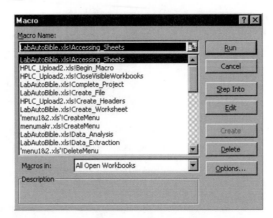

FIGURE 2.7 Should users be required to wade through all these subroutines to find the one that starts the macro in which they are interested?

	A	B	C	D	E	F	G
1	Level	Caption	Position/Macro	Divider	FaceID		Click the link below
2	1	&Automation	6				identify the FaceI
3	2	&Section 1					http://www.j-walk.com/ss/
4	3	Load_Data	LabAutoBible.xls!Load_Data				
5	3	Row_Analysis	LabAutoBible.xls!Row_Analysis				
6	3	Accessing_Sheets	LabAutoBible.xls!Accessing_Sheets				
7	3	Create_Worksheet	LabAutoBible.xls!SS_Create_Worksheet			Create Menu	
8	3	Save Worksheet	LabAutoBible.xls!SS_SaveWorksheet				
9	3	Create_Files	LabAutoBible.xls!SS_Create_Files				
10	3	Completed Project	LabAutoBible.xls!Complete_Project			Delete Menu	
11	2	&Section 2					
12							

FIGURE 2.8 The menumakr shareware program by John Walkenbach & Associates. (From *The SpreadSheet Page*, menumakr.xls utility, JWalk & Associates, Inc., La Jolla, CA, http://www.j-walk.com, 2002. With permission.)

- Level — the level of the particular item. Valid values are 1, 2, and 3. A level of 1 is for a menu; 2 is for a menu item and 3 is for a submenu item. Normally, a user will have one level 1 item, with level 2 items below it. A level 2 item may or may not have level 3 (submenu) items.
- Caption — the text that appears in the menu, menu item, or submenu. Use an ampersand (&) to specify a character that will be underlined.
- Position/Macro — for level 1 items, this should be an integer that represents the position in the menubar. For level 2 or level 3 items, this will be the macro that is executed when the item is selected. If a level 2 item has one or more level 3 items, the level 2 item may not have a macro associated with it.
- Divider — displays true if a divider should be placed before the menu item or submenu item.
- FaceID — an optional code number that represents the built-in graphic images displayed next to an item. Follow this link for more information about determining FaceID code numbers. This link provides an easy method to identify FaceID codes: http://www.j-walk.com/ss/excel/tips/tip40.htm.

This workbook will then autoload menu items (in the position of the developer's choosing) to execute the macros required and can then be hidden and saved to the XLStart directory. The menus will then be automatically loaded behind the scenes. The developer can use as many of these workbooks as desired as long as each workbook is given a unique name. It would be wise to use a different workbook for the menu items in different projects, thereby keeping the menu items for a particular project managed locally by one workbook.

One point that bears mentioning is that the menu item may sporadically lose its link with the macro it is supposed to run. The user will choose the menu item and then nothing will happen. The first thing to check is to make sure that the worksheet that contains the macro is accessible (the macro is listed in the macro dialog box). Once it has been determined that the link is accessible, the next thought on any sane developer's mind is that it is once again time to shut down Windows® and reboot. The reboot, however, does not fix the problem. The problem is fixed by running the macro manually by selecting it from the macro dialog box. This will somehow reestablish the link from the menu item to the macro. It is unknown why this phenomenon occurs, but it does, and this is how it can be fixed. It usually occurs when a developer has been modifying the workbook to which the link exists and has been opening, changing, and saving the workbook repeatedly.

A more permanent fix to this problem can be found in Microsoft's article (Q213463) "Cannot Find Macro Error Message When Running Macro from Object". To ensure that this error message does not occur, it suggests: (1) using unique names for macros as well as for names on worksheets or (2) specifying the complete path to the macro, as in the following example:

```
module name.macro name
```

For the macro Load_Data included in Chapter 1, this command would be:

```
LabAutoBible.xls!Module1.Load_Data
```

and would be placed in the position/macro column of the menumakr application worksheet.

2.3 KEEPING A MACRO FROM CRASHING WHEN A CELL HAS INVALID DATA

Sooner or later a macro with any degree of complexity will produce a result that is invalid. A very common example of this is when a number is divided by 0 and the result is put into a cell and produces the text #DIV/0! in the cell (Figure 2.9). Although this is correct, the problem arises when the value in this cell is accessed by the macro and an attempt is made to perform a mathematical operation on

	A	B	C	D
1	Output from Photons striking incident detector			
2	Detector Type	Photons	voltage out (mV)	Photon/mV
3	Si	300	3	100
4	Ge	74	0	#DIV/0!
5	GaAs	500	5	100

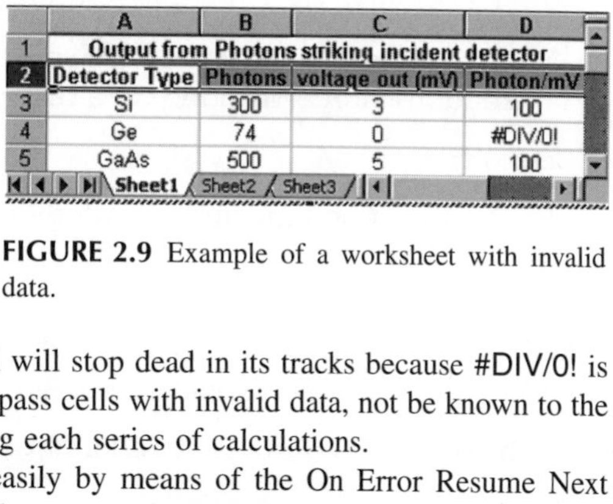

FIGURE 2.9 Example of a worksheet with invalid data.

#DIV/0!. If such an operation is attempted, Excel will stop dead in its tracks because #DIV/0! is not a valid value. A method is needed that will bypass cells with invalid data, not be known to the user until the macro is run, and be different during each series of calculations.

Such erroneous data can be bypassed very easily by means of the On Error Resume Next statement. Suppose, for example, it was desired to have an additional column of data that contained the ratio of photon/volts. This code would need the ability to run and not choke when it encountered a cell with a value of #DIV/0!. The following macro illustrates the use of the On Error Resume Next statement. It can be run by selecting Automation->Section 2->Ignore from the menu. The user can also manually run the macro "Begin_IgnoreMacro." When prompted for a file, use "#DIV0 Example.xls" provided on the website, http://www.pharmalabauto.com.

```
Sub Begin_IgnoreMacro()

Call Load_PhotonData

Call Calc_PV

End Sub

Sub Load_PhotonData()
```

```
'Subroutine to Load text or comma delimited file into a
worksheet

Dim vFileName As String

Dim nochoice As Integer

Dim WindowCaption$

Dim ErrorBoxCaption$, ErrorBoxMsg$

On Error GoTo FileOpenErr

'Set Up Captions and Messages

WindowCaption$ = "Load Photonic Data"

ErrorBoxCaption$ = "Error Loading Photonic Data"

ErrorBoxMsg$ = "Photonic Data NOT Loaded !!!"

vFileName = Application.GetOpenFilename("Excel Files (*.xls),
*.xls",, WindowCaption$)

If vFileName <> "" Then

    Workbooks.Open FileName:=vFileName

    'ActiveSheet.Name = "HPLCDATA"

    'ActiveWorkbook.SaveAs FileName:=
    "C:\TEMP\HPLC\HPLCDATA.CSV", FileFormat:=xlCSV, _

    'Password:="", WriteResPassword:="",
    ReadOnlyRecommended:=False, CreateBackup:=False

End If

'File was chosen and loaded into worksheet - now exit

Exit Sub

FileOpenErr:

'If no file is chosen you end up here

nochoice = msgbox(ErrorBoxMsg$, vbOKOnly, ErrorBoxCaption$)

End

End Sub

Sub Calc_PV()

'Create Additional Column with Photons/Volts Ratio

Dim row As Integer 'Rows

Dim retval As Integer

'Should we Ignore the error message that appears on blank cells?

retval = msgbox("Do you wish to ignore Errors?", vbQuestion
+ vbYesNo, "Process Method")

If retval = vbYes Then
```

```
    On Error Resume Next

End If

'Column Header

Worksheets(1).Cells(2, 5).Value = "Photon/Volts"

For row = 2 To 5

    Worksheets(1).Cells(row, 5).Value =
    Worksheets(1).Cells(row, 4).Value * 1000

Next row

End Sub
```

When the macro is run a message box appears asking the user if errors are to be ignored. If yes is chosen, the macro will execute without errors and the cell that was to be calculated using #DIV/0! is ignored and left blank. The program then continues calculating the rest of the elements in the worksheet (see Figure 2.10). However, if no is chosen, the macro will stop and the error message shown in Figure 2.11 will be presented.

If the debug button is pushed, the program will point to the area in which the error was encountered. Figure 2.12 illustrates a mathematical operation performed on a cell that had #DIV/0! for a value.

From an automation standpoint, it is far better to have the program ignore an error and leave a cell value blank than for it to crash and require user intervention. The only drawback to this method is that if the program is processing hundreds or even thousands of calculations, a method of keeping track of which ones failed must be instituted. This can easily be done by having the program write "Sample No. X was not calculated" to an error log file. Then the user knows where to begin looking for errors.

	A	B	C	D	E
1	Output from Photons striking incident detector				
2	Detector Type	Photons	voltage out (mV)	Photon/mV	Photon/Volts
3	Si	300	3	100	100000
4	Ge	74	0	#DIV/0!	
5	GaAs	500	5	100	100000

Sheet1 / Sheet2 / Sheet3 /

FIGURE 2.10 Choosing "yes" results in a worksheet modification.

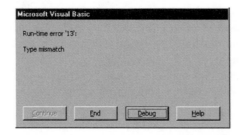

FIGURE 2.11 Calculations performed on a cell with an invalid value will produce this error.

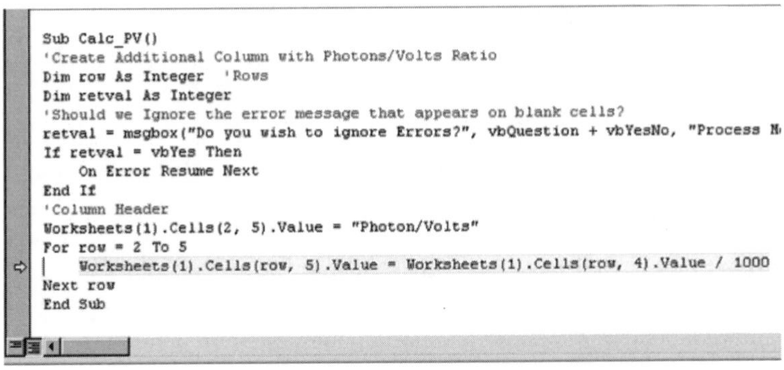

```
Sub Calc_PV()
'Create Additional Column with Photons/Volts Ratio
Dim row As Integer   'Rows
Dim retval As Integer
'Should we Ignore the error message that appears on blank cells?
retval = msgbox("Do you wish to ignore Errors?", vbQuestion + vbYesNo, "Process M
If retval = vbYes Then
    On Error Resume Next
End If
'Column Header
Worksheets(1).Cells(2, 5).Value = "Photon/Volts"
For row = 2 To 5
    Worksheets(1).Cells(row, 5).Value = Worksheets(1).Cells(row, 4).Value / 1000
Next row
End Sub
```

FIGURE 2.12 The debugger freezes at the point where an invalid cell value is accessed for calculation purposes.

2.4 NOT ALL INPUTBOXES ARE EQUAL

InputBoxes are a very convenient way to obtain input from a user. However, unknown to many users, there are actually two types of InputBoxes and they are not equal. The first is the InputBox function that is most commonly used. The second is the InputBox method by the application object. There are advantages and pitfalls to each type of InputBox.

The InputBox method has the advantage that it can determine the types of data entered into it and restrict the user to entering data of only a certain type. This saves the developer the chore of writing validation code to determine if the right kinds of data were entered into the InputBox.

This method has an additional advantage in that it can determine when the cancel button has been pressed, and take appropriate action on this event. The disadvantage to the InputBox function is that it is unable to qualify what types of data it receives automatically. If the macro must have the ability to determine if the cancel button is pressed, the developer has no option but to use the InputBox method.

The pitfall to using the InputBox function is that, when the cancel button is pressed, this function returns an empty ("") string. If the user is always supposed to enter *something* into the InputBox then this is not a problem. But suppose the InputBox is used to add something such as comments. The user may or may not choose to add comments; however, if the user were to press cancel, the macro should exit out of the routine that it is in, and go back to another point in the macro to perform some other function. The problem here is that the program cannot determine if the user did not wish to enter any comments or if he hit cancel to return to a previous portion of the macro. This is a big problem, and there is no workaround for it.

A second disadvantage to using the InputBox function is that it cannot qualify what kinds of data are entered into it (Figure 2.13). The developer must therefore write validation code (see Table 2.1) to determine if the right kinds of data were entered into the InputBox. Bottom line: always use the InputBox method because it is much more flexible than the InputBox function.

Two sample macros that illustrate the use of both types of InputBoxes can be run from the menu. The first, Automation->Section 2->Good InputBox, demonstrates the use of the more flexible InputBox method. The second, Automation->Section 2->Crummy InputBox, demonstrates the less flexible InputBox function. The code for both macros follows:

```
Sub AppIBoxExample()

'Example Application Input Box Method

Dim Response As String, Number As Single

'Run the Input Box.
```

FIGURE 2.13 The application InputBox method can automatically check what kinds of data are entered.

TABLE 2.1
InputBox Method Codes

Type Data Accepted

0	Formula
1	Number
2	Text
4	Boolean (true/false)
8	Array of data from a range of cells or reference to cells
16	Error value

```
'Notice that the type qualification (Type:=2)

'Response = Application.InputBox(prompt:="Please Enter Some
Text",_

'Title:="Application InputBox Example", Default:="John Doe",
Type:=2)

'Run the Input Box.

'Notice that the type qualification (Type:=1) Number

Number = Application.InputBox(prompt:="Please Enter one
Number",_

Title:="Application InputBox Example", Default:="10", Type:=1)

'This Type of Input Box CAN Determine if the Cancel Button
was Pressed

Debug.Print Response

' Check to see if Cancel was pressed.

If Response = "False" Then

    msgbox "You Pressed Cancel," vbInformation + vbOKCancel,
    "User Canceled Action"

    Exit Sub
```

```
    Else

    msgbox "This is the Number you entered: " & Str(Number),
    vbInformation + vbOKOnly, "Input Number"

End If

End Sub

Sub FuncIBoxExample()

'Example of Function Type Input Box

Dim Response As Variant

'Run Input Box

'This type of Input Box CANNOT Tell if Cancel was pressed

'Nor Can it Reject Data which is not of a specified type

Response = InputBox("Enter Only a Number", "Function InputBox
Example", 5)

If Response = "" Then

    msgbox "Did you press cancel or enter nothing, nobody
    knows!", vbInformation + vbOKCancel, "What Happened"

    Exit Sub

    Else

    msgbox "This is what you entered: " & Response, vbInformation
    + vbOKOnly, "Who knows what you input?"

End If

End Sub
```

One last note: the InputBox method type 2 (Text) does not buy the programmer much in terms of qualifying data. Because nearly everything can be considered text, numbers, punctuation, etc., type 2 will rarely if ever return a "Text is not valid." messagebox.

2.5 USING TEMPLATES AND GUIs

Templates are a wonderful mechanism by which a developer can save a great deal of time by not having to code trivial things such as headers or format the number of decimal places, etc. Templates should be used whenever possible to minimize the amount of coding needed. Like all things, however, templates have their limitations. It will be impossible to use a template if the features in the report are not going to remain consistent from run to run. In such an instance, hardcoding labels and formatting to an individual worksheet template would make no sense. Depending upon the circumstances, sometimes the top portion of a report can be formatted in a template and the bottom portion must be coded into the macro. This is often the case when the number of rows of data is different from run to run.

Templates can be loaded into Excel in exactly the same manner that worksheets are. In fact, because the exact same template will usually be loaded each time, it is often unnecessary to have the user pick the file from a dialog box; it can be loaded automatically. The following subroutine automatically loads a template named "LAB_SampTots.xlt". (Be sure to copy this file from the website, http://www.pharmalabauto.com, before running this subroutine.)

```
Sub Open_Template()

'Automatically open a TEMPLATE WORKBOOK in Excel

'Open the Specified Template

Workbooks.Open FileName:="C:\Program Files\Microsoft
Office\Templates\LAB_SampTots.xlt"

'Save Template as a Workbook with a New Name to avoid corrupting
the Template

ActiveWorkbook.SaveAs FileName:="C:\TEMP\LAB_SampTots.xls",
FileFormat:=xlNormal_

, Password:="", WriteResPassword:="", ReadOnlyRecommended:
=False,_

CreateBackup:=False

End Sub
```

Of course the subroutine could also be set up to load files in by means of a dialog box as Excel worksheets and comma-delimited files were in Chapter 1. Remember to change the file extension from (*.xls or *.csv) to (*.xlt).

Notice that as soon as the template is opened it is immediately saved as a workbook to the temp directory. This is done to change the status of the template to that of a workbook and to keep the template from being corrupted. Once the analysis on the workbook has been finished, it will then be saved again under its final filename to another directory.

If the preceding macro is run, a template with the header shown in Figure 2.14 will appear in Excel. This template will be used in macros later in this chapter to demonstrate some of the other techniques covered here.

Although the message boxes offer a "quick and dirty" way to get information from the end user, chances are most programs will be best served by having custom forms that have labels for very specific data to be obtained. In previous versions of VBA, this was accomplished by means of dialog sheets. Although it is still possible to use dialog sheets, they are outdated and have been replaced by user forms. To insert a user form into a workbook, choose Insert->UserForm from the menu (see Figure 2.15).

Notice the controls toolbar on the right-hand side of the user form. It is possible to add even more controls to this toolbar by right clicking on it. However, a word of warning is in order here. VBX and OLE controls meant to be used with VB may not necessarily work under VBA. Many times the controls can be added and are accessible, but for some reason certain properties and methods do not work. Bottom line: using third party controls in VBA is not recommended. In fact, if the user is running MSOffice 2000 and has installed service pack 1, no third party controls will work under VBA.

While most third party controls are far more robust than the standard Microsoft controls, often they may not follow Microsoft's conventions for coding VBA and OLE controls, which leads to bugs and incompatibility in some situations. Using the controls under VBA as opposed to standalone VB seems to aggravate this situation. Before using any third party control in VBA, it would be

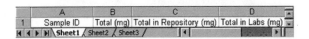

FIGURE 2.14 The LAB_SampTots.xlt template header.

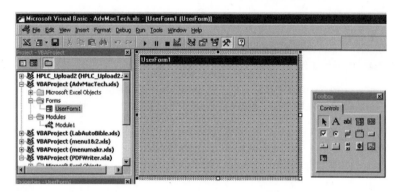

FIGURE 2.15 A user form within a workbook (AdvMacTech.xls).

wise to call the manufacturer and ask if it is guaranteed VBA compatible, and under what conditions. For 99.0% of applications, the controls provided within Excel will accomplish most tasks adequately.

One last consideration about third party controls is deployment. Deployment of an application is where all the unforeseeable problems pop up. The application will work fine on the machine it was developed on, but will hang on other users' machines. One reliable way to make an application hang is to distribute a program that utilizes a third party control and forget to include the licensing information for use with that control. Sometimes distributing the licensing information with the application is easy, and sometimes it is not. A genius can think of half the potential problems an individual may run into during deployment, but most people are not geniuses. It is best to minimize potential problems at the beginning of any potential project.

A sample macro has been provided to demonstrate the utility of user forms. It can be run by selecting Automation->Section 2->Form Demo from the menu. Doing so will produce the GUI (user form) illustrated in Figure 2.16.

This is a nice example because it puts together a lot of concepts that are essential for an effective GUI. A value can be selected and extracted from any visible workbook by selecting the row and column and pressing the "get value" button. The workbook from which the selection is to be made as well as the worksheet is selectable at the bottom of the GUI. Only visible workbooks are shown in the "Available Workbooks" drop down list. The extracted value (or any value typed in the textbox, for that matter) can then be written to any workbook and worksheet using the "Set Value" button. The large button in the upper right corner dismisses (or closes) the user form. This is of particular significance because many developers do not get rid of user forms properly (see Figure 2.17).

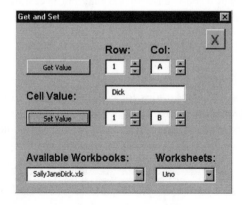

FIGURE 2.16 Example of "get and set" GUI.

The following subroutine shows how to load and display a user form:

```
Sub Form_Demo()
'This Loads the Form into Memory
Load GSForm
'The allows the form to be seen
GSForm.Show
End Sub
```

The load statement is what loads the contents of the form into the computer's memory. Once the form has been loaded it can be displayed using the (.Show) method. When the form does not need to be seen by the user it can be hidden using the (.Hide) method. The problem that arises here is that out of sight is also out of mind. Even though the user form is not visible when the (.Hide) method is employed, the form still resides in memory and is allocating resources that could be utilized somewhere else. When the macro is finished with the form it should not only be hidden with the (.Hide) method but unloaded as well. This will free up all the resources (memory) that the form allocated when it was loaded.

Visual Basic is such a high-level language that developers do not need to keep track of memory allocations, pointers to memory locations, or memory usage in general. As a result, developers are often wasteful of memory by utilizing poor programming practices such as not unloading forms when they are needed, or declaring arrays larger than they need to be. For small applications this is not a problem. In larger applications, such practices can cause real performance issues and may even crash the program should it run out of useable memory. The situation is analogous to a renter whose utilities are included in the rent and thus has little incentive to conserve energy. Even though memory is not managed by the developer, it is not a "free" commodity. It should not be wasted, but utilized as efficiently as possible.

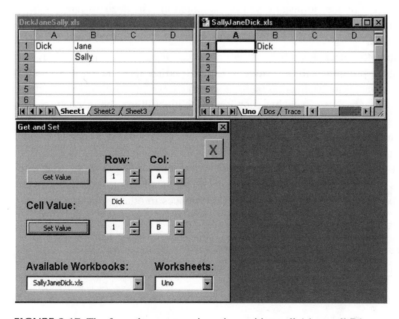

FIGURE 2.17 The form demo macro in action writing cell A1 to cell B1.

The following code is used to unload the user form and free up the resources it allocated from the system:

```
Private Sub Close_Button_Click()

'The Proper Way to Close a User Form

'This Simply Hides the Form

GSForm.Hide

'This Releases the Memory allocated by the form !!!

Unload GSForm

End Sub
```

One last word on user forms: message boxes are examples of *modal forms*, which are forms displayed until a user completes an action on the form. The program will not continue until the user completes this action and closes the form. A *modeless form* can be present while code is executing in the background. A user form is really quasi modeless. Like a modeless form, a user form does not have to be closed for code to execute in the background. However, like a modal form, the user form must be closed before a user can access and change values in a worksheet.

2.6 OPENING AND USING OTHER APPLICATIONS WITHIN EXCEL

Without a doubt Excel has a lot of functionality built into it; in areas where it is lacking, many third-party vendors offer libraries for purchase. Despite this, there will still be instances when another piece of software is better suited to perform some calculations or parse out some data. An individual could easily be in a situation where a software package for a laboratory instrument performs a series of proprietary calculations on a set of data. Duplication of these calculations in Excel could be costly or even impossible if the vendor will not disclose the algorithm used to make the calculations. Anyone doing any serious automation will need to harness the computational power of another application at one time or another and return the calculated results to Excel. Here is how it is done.

To illustrate the mechanism behind starting, accessing, and returning values from other applications, a simple program will be written in Excel to utilize the calculator program (provided with Windows) to create a multiplication table in a worksheet. Although this application is trivial (and really unnecessary because Excel has such functionality built in), the mechanism here is important. In addition, an application had to be chosen that all individuals would have on their computers in order to learn the concepts. The application being accessed could be changed to one that calculates fast Fourier transforms whose functionality is *not* included as part of standard Excel.

The first thing to do is to get the calculator program running and accessible to Excel's VBA code. This is accomplished by means of the shell command. (Note: examples under Windows NT 4.0. Paths may need to be changed to reflect differences in other operating systems.)

```
Sub CreateTimesTable()

CalcTask = Task_ID("c:\winnt\system32\calc.exe")

End Sub

Function Task_ID(fullpath$) As Variant

Task_ID = Shell(fullpath$, vbNormalFocus)

Debug.Print "Task ID: "; Task_ID
```

```
End Function
```

The function Task_ID starts the program whose full path is passed by means of the fullpath$ variable (in this case the calculator program). The shell command starts the program located at the said path and will place it in any number of conditions specified by the second variable, which determines the focus of the activated program. The program can be minimized, maximized, hidden, etc. at startup depending upon the value of the second variable. Most importantly, when the shell function executes, it returns a unique variant (or double) variable number, which is called the task ID. The task ID is a unique number used to reference the started application. It should therefore be stored to a global variable for easy reference in all modules; in this case the variable is named CalcTask.

2.7 THE IMPORTANCE OF "IDLE TIME"

Unfortunately, the shell function runs other programs in an asynchronous manner, meaning that the shell function may be called upon to start another program, but before that program is up and ready to receive commands, the statements following the shell command can be executed. If the statements immediately following the shell function should reference the started application (highly likely), and the application is not ready to service them, an error will result. For this reason, an idling mechanism should be called following the shell function, which will pause the program until the application executed by the shell statement is up and running.

Further complicating matters is the fact that the amount of time the program will need to idle following execution of the shell function can vary widely depending upon such things as processor speed, available memory, processor utilization, etc. Therefore, it is best to be conservative and allow more than enough time for the program to start and enter a "ready" condition. How much time is enough? One way to find out is to measure the amount of time it takes the program to open up on the slowest machine upon which the program will be run and then to add a safety factor to this of a few seconds.

Another complication is the fact that the program not only has to pause, but must also allow background tasks to continue processing so they will be complete once the idling period is finished. Fortunately, a built-in function called DoEvents does that. The following subroutine can be used to idle a macro and allow background processes to continue:

```
Public Sub Idle_Time(ByVal nSecond As Single)

Dim t0 As Single, tn As Single

Dim dummy As Integer, temp$

t0 = Timer

Do While Timer - t0 < nSecond

dummy = DoEvents()

    'If we cross Midnight, back up one day

    If Timer < t0 Then

        t0 = t0 - CLng(24) * CLng(60) * CLng(60)

    End If

Loop

tn = Timer

End Sub
```

The value passed to the Idle_Time subroutine is the number of seconds the program is to idle and can be expressed in decimals. A half second (0.5) is excellent to allow a window to be redrawn after some processing has occurred. When using the shell function, chances are the program will need to be paused for several seconds, maybe even longer. When the Agilent Chemstation software is loaded, it is necessary to allow a half a minute (30 seconds) on slower machines for it to load up fully.

The reason idling periods are necessary is lack of synchronicity. When two events (programs) are synchronous, they have events that occur on the rising edge of the same clock. This is very easily accomplished in hardware where one master system clock is used on a board, and other clocks are derived from the master clock. In software, however, the rate at which events occur can depend upon many factors, so programs are rarely in synch with each other.

Synchronicity can be thought of in terms of runners. Rarely can two people run together side by side for any length of time without one person holding the other up. A synchronous process is like two runners who keep a perfect stride with each other throughout a run. From an engineering standpoint, synchronicity is desirable because each process knows at what point of execution the other process is. In hardware it is easy to design processes to occur in a synchronous manner. However, processes must be coerced into being synchronous in software, often at the expense of holding another process up; this is why software for the most part functions in an asynchronous manner. From a design standpoint, the speed differentials inherent in asynchronous systems must be accounted for when constructing any software process that utilizes multiple programs.

2.8 THE MAGIC OF THE "SENDKEYS" STATEMENT

One of the most frustrating things about off the shelf software is the lack of customization available to the end user. It seems that every end user always wants to accomplish something above and beyond the software's capability; oftentimes such a capability will lead to a great competitive advantage. Unfortunately, it is very difficult to get a company to customize software for a user. Even if the user is from a large corporation with the financial prowess to pay their programmers to make the modifications, often the companies are not interested in doing this unless: (1) the capability will be utilized by a variety of their clients or (2) the company's programmers can spare the time from other tasks that so desperately require their attention. Not to mention that, it will be impossible to keep the modification secret should it be disclosed to another entity (no matter how many confidentiality agreements are signed).

However, almost all programs have "shortcut keys" for every menu item. For example in many applications, pressing Alt+F+S will save the current document. Even if the application being worked with does not have shortcut keys, they can be added very easily by the owner of the software. (It will take a programmer less than an hour to add the shortcut keys and recompile the program.) Most companies are willing to do this for a fee because it will not place an undue burden on their resources, as writing custom software for a client would. The significance of having shortcut keys in an application is that if a process can be accomplished by means of keystrokes (rather than using the mouse), this process can be duplicated by using the SendKeys statement from code.

Syntax:

```
SendKeys string[, wait]
```

String (required) — string expression specifying the keystrokes to send.
Wait (optional) — Boolean value specifying the wait mode. If False (default), control is
    returned to the procedure immediately after the keys are sent. If True, keystrokes must be
    processed before control is returned to the procedure. Best programming practice is to use
    True.

TABLE 2.2
Codes for Sending Special Key
Characters That Do Not Require {}

Key	Code
SHIFT	+
CTRL	^
ALT	%

Not only can regular keyboard characters be sent via the SendKeys command but special characters such as CTRL and ALT can be sent as well.

To send a special character it must be enclosed in curly cue braces {}. The only special characters that do not need to be enclosed in curly cue braces are those listed in Table 2.2. For example, to send ALT, it is only necessary to send the string "%". To send the actual characters "+", "^", or "%", it is necessary to enclose them in curly cue brackets like this: {+}, {^}, {%}.

To send curly cue braces enclose them in curly cue braces: {{} sends {. To specify that any combination of SHIFT, CTRL, and ALT should be held down while several other keys are pressed, enclose the code for those keys in parentheses. For example, to specify to hold down ALT key while F and S keys are pressed, use "+(FS)". To specify to hold down SHIFT while A is pressed, followed by B without SHIFT, use "+AB". It is also possible to specify that a key be pressed several times using the form {key number}. A space must be present between the key and the number. For example, {B 10} means press the B key 10 times. Table 2.3 lists additional special key codes.

TABLE 2.3
Additional Special Key Codes That Require {}

Key	Code
Backspace	(BACKSPACE), (BS), or (BKSP)
Break	(BREAK)
Caps Lock	(CAPSLOCK)
Del or Delete	(DELETE) or (DEL)
Down Arrow	(DOWN)
End	(END)
Enter	(ENTER)
Esc	(ESC)
Help	(HELP)
Home	(HOME)
Ins or Insert	(INSERT) or (INS)
Left Arrow	(LEFT)
Num Lock	(NUMLOCK)
Page Down	(PGDN)
Page Up	(PGUP)
Print Screen	(PRTSC)
Right Arrow	(RIGHT)
Scroll Lock	(SCROLLLOCK)
Tab	(TAB)
Up Arrow	(UP)

Keystrokes are being sent — but where? Now the task ID returned by the shell statement that started the program comes into action. By using the AppActivate statement, the macro will specify what application has focus, or to which application the commands issued by SendKeys will be sent:

```
AppActivate title[, wait]
```

Title (required) — string expression specifying the title in the title bar of the application to be activated. The task ID returned by the Shell function can be used in place of title to activate an application (this method is preferable as well).

Wait (optional) — Boolean value specifying whether the calling application has the focus before activating another. If False (default), the specified application is immediately activated, even if the calling application does not have the focus. If True, the calling application waits until it has the focus, then activates the specified application. Best programming practice is to use True.

The following example macros will start the calculator application, calculate the value of 3*9, and copy the result to the clipboard:

```
Sub MultExamp()

'A simple example of how to multiply 3*9 remotely

'Start the Calculator Application

CalcTask = Task_ID("c:\winnt\system32\calc.exe")

'Asynchronous - Wait for Application to Open!

Call Idle_Time(5)

'Call Attention to Calculator Program

AppActivate CalcTask

'Send Commands to Multiply 3*9

SendKeys "3*9=", True

'Now Copy the Result to the Clipboard

Call CopyResult

End Sub

Public Sub CopyResult()

'Subroutine Copies the Current Value in the Calculator Display

'Call Attention to Calculator Program

AppActivate CalcTask

'Send the String Alt+E+C

SendKeys "%(EC)", True

End Sub
```

Notice that the preceding example gives the calculator application plenty of time (5 seconds) to open and get ready to receive commands. Once the results have been calculated, the CopyResult subroutine copies the results to the Windows clipboard. From there they can be pasted into Excel, or any other application for that matter. In fact, several applications could be opened using shell (each with a unique task ID) and each could calculate a value and pass it to the next application, until the final result is calculated. In this way, anything that can be calculated in an application other than Excel can be accessible to Excel by harnessing the calculating power of that application remotely. Although this application is trivial, the methodology employed could be harnessed to calculate everything from the strength of ionic bonds to the stiffness of breast implants, the results of which can all be returned to Excel.

2.9 EXCHANGING DATA BETWEEN EXCEL AND OTHER APPLICATIONS

A typical application harnessing the SendKeys statement would involve copying numbers from various cells in an Excel worksheet and pasting them into appropriate textboxes in another application. The remote application could then calculate a value that would be copied and pasted into Excel. This process will be demonstrated using the calculator program that comes with Windows. The following macro will construct a workbook that has a multiplication table for the integer 5:

```
Sub CreateTimesTable()

Dim ii As Integer

'Start the Calculator Application

CalcTask = Task_ID("c:\winnt\system32\calc.exe")

'Asynchronous - Wait for Application to Open!

Call Idle_Time(5)

'Add a new Workbook

Workbooks.Add

'Create 5X Times Table

For ii = 1 To 12

    'Fixed Labels

    Worksheets(1).Cells(ii, 1).Value = "5"

    Worksheets(1).Cells(ii, 2).Value = "X"

    Worksheets(1).Cells(ii, 4).Value = "="

    'Variable Label

    Worksheets(1).Cells(ii, 3).Value = ii

Next ii

'Make Pretty

Columns("A:A").EntireColumn.AutoFit

Columns("B:B").EntireColumn.AutoFit

Columns("C:C").EntireColumn.AutoFit

Columns("D:D").EntireColumn.AutoFit

'Now Use Calculator Program to Get Results
```

```
'Make the Calculator the "Active" Application
AppActivate CalcTask
For ii = 1 To 12
    'Copy 1st col (always "5")
    Range(Cells(ii, 1), Cells(ii, 1)).Select
    Selection.Copy
    'Now Paste in Calculator window
    SendKeys "%(EP)", True
    Call Idle_Time(0.1)
    'Hit * Button
    SendKeys "*", True
    Call Idle_Time(0.1)
    'Copy 3rd col (varies {1-12})
    Range(Cells(ii, 3), Cells(ii, 3)).Select
    Selection.Copy
    'Now Paste in Calculator window
    SendKeys "%(EP)", True
    Call Idle_Time(0.1)
    'Hit * Button
    SendKeys " = ", True
    Call Idle_Time(0.1)
    'Now Copy Results from Calculator
    SendKeys "%(EC)," True
    Call Idle_Time(0.1)
    'Now Paste Results into Excel Spreadsheet
    Range(Cells(ii, 5), Cells(ii, 5)).Select
    ActiveSheet.Paste
Next ii
'Make Results Pretty
Columns("E:E").EntireColumn.AutoFit
'Close Calculator Alt+F4
SendKeys "%{F4}", True
End Sub
```

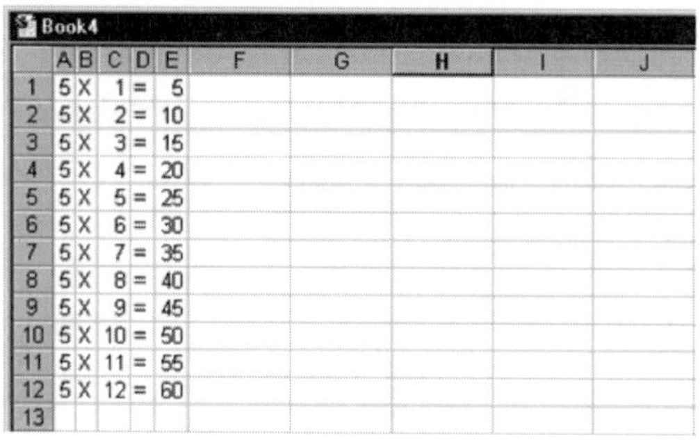

FIGURE 2.18 Worksheet produced by the times table macro.

This particular script has several important features. The Idle_Time statement is immediately called after each SendKeys statement providing an interval for the remote application to process the commands sent to it. Also note that the application could very easily pass the values in the times table directly to the calculator but instead chooses to copy them from the worksheet. While in this application a simple multiplication is being created, in reality an application of sophistication would be calculating results from unevenly spaced nonintegers, which could not be reproduced by means of a loop. Last, notice that when the results are ready to be pasted into Excel, the range is chosen but the command "ActiveSheet.Paste" is used rather than "Selection.Paste". This process is counterintuitive because "Selection.Copy" is used to copy the contents within a cell. However, the application is pasting to the ActiveSheet, not to the selection. The selection merely *points* to where the information will be copied to on the worksheet.

In this instance, it is very easy to paste the data into the calculator program because it has only one textbox. Rarely will this be the case. Many times the user will have to open a form and select from a number of textboxes in order to place data in the right area. Most often, when multiple objects coexist on a form, focus can be switched among objects using the tab key. For example, it may be necessary to open a form and hit the tab key 12 times to activate the textbox that data should be placed in. This can be done with the command: SendKeys {TAB 12}, True — which will push the tab button 12 times.

The previous application can be run by selecting Automation->Section 2->Times Table from the menu. When finished, the application will produce a worksheet like that in Figure 2.18, which utilizes the calculator to calculate the values (see Table 2.4).

2.10 PROGRAMMING FUNCTIONS INTO WORKBOOK CELLS

Although the times table macro produces an accurate report, it does have one major deficiency. If a user should change a value within the worksheet, the values within the worksheet are not recalculated based on the changed value. This is because the values are calculated within the macro, and only their values are written into the worksheet cells. While for some applications this may be sufficient, in many instances the worksheet produced by the macro may need to have its values adjusted, and cells dependent upon the adjusted values should be automatically updated.

For example, in the times table macro, if the user changes cell C2 from "2" to "2.5" in the final worksheet, row 2 will look like "5 × 2.5 = 10." This is, of course, incorrect and, because cell E2 is a value and not a function (or formula), this cell will not be updated to reflect the new equation the user adjusted by changing cell C2 from "2" to "2.5."

Whenever a report is created, it is always preferable to program formulas into cells if the values contained within those cells are dependent upon other values in the worksheet. Should the final worksheet then undergo any change, the worksheet will automatically be updated to reflect any changes that the new value added.

TABLE 2.4
Keyboard Equivalents for the Windows Calculator Program

Button = Key	Button = Key	Button = Key
% **%**	Back **BACKSP**	Hyp **h**
(**(**	Bin **F8**	In **n**
) **)**	Byte **F4**	Int **;**
* *****	C **ESC**	Inv **i**
+ **+**	CE **DEL**	log **l**
+/- **F9**	cos **o**	Lsh **<**
- **-**	Dat **INS**	M+ **CTRL+P**
.. **or ,**	Dec **F6**	MC **CTRL+L**
/ **/**	Deg **F2**	Mod **%**
0-9 **0-9**	dms **m**	MR **CTRL+R**
1/x **r**	Dword **F2**	MS **CTRL+M**
= **ENTER**	Exp **x**	n! **!**
A-F **A-F**	F-E **v**	Not **~**
And **&**	Grad **F4**	Oct **F7**
Ave **CTRL+A**	Hex **F5**	Or **|**

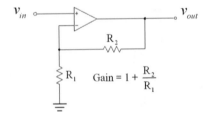

FIGURE 2.19 A noninverting operational amplifier.

To illustrate this point, a macro will be created to calculate the gain and the thermal noise for a noninverting amplifier (see Figure 2.19).
The gain for such a system is:

$$Gain = 1 + (R_2/R_1)$$

The thermal or Johnson noise for such a system would be:

$$e_n^2 = 4k(R_1 + R_2)T(f_2 - f_1)$$

where:

k = Boltzmann's constant (1.38E − 23)
T = resistor temperature in °K (300 K)
$(f_2 - f_1)$ = bandwidth in Hertz (50 kHz)

The following macro will create a worksheet that will calculate the gain and the noise for a noninverting amplifier. The macro will prompt the user to write functions into the worksheet cells or to write values into the worksheet cells. The difference is that if functions are written into the worksheet cells and the user should change the values of R_1 or R_2 at the conclusion of the macro, the values for the gain and the noise will be updated automatically within the worksheet. If the values are written to the worksheet, the gain and the noise values will need to be recalculated and reentered by the user if R_1 or R_2 are changed in the worksheet. This macro can be run by choosing Automation->Section 2-> Cell Funtions from the menu bar. The code for this macro is

```
Sub CellFunc()
Dim ii As Integer, retval As Integer
'Add a new Workbook
Workbooks.Add
'Create Labels
    Worksheets(1).Cells(1, 1).Value = "R1 (K)"
    Worksheets(1).Cells(1, 2).Value = "R2 (K)"
    Worksheets(1).Cells(1, 3).Value = "Gain"
    Worksheets(1).Cells(1, 4).Value = "Noise"
'Populate Resistance Values
For ii = 10 To 50 Step 10
    Worksheets(1).Cells(1 + ii/10, 1).Value = "5"
    Worksheets(1).Cells(1 + ii/10, 2).Value = ii
Next ii
retval = msgbox("Use Functions", vbQuestion + vbYesNo, "Cell
Calculation Method")
If retval = vbYes Then
    'Assign Functions to Cells to Make Worksheet Calculations
    Call FuncCalc
    Else
    'Calculate Values in Code and assign Values to Cells
    Call HardCalc
End If
End Sub
Sub FuncCalc()
'Assign Functions to Cells to Make Worksheet Calculations
Dim ii As Integer
'Add Gain Formula
For ii = 2 To 6
    Worksheets(1).Cells(ii, 3).Formula = "=(1+B" &
    Trim(Str(ii)) & "/A" & Trim(Str(ii)) & ")"
```

```
Next ii

'Add Noise Formula

For ii = 2 To 6

    Worksheets(1).Cells(ii, 4).Formula = "=SQRT(4*1.38E-
    23*300*((A" & Trim(Str(ii)) & "+ B"_

& Trim(Str(ii)) & ")*1000)*50000)"

Next ii

End Sub

Sub HardCalc()

'Define Constants (for use in cell functions demo)

Const Boltz As Double = 1.38E-23, KelvTemp As Double = 300

Dim ii As Integer

'Place Gain VALUES ONLY into Worksheet Cells

For ii = 2 To 6

    Worksheets(1).Cells(ii, 3).Value = 1 +
    (Worksheets(1).Cells(ii, 2).Value/Worksheets(1).Cells(ii,
    1).Value)

Next ii

'Place Noise VALUES ONLY into Worksheet Cells

For ii = 2 To 6

    Worksheets(1).Cells(ii, 4).Value = ((4 * Boltz * KelvTemp
    * ((Worksheets(1).Cells(ii, 2).Value + _

    Worksheets(1).Cells(ii, 1).Value) * 1000)) * 50000) ^ 0.5

Next ii

End Sub
```

Looking at this code it should be pointed out how to program a function into a cell. Unfortunately, there is no direct way to write the formula in code. The formula must be written in as a string. This restriction is very limiting because it precludes using variables in the macro code, even such references to cells using the (.Cells(row,col)) property. This is because the macro code will not be able to distinguish between what is part of the formula and what is a macro variable. Furthermore, the cell will not be able to reference the macro variable once the macro has terminated; even if it does reference a cell value, the reference will be lost. When programming functions into cells, the functions must reference cells using the R1 and C1 type notation. This can be somewhat troublesome to program.

Looking at the code:

```
Worksheets(1).Cells(ii, 3).Formula = "=(1+B" & Trim(Str(ii))
& "/A" & Trim(Str(ii)) & ")"
```

Here the gain formula "= $1 + (R_2/R_1)$" (where R_1 is in cell A*n* and R_2 is in cell B*n*) is added programmatically to the appropriate cell. Notice that a cell can be referenced without using the

	A	B	C	D
1	R1 (K)	R2 (K)	Gain	Noise
2	5	10	3	3.52E-06
3	5	20	5	4.55E-06
4	5	30	7	5.38E-06
5	5	40	9	6.1E-06
6	5	50	11	6.75E-06

FIGURE 2.20 The cell function macro worksheet.

(.Cells(row,col)) property. Here the cell row is provided by means of the loop variable ii that is converted to a string and stripped of any leading spaces and added to the formula. A more complicated example that calculates the thermal noise using the formula $e_n = \sqrt{4k(R_1 + R_2)T(f_2 - f_1)}$, follows:

```
Worksheets(1).Cells(ii, 4).Formula = " = SQRT(4*1.38E-
23*300*((A" & Trim(Str(ii)) & "+ B"_ & Trim(Str(ii)) &
")*1000)*50000)"
```

Also notice that the equal sign must be the first thing in the string that contains the formula. For a really long formula, an individual might want to break the formula up into several substrings, and assemble it into the final formula string, to which the .Formula property can be set equal. The previous application can be run by selecting Automation->Section 2->Cell function from the menu and the code will produce a worksheet that looks like the one depicted in Figure 2.20.

At some point in time it may become necessary to create a very sophisticated report with multiple worksheets. Although each worksheet would be its own entity, there could be interdependence among worksheets as they may well share some of the same parameters. Ideally, when an individual alters a parameter, the change should be reflected in all reports that utilize this parameter. If the user must go in and edit the reports by hand, the process is prone to errors because the user might easily miss an individual instance of the parameter that needs to be changed. The solution to this problem is to have a "Master" worksheet that has all of the parameters to be utilized in the report. If a value on the master sheet is changed, the change will then propagate to all other reports in the workbook.

This is accomplished by setting up a master worksheet whereby each column represents a single parameter that will be utilized in the other report worksheets. Typically, the first row of the column will contain a header ID that will identify the parameter stored in row two of the column. Conversely, the master worksheet could also be set up as rows, where each row represents a single parameter with the first column containing the parameter headers. The example files in this text use the former column method.

To extract a value from the master worksheet, a report worksheet must use a formula that references the cell of the master worksheet containing the parameter to be extracted. An example of such a formula reference would be:

```
=Master!B3
```

Here the value of cell B3 in the worksheet named "Master" is transferred to the cell of any worksheet that has this formula attached to it. The problem with this formula is that, if the cell (in this case B3) has nothing in it, a zero will be written into the cell containing the formula. To get around this, an "if" statement must be used in the formula. An example of how to accomplish this is:

```
IF(Master!B2<>"",Master!B2,"")
```

Here, if cell B2 in the master worksheet contains anything, the contents of cell B2 in the master worksheet will be written into the cell that has this formula reference. However, if cell B2 in the master worksheet contains nothing, the cell that has this formula reference will be left blank. This is extremely important because a null string "" and a 0 are not equivalent expressions. A zero present where a null string should be will cause errors such as division by 0, etc.

An example of this can be run by selecting Automation->Section 2-> Master Worksheet from the menu bar. This macro will create two worksheets, master and report. The master worksheet contains two resistor values, R_1 and R_2. Whenever the values of R_1 or R_2 are changed on the master worksheet, these changes are instantaneously made on the report worksheet as well. In addition, the gain and noise calculations are updated to reflect the new values of R_1 and R_2 on the master worksheet. The important thing to keep in mind here is that if 10 reports are in a workbook, this is a method of updating all reports at once, as opposed to a piecemeal operation. The code for this example follows:

```
Sub MasterExample()

'How to Create a Master Value Worksheet which propagates changes to all

'subsequent report worksheets

Dim ii As Integer, retval As Integer

'Add a new Workbook

Workbooks.Add

'Name "Master" Worksheet

Worksheets(1).Name = "Master"

'Create Labels

    Worksheets(1).Cells(1, 1).Value = "R1 (K)"

    Worksheets(1).Cells(1, 2).Value = "R2 (K)"

'Populate 2 Resistance Values

For ii = 10 To 10

    Worksheets(1).Cells(1 + ii/10, 1).Value = "5"

    Worksheets(1).Cells(1 + ii/10, 2).Value = ii

Next ii

'Now Create Report Based on Values in Master

Worksheets(2).Name = "Report"

'Create Labels

    Worksheets(2).Cells(1, 1).Value = "R1 (K)"

    Worksheets(2).Cells(1, 2).Value = "R2 (K)"

    Worksheets(2).Cells(1, 3).Value = "Gain"

    Worksheets(2).Cells(1, 4).Value = "Noise"

'Extract Resistance Parameters from MASTER WORKSHEET

'Add Resistances from Master Template, if cell = "" then write nothing, not zero

Worksheets(2).Cells(2, 1) = BuildWBFunction(2, 1)

Worksheets(2).Cells(2, 2) = BuildWBFunction(2, 2)
```

```
'Calculate Gain using Formula

Worksheets(2).Cells(2, 3).Formula = "=(1+B" & Trim(Str(2))
& "/A" & Trim(Str(2)) & ")"

'Calculate Noise using Formula

Worksheets(2).Cells(2, 4).Formula = "=SQRT(4*1.38E-23*300*((A"
& Trim(Str(2)) & "+ B"_

& Trim(Str(2)) & ")*1000)*50000)"

End Sub

Function BuildWBFunction(Row, Col) As String

'Build the Appropriate Worksheet Function to Replicate Data

BuildWBFunction = "=IF(Master!" & Trim$(Chr(64 + Col)) &
Trim$(Str(Row)) & "<>" & Chr(34) & Chr(34) &_

",Master!" & Trim$(Chr(64 + Col)) & Trim$(Str(Row)) & "," &
Chr(34) & Chr(34) & ")"

End Function
```

2.11 ADDING DROP DOWN BOXES TO WORKSHEET CELLS

Very often it is desirable to have a set of standardized choices for a user to pick from when creating a report worksheet. This is especially true when the report being generated will be uploaded to a corporate database. For example, if a report has a comments field for running a method using an HPLC, it would be prudent to have a set of standard comments such as: "No Peaks Detected," "Multiple Peaks Detected", and "Not Amenable to Measurement".

Having the user type such comments in by hand is a bad idea because he will invariably fail to word the comments consistently. If the comments are not uniform throughout all reports uploaded to the database, then the comments field cannot be effectively queried. For instance, at some point in time a user might wish to see how many compounds were assayed in which multiple peaks were detected when run on a particular HPLC. Such a query would not be effective if the comments fields are not identical across the playing field.

The solution to such a problem is to create a report in which a user can select a choice from a set of standard fields located within a drop down box. Such a method will protect the consistency of the report.

Suppose that in the gain and noise example given earlier, one additional column was added to the end of the worksheet and called "Application." In the application column it is desired to have the name of the device in which the amplifier is to be utilized, for example, spectraphotometer, nephelometer, diode array, turbidimeter, or electrode.

First, a line of code would be added to create an additional column header called "Application":

```
Worksheets(1).Cells(1, 5).Value = "Application"
```

Next, a subroutine would be developed to add the appropriate comments to a specific range of cells. The following subroutine will accomplish this nicely:

```
Sub Add_Comments(cellrng$)

'Adds Drop Down List of Comments for the cells passed in
cellrng$
```

```
Dim Cmt(10) As String, ii As Integer, CmtList$
'Create a List of "Standard" Comments
Cmt(1) = "spectraphotometer"
Cmt(2) = "nephelometer"
Cmt(3) = "diode array"
Cmt(4) = "turbidimeter"
Cmt(5) = "electrode"
For ii = 1 To UBound(Cmt())
    If ii < UBound(Cmt()) Then
        CmtList$ = CmtList$ & Cmt(ii) & Chr(44)
    Else
        CmtList$ = CmtList$ & Cmt(ii)
    End If
Next ii
'Select the Range in which the List is to appear
Range(cellrng$).Select
    Selection.HorizontalAlignment = xlLeft
        With Selection.Validation
        .Delete
        .Add Type:=xlValidateList, AlertStyle:=
        xlValidAlertStop, Operator:=_
        xlBetween, Formula1:=CmtList$
        .IgnoreBlank = True
        .InCellDropdown = True
        .ShowInput = True
        '.ShowError = True
        End With
  End Sub
```

Notice in the preceding subroutine that the comments an individual wishes to add are stored in the array Cmt, which is set to have 11 array elements (0 to 10) by default. Additional comments can be hard coded into the subroutine or can even be read in by means of an *.ini file or through use of the Windows registry (covered in another section).

   To use the subroutine, an individual must pass a range of cells in which the comment list is to be added. The following snippet of code can be added to any subroutine to accomplish this:

```
'Add a Drop Down List of Potential Applications which the user
may Select

cellrange$ = cpt$ & Trim$(Str(startpt)) & ":" & cpt$ &
Trim$(Str(endpt))
```

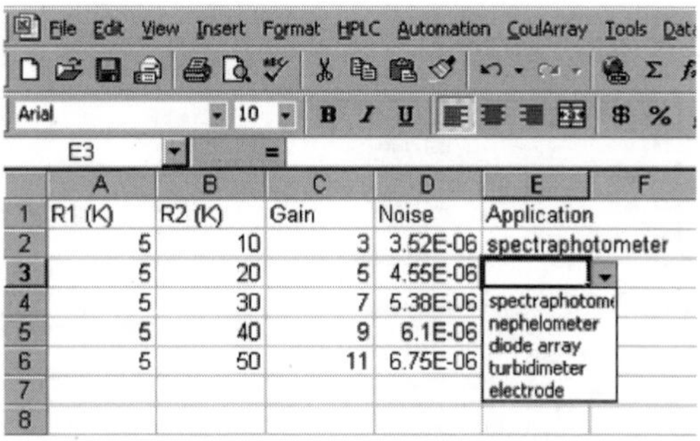

FIGURE 2.21 Generating a drop down list within a range of cells in a report.

```
cellrange$ = "E" & Trim$(Str(2)) & ":" & "E" & Trim$(Str(6))

Call Add_Comments(cellrange$)
```

While in this instance the range is hard coded into the variable cellrange$, notice the commented statement above it. This illustrates how to create a range using variables (in this case cpt$, startpt, endpt), which can be extremely useful in cases where a particular range might be defined by elements of a loop or bounded by particular variables. The ability to create such a range on demand using system variables adds great flexibility to the procedure.

The revised application can be run by selecting Automation->Section 2->Cell Function Cmmt from the menu. The macro will produce a worksheet that looks like the one shown in Figure 2.21.

2.12 OPENING FILES IN OTHER APPLICATIONS USING MACROS

Once it is known how to manipulate applications using Excel macros, the next logical pitfall is that, to make use of application X, data file Y must first be loaded. To the novice this may seem like an insurmountable task, but in most cases it is not that difficult. The following example code opens the notepad.exe application and automatically loads the setuplog.txt file present in Windows NT. If the system used to run the macro is not Windows NT, the file and/or path will need to be changed.

```
Sub FileLoader()

'Demo on how to Load Files into Notepad Using Macros

Dim fullpath$

'Start the Calculator Application

CalcTask = Task_ID("c:\winnt\system32\notepad.exe")

'Open File Box

SendKeys "%(FO)", True

Call Idle_Time(0.1)

'If system is Win NT this file will be present

'Can be changed to any filename & path !

fullpath$ = "c:\winnt\setuplog.txt"

SendKeys fullpath$, True

Call Idle_Time(0.1)

'Hit Return to Load File

SendKeys "{ENTER}", True
```

FIGURE 2.22 The file open dialog box in notepad.exe — tab order marked with numbers in bold.

```
Call Idle_Time(0.1)

End Sub
```

One thing that the programmer should pay special attention to is the File Open dialog box. Focus is first given to the FileName textbox. However, by striking the tab key, focus is moved to other objects in the form in the following order: Filename (textbox) -> FileType (drop down listbox) -> Open (button) -> Cancel (button) -> Look In (drop down listbox) -> Folder (listbox).

Also notice that if the return key is struck after entering a filename in the filename textbox, the file will then load without pressing the open button. Further notice that in the folder listbox, when the up or down arrows are pressed, files will be selected relative to their position in the listbox. Also note that when a file is selected, its name appears in the filename textbox.

These features are important because a loop can be written to open all or any datafiles in a particular directory. Upon each rotation of the loop, the down arrow would be pressed and the name of that file would appear in the file name textbox. The values that appear in this box could be monitored. When these values begin to repeat, all the files in that particular directory have been opened. This is illustrated in Figure 2.22.

The bottom line is that by manually playing around with different dialog boxes, it is possible to discern which keystrokes produce what results. These keystrokes can then be sent via a macro to produce the results desired. Often it is possible to automate many tasks that the designers of the program never intended to have automated.

2.13 INTERWORKBOOK RECOGNITION OF VARIABLES

A variable declared public has the broadest scope of all variables; it is recognized by every module in the active workbook. They also retain their last value until Excel is shut down. But what happens when workbook A runs a macro that should have access to a public variable in workbook B? Unfortunately, public declarations of variables do not allow access to those variables via other workbooks without the use of something called a reference. Boundaries have been put in place between workbooks that do not allow workbooks to gain access to other workbooks' public variables, unless they are given explicit permission to do so. This was probably a smart thing to do because, if many workbooks have been created, each with unique code, the likelihood of their having some public variables with the same name is fairly high.

Creating a reference to another workbook is not difficult; however, one major pitfall is waiting to snare users of Excel 97. In Figure 2.23, there are quite a few projects, each of which is contained within a separate workbook. The workbook's filename is in (). The problem arises in that for every new project started in Excel, a default name of "VBAProject" is provided. If one workbook (project) with the default name "VBAProject" tries to reference another workbook (project) with the default name "VBAProject", an error message will occur that says "Name Conflicts with Existing Module".

FIGURE 2.23 The project explorer window in Microsoft Excel VBA.

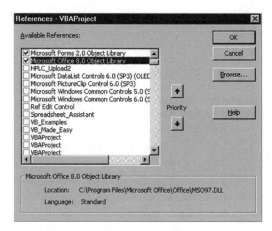

FIGURE 2.24 The references dialog box.

On one level, an individual would think that because the workbook filenames are different, that should be enough to differentiate between the different projects (Workbooks). Unfortunately, that is not the case. It is good programming practice always to rename a workbook's project name when beginning to program macros in a workbook. To change the default name of a project, go to: Tools->VBAProject Properties. From the general tab change project name to any name desired.

Once two projects have unique names, an individual can make a reference to another. This is done by selecting Tools->References while the workbook *from* which one wishes to make a reference is selected in the project explorer. From the dialog box the user then selects the workbook *to* which he would like to make a reference.

Notice in Figure 2.24 that numerous projects do not have a unique project name and are referenced as VBAProject. The difficulty arises when one project has no idea in which "VBAProject" it should be looking for the public variable.

2.14 WAITING FOR USER SELECTIONS

The question a person might be asking now is why a macro running in one workbook would need to know the status of a variable contained within a macro of another workbook. One very common

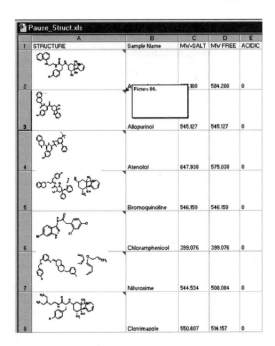

FIGURE 2.25 The Pause_Struct.xls file.

occurrence is for hidden workbook A, which contains the macro code, to wait for an event to occur in visible workbook B, in which the macro code is manipulating data. How can the hidden workbook A that is running the macro code determine if an event has transpired in visible workbook B? One very common way to accomplish this task is to have a variable that is recognized by both workbooks change value when a specific task has been completed. Such variables are referred to as *flags*, and will often be Boolean or integer in type. Flags will usually change from true to false to determine the passing of an event, or return an integer number to determine a type of selection.

The following program is an excellent example of how to allow users to select choices from various groupings of data. It can be run from the Automation->Section 2->Pick Demo. This macro will prompt the user for two files, the first of which is Pause_Struct.xls. This workbook, shown in Figure 2.25, contains the sample names along with structure jpgs, molecular weight, and acid information.

Notice the triangle at the upper right-hand corner of every cell with a structure jpg. This indicates that a comment is present for that cell. The comments text can be viewed by placing the mouse over this triangle as shown in the illustration. In this case the comment contains the name of the shape object, which contains the picture of the structure. This information is critical to have if the structures are to be manipulated and used between workbooks and worksheets.

The second file that the macro will prompt the user for is "Pause_pkas.xls". This file contains experimental pKa values for each sample. (Note: these are not real pKa values; they were all made up for the purpose of this example.) This file contains one or more rows of results for each sample present in the structure workbook. What would be ideal in this situation is to create a final listing of results. Samples with only one result would automatically have that sample placed in the list. Samples with more than one result would place all the results on a worksheet along with the sample name and structure and allow the scientist to pick the most reasonable result (given the structure) or reject all the results. This macro accomplishes exactly that by using a template file, "Pause_Temp.xlt".

The worksheet page shown in Figure 2.26 is from the "Pause_Temp.xlt" template. The first row contains the sample name. The second row contains labels for up to three pKa values. Rows

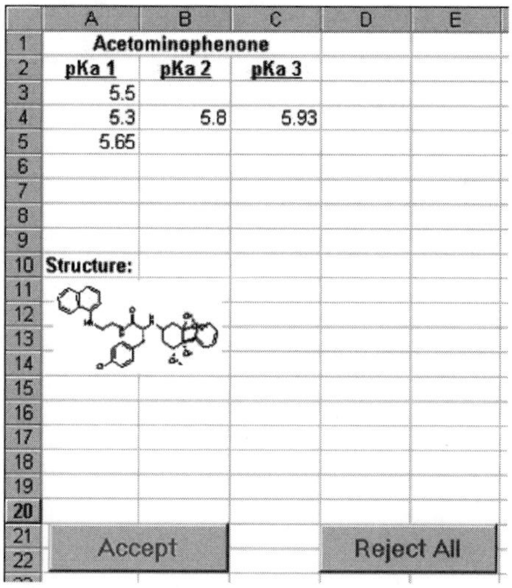

FIGURE 2.26 Template worksheet page allows a scientist to select multiple results (across rows).

3 to 9 contain the results loaded from the "Pause_pkas.xls" file. To select the results to be accepted, the user clicks on the row that they are located in and then presses the "Accept" button. If no reasonable results are found within the data, the user can press the "Reject All" button. The sample name will be written to the upload file with no pKa results but a comment will be inserted that says "Not Amenable to pKa Analysis." Two sheets are on the template workbook; "Select" is the sheet pictured in Figure 2.26 that allows selection of multiple results. The second worksheet, "FinalList", is a compilation of all the selected results, which will be uploaded to a database.

To create an application such as this, many determinations need to be made and acted upon. The number of results must be counted. Multiple results and their corresponding structures must be moved to the selection worksheet. Once a selection has been made, the selection worksheet must be "cleaned up"; that is, the old structure and results must be deleted and made ready for the next compound with multiple results. The simple list in Figure 2.26 seems trivial in nature. However, in practice unforeseen complications will always increase the level of difficulty involved in putting an idea into practice.

Subroutines and functions that make this application work will be discussed at length later as well as potential pitfalls, which can bring any project to a grinding halt. All the code snippets for this section are located in module 3 in the AdvMacTech.xls Workbook that can be downloaded from http://www.pharmalabauto.com.

The following two variables (WaitForSelect and pKaSelectRow) are available to the template workbook (Pause_Temp.xlt) because it makes a reference to the workbook with the macro code in it (AdvMacTech.xls). The template is saved with this reference as well as buttons with attached code in the "Select" worksheet that set the values of the two public variables. The WaitForSelect variable tells when a selection has been made. The pKaSelectRow is set to the row in which the data should be extracted.

```
'Intra Workbook Variables (Templates Reference this Workbook)

Public WaitForSelect As Integer

Public pKaSelectRow As Integer
```

This macro is called when the Pick Demo menu item is selected:

```
Sub Pause_Demo()
```

```
'Suppress Excel Warning Dialog Boxes

Application.DisplayAlerts = False

'Load Structure Worksheet

Call Load_StructData

'Load pKa Data

Call Load_pKaData

'Load Template to Choose Data

Call OpenPauseTemplate

'Choose Data

Call Data_Selector

End Sub
```

The subroutines used to load data file workbooks (Load_StructData, Load_pKaData) are not listed because they are nearly identical to the data loading subroutines covered in Chapter 1. (Their code can, however, be viewed in module 3.)

The Data_Selector subroutine is the heart of this application. It is here that the user is allowed to select from multiple data results. Of particular interest in this subroutine is the code snippet:

```
Do

    If WaitForSelect <> 0 Then

        Exit Do

    End If

    DoEvents

Loop
```

This is the mechanism that allows the program to wait for user input. Here the program enters into a loop that will be infinite if the WaitForSelect variable does not change. However, the WaitForSelect variable can be altered by *both* the template workbook and the hidden workbook with the code (see Figure 2.27) because the template workbook has a *reference* to the hidden

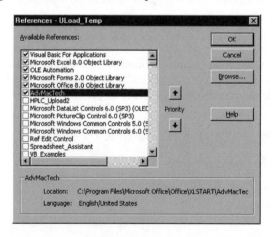

FIGURE 2.27 This reference makes it possible for the two variables to be recognized by both workbooks.

workbook. Then, when the user selects a row and pushes a button on the "Select" worksheet to finalize the selection, the template Workbook tests the validity of the selection made. If the selection is valid, the value of the WaitForSelect variable is changed (to anything but 0), which allows the program to exit the loop. The WaitForSelect variable can be set to different integer values to differentiate which action should be taken upon termination of the "Do" loop. Typically a Select Case statement would follow the "Do" loop to implement the course of action to be taken:

```
Sub Data_Selector()

Dim ii As Integer, TRows As Integer

Dim prev$, this$

Dim line_no As Integer, SR As Integer, ER As Integer

Dim startpt As Integer, endpt As Integer

Dim view As Boolean, entries As Integer, dummy As Integer

Dim rowptr As Integer, msg$, temp$, PictureName$

Call ActivateWorkbook(pkaWrkBk)

TRows = TotalRows(pkaWrkBk, ActiveSheet.Name)

rowptr = 1

For ii = 2 To TRows

    'Reset Intra Workbook Variables

    pKaSelectRow = 0: WaitForSelect = 0

    'Reset View

    view = False

    Select Case ii

        Case 2

            startpt = ii

            prev$ = Workbooks(pkaWrkBk).Worksheets(1).Cells(ii,
            1).Value

        Case 3 To TRows - 1

            'The Current cell is this$ value

            this$ = Workbooks(pkaWrkBk).Worksheets(1).Cells(ii,
            1).Value

            If this$ <> prev$ Then

                'End of Results for Compound X

                endpt = ii - 1 'End Point Marker for this stream

                entries = endpt - startpt 'Differential of Rows
                for this Stream

                'Save Row Range of Possible Selections before
                reassigning Start Point
```

```
            SR = startpt: ER = endpt

            startpt = ii 'Starting Point for next (new) Stream

            prev$ = Workbooks(pkaWrkBk).Worksheets(1).
            Cells(ii, 1).Value

            view = True

        End If

    Case TRows

        'Last Compound will always be the end of the Results

        endpt = ii '- 1 'End Point Marker for this stream

        entries = endpt - startpt 'Differential of Rows for
        this Stream

        startpt = ii 'Starting Point for next (new) Stream

        view = True

End Select

If view = True Then

    'All the results for Compound X have been tabulated

    rowptr = rowptr + 1 'Increment Row Pointer

    Select Case entries

        Case 0

        'Only one Row of Data, Copy the single Result over

        'Same Compound - Write info to Selection Sheet

        'Write Compound Number

        Workbooks(pickWrkBk).Worksheets("FinalList").Cells
        (rowptr, 1) =_

        Workbooks(pkaWrkBk).Worksheets(1).Cells(ii - 1, 1)

        'Write pKa 1

        Workbooks(pickWrkBk).Worksheets("FinalList").Cells
        (rowptr, 2) = _

        Workbooks(pkaWrkBk).Worksheets(1).Cells(ii - 1, 2)

        'Write pKa 2

        Workbooks(pickWrkBk).Worksheets("FinalList").Cells
        (rowptr, 3) =_

        Workbooks(pkaWrkBk).Worksheets(1).Cells(ii - 1, 3)

        'Write pKa 3

        Workbooks(pickWrkBk).Worksheets("FinalList").Cells
        (rowptr, 4) =_
```

```
Workbooks(pkaWrkBk).Worksheets(1).Cells(ii - 1, 4)
Case Else
'Multiple Rows of Data - Must Manually Choose Result
'Paste Information Into Selection Worksheet
'Write Compound Number
Workbooks(pickWrkBk).Worksheets("Select").Cells(1, 1)
=_
Workbooks(pkaWrkBk).Worksheets(1).Cells(ii - 1, 1)
'Place Structure in Select Sheet
'Capture Image Name (in comments field)
PictureName$ = ImageName(Workbooks(pickWrkBk).
Worksheets("Select").Cells(1, 1))
Call Fetch_Structure(PictureName$)
Call ActivateWorkbook(pkaWrkBk)
Range(Cells(SR, 2), Cells(ER, 4)).Select
Selection.Copy·
'Paste Common Info to LArea
'Workbooks("HPLC_Macro").Activate
Call ActivateWorkbook(pickWrkBk)
Worksheets("Select").Select
Range(Cells(3, 1), Cells(3, 1)).Select
ActiveSheet.Paste
Rows(1).Select 'Ensure User Selects Row
Workbooks(pickWrkBk).Worksheets("Select").Select
    Call Idle_Time(0.05)
    Do
        If WaitForSelect <> 0 Then
            Exit Do
        End If
        DoEvents
    Loop
Select Case WaitForSelect
    Case 1
    'A pKa from the list was chosen - Paste row to
    final sheet
```

```
'Write Compound Number

Workbooks(pickWrkBk).Worksheets("FinalList").Cells
(rowptr, 1) =_

Workbooks(pickWrkBk).Worksheets("Select").Cells(1,
1)

'Write pKa 1

Workbooks(pickWrkBk).Worksheets("FinalList").Cells
(rowptr, 2) =_

Workbooks(pickWrkBk).Worksheets("Select").Cells(pK
aSelectRow, 1)

'Write pKa 2

Workbooks(pickWrkBk).Worksheets("FinalList").Cells
(rowptr, 3) =_

Workbooks(pickWrkBk).Worksheets("Select").Cells(pK
aSelectRow, 2)

'Write pKa 3

Workbooks(pickWrkBk).Worksheets("FinalList").Cells
(rowptr, 4) =_

Workbooks(pickWrkBk).Worksheets("Select").Cells(pK
aSelectRow, 3)

Case 2

'No viable pKa

'Write Compound Number

Workbooks(pickWrkBk).Worksheets("FinalList").Cells
(rowptr, 1) =_

Workbooks(pickWrkBk).Worksheets("Select").Cells(1,
1)

'Write in Comment

Workbooks(pickWrkBk).Worksheets("FinalList").Cells
(rowptr, 5) =_

"Not Amenable to pKa Measurements"

    End Select

End Select

'Clear cells of old pKa's

Call ActivateWorkbook(pickWrkBk)

Range("A3:C9").Select

Selection.ClearContents

'Remove Structure Image
```

```
        Call Remove_Structure

    End If

Next ii

Workbooks(pickWrkBk).Worksheets("FinalList").Select

End Sub
```

Copying the sample information to the "Select" worksheet is a fairly straightforward process. However, copying the image of the compound structure is not. The images are stored in shape objects and in order to copy them an individual must know the name of the particular image. Fortunately, the name of the image to be copied is stored as a comment within the cell in which the image resides. The following function returns the name of the image associated with a particular sample.

```
Function ImageName(sampname$) As String

'Returns the Name of the Image Associated with Sample X

Dim TRows As Integer, ii As Integer

Dim C As Range

TRows = TotalRows(structWrkBk, "Sheet1")

For ii = 2 To TRows

    If Workbooks(structWrkBk).Worksheets(1).Cells(ii, 2) =
    sampname$ Then

        Workbooks(structWrkBk).Worksheets(1).Cells(ii, 1).Select

        Set C = ActiveCell

        ImageName = C.Comment.Text

        Exit Function

    End If

Next ii

End Function
```

The comment can be extracted by using a variable set to a range and setting the range to a single active cell. Once the range is set to the cell containing the image, the comment can be extracted using the .Comment and .Text properties. Unfortunately not a great deal of documentation is available on how to use ranges and objects within Visual Basic. However, the AutoComplete feature of VB is a lifesaver because it lets the user know what properties and methods can be acted upon (see Figure 2.28).

The next subroutine actually copies and pastes the structure image into the "Select" worksheet. This is fairly straightforward. Once the name of the shape that contains the image is known, it can be selected and copied just like the contents of any cell. To determine placement of the image to be pasted select the cell where the upper left corner of the image is to start.

```
Sub Fetch_Structure(image$)

'Copy the Image associated with Sample X to the Selection

'Worksheet, its name is in the comments field of the cell
```

```
Function ImageName(sampname$) As String
'Returns the Name of the Image Associated with Sample X
Dim TRows As Integer, ii As Integer
Dim C As Range
TRows = TotalRows(structWrkBk, "Sheet1")
For ii = 2 To TRows
    If Workbooks(structWrkBk).Worksheets(1).Cells(ii, 2) = sampname$ Then
        Workbooks(structWrkBk).Worksheets(1).Cells(ii, 1).Select
        Set C = ActiveCell
        ImageName = C.Comment.Text
        C.
End
```

FIGURE 2.28 The autocomplete feature shows the properties and methods available for use with range and worksheet type variables that are poorly documented.

```
If image$ = "" Then Exit Sub

Call ActivateWorkbook(structWrkBk)

ActiveSheet.Shapes(image$).Select

Selection.Copy

Call ActivateWorkbook(pickWrkBk)

Sheets("Select").Activate

Range("A11").Select

ActiveSheet.Paste

End Sub
```

Unfortunately, removing the structure from the "Select" worksheet is not nearly as easy as adding it. For some reason, when the shape is pasted into this worksheet, it may not retain the same name from the worksheet from which it was copied. However, because only one picture (the structure image) is on the "Select" worksheet, the structure can be deleted by looping through every shape in the worksheet and deleting any that are pictures. If this snippet of code was not present:

```
If ws.Shapes.Item(ii).Name Like "*Picture*" Then

    ws.Shapes.Item(ii).Delete

    Exit Sub

End If
```

then every shape on the worksheet would be deleted — including the command buttons. This would make the selection worksheet useless for the next batch of samples from which a result needed to be chosen.

```
Sub Remove_Structure()

Dim ws As Worksheet

'When Copied Shapes Do Not retain their Original Image Name!
```

```
'Since there is only 1 image on worksheet find it and delete it!

For Each ws In Worksheets

    For ii = 1 To ws.Shapes.Count

        If ws.Shapes.Item(ii).Name = "" Then Exit Sub

        If ws.Shapes.Item(ii).Name Like "*Picture*" Then

            ws.Shapes.Item(ii).Delete

            Exit Sub

        End If

        'Debug.Print ii, ws.Shapes.Item(ii).Name

    Next ii

Next ws

End Sub
```

One property that was not utilized in this macro but has proven to be very valuable is the (.Row) property. Many times in files generated from relational databases an entire row will be dedicated to data relating to a single compound. In such instances it might be advantageous to have the ability to select and extract an entire row of data. This can be accomplished by using the command ActiveCell.Row property. This property will return an index to the row of the cell that the user has selected or has currently active. For example, if cell C7 is currently active, the following will be returned using the ActiveCell methods:

```
ActiveCell.Row = 7

ActiveCell.Col = 3

ActiveCell.Address = $C$7
```

2.15 THE END: CLOSING ALL NONHIDDEN WORKBOOKS

When a macro completes its analyses, and all tasks have been completed, a natural progression is to close all visible workbooks and leave Excel in a "clean state." The problem here is how to close only the visible workbooks. Chances are that any user who really harnesses the power of Excel will have a number of hidden workbooks in the background for running various macros. If the hidden workbooks are closed, then the Excel session will lose the capability of running those macros that are stored in the hidden workbooks. The following subroutine accomplishes this task quite well:

```
Sub CloseVisibleWorkbooks()

'Subroutine to Close ALL VISIBLE Workbooks

Dim NumberofVisibleWorkbooks As Integer

Dim bookCounter As Workbook

For Each bookCounter In Application.Workbooks

    If Windows(bookCounter.Name).Visible = True Then

        bookCounter.Close (False)
```

```
    End If

Next bookCounter

End Sub
```

Suppose the user would like to leave open a workbook called "LeaveOpen.xls" at the end of all operations. The following modification could be made, which would close all workbooks except "LeaveOpen.xls":

```
If Windows(bookCounter.Name).Visible = True And_

    bookCounter.Name <> "LeaveOpen.xls" Then

    bookCounter.Close (False)

End If
```

2.16 SAVING SETTINGS TO THE WINDOWS REGISTRY

In any program of any complexity there are usually settings that should be stored and reloaded in between executions of the programs. The first method developed to accomplish this task was the *.ini (for initialization) file. Once a program was started it would immediately read this file and adjust its default settings based on the data contained in the file. Initialization files were used quite extensively in older operating systems such as Windows 3.1. Although *.ini files are still used today under some circumstances, a better method for allowing programs to "remember" their settings has been developed.

The Windows Registry is essentially a large database of settings used for the Windows Operating System and the software programs that operate under its auspices. Writing and reading values from the registry are possible using Visual Basic; however, the consequences for writing an invalid value or writing a value to an improper place can be horrific. They typically range from taking down the operating system to the point where it will no longer function or boot, showing the "blue screen of death" when a program is run, or having the program run with an incorrect parameter setting.

For this reason, many programmers avoid utilizing the registry entirely, and choose to continue utilizing *.ini files. Utilizing the registry, however, gives programs a touch of professionalism, and is also inherently faster when large numbers of settings need to be stored.

This brings on the question of how to get to the registry. Windows has a registry editor program that allows a manual edit and view of individual settings within the registry. The program does not have an icon on the desktop or the start menu to run the program, however, because Microsoft does not want users playing with the system settings in the registry. This greatly reduces unnecessary technical support calls. The registry editor program can be run from c:\winnt\regedit.exe. This program is good for viewing values a program has written to the registry *after the fact*. The program is not utilized by Visual Basic to *write* the values to the registry.

Values are written to the Windows registry as "Keys," which are objects that have a name and a value. For example, a key might have a name of "TempFilePath" and a value of "c:\temp". Notice that the value does not need to be numeric; it can consist of text, numbers, and symbols. Keys are stored within the registry by application name and section. Figure 2.29 illustrates the hierarchy of stored keys within the Windows registry. Two key points to note from the figure: first, all settings programmatically saved from VB and VBA are stored under the folder "VB and VBA Programming Settings." Second, notice that the application name (AppName) and section are subfolders under the "VB and VBA Programming Settings" folder.

A value is saved to the Windows registry using the SaveSetting function:

```
SaveSetting AppName, Section, Key, Value
```

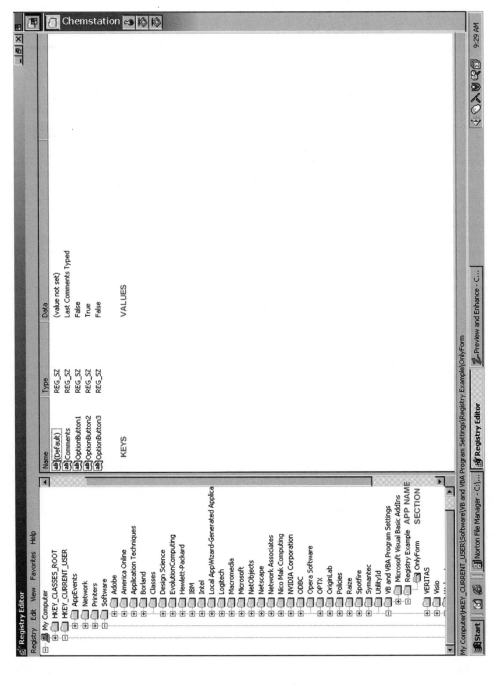

FIGURE 2.29 Key storage within the Windows system registry.

- AppName — the application name. The programmer creates this.
- Section — the folder under the application name. It might be advantageous to store the default values for each user form in a separate subfolder (section) under the application name (folder).
- Key —the name for the key. The programmer creates the name for the key; it is generally descriptive of the contents of the key.
- Value —the value held within the key (what the programmer wishes to store.) It can be numerical, text, symbolic, or any combination thereof.

A value can be read from the Windows registry using the GetSetting function:

```
GetSetting AppName, Section, Key
```

The parameters here are the same as for the SaveSetting function but the value is returned.

An application has been constructed to demonstrate proper usage of the registry to save user settings in a form. It can be run by selecting Automation->Section 2->Registry Example from the menu bar. The first time the application is run the user will see a message box like the one illustrated in Figure 2.30.

FIGURE 2.30 The first time the program is run this warning message will appear.

This warning message will appear the first time the program is run because no settings (keys) in the registry correspond to those the application is looking to load. The loading of the form parameters takes place in the UserForm_Initialize subroutine. Notice that an On Error statement is provided in the event that the registry information cannot be found.

```
Private Sub UserForm_Initialize()

On Error GoTo Reg_Err

RegistryExample!Comments.Text = GetSetting("Registry Example",
"OnlyForm", "Comments")

OptionButton1.Value = GetSetting("Registry Example",
"OnlyForm", "OptionButton1")

OptionButton2.Value = GetSetting("Registry Example",
"OnlyForm", "OptionButton2")

OptionButton3.Value = GetSetting("Registry Example",
"OnlyForm", "OptionButton3")

Exit Sub

Reg_Err:

msgbox "Errors Loading Registry Data!", vbOKOnly +
vbExclamation, "Registry Error"

Exit Sub

End Sub
```

Once the program has been run and the form has been closed, the last settings of the objects in the form will be stored to the registry. The UserForm_Terminate subroutine writes the values of all the objects in the form to the system registry. The next time the program is run and the form is loaded, the object settings are read from the registry and the objects are initialized based on the values read from the registry.

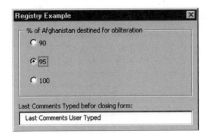

FIGURE 2.31 Example choices for the registry example macro.

```
Private Sub UserForm_Terminate()

SaveSetting "Registry Example", "OnlyForm", "Comments",
RegistryExample!Comments.Text

SaveSetting "Registry Example", "OnlyForm", "OptionButton1",
RegistryExample!OptionButton1.Value

SaveSetting "Registry Example", "OnlyForm", "OptionButton2",
RegistryExample!OptionButton2.Value

SaveSetting "Registry Example", "OnlyForm", "OptionButton3",
RegistryExample!OptionButton3.Value

End Sub
```

The objects saved and the values of the keys that store the object settings can easily be viewed using the registry editor (c:\winnt\regedit.exe). Depending upon the last settings of the option buttons and the text in the text box, if the last user who ran the program made the selections and comments shown in Figure 2.31, then the registry editor would look like that shown in Figure 2.32.

2.17 SELF-WRITING MACROS

Self-writing macros are a very useful feature, especially for those new to programming in VBA. Many times a user will know how to accomplish a task manually using mouse clicks and keystrokes, but not know what properties and methods are required to accomplish this same task using macro code. In such an instance, it is possible to record a task being performed and later look at the code required to complete such a task. For instance, suppose someone wanted to create a worksheet with the text "Title of this Report" written in bold font centered across columns A to F. Furthermore, the user would like the numbers 1 to 10 written in cells A2 to A11.

To record a macro, select Tools->Macro->Record New Macro from the menu. A dialog box like that in Figure 2.33 will appear. Type the name of the macro to be stored (in this case, "SelfWriteTest") and choose the workbook in which the macro code is to be stored. After the "OK" button is pressed, a symbol like the one in Figure 2.34 will appear. It may appear on the toolbar if View->Toolbars->Visual Basic is checked. Excel is now recording every action the user takes and placing the code required to repeat this action in a VB module. To stop the recording process, the user pushes the square button on the toolbar (Figure 2.34).

Once a macro has been recorded, the code generated during the recording process can be viewed by choosing Tools->Macro->Macros, which will bring up the dialog box depicted in Figure 2.35. The user now needs to highlight the macro of interest and press the "Edit" button. If the macro is not present, the user can try changing the "Macros in:" drop down box selection to "All Open Workbooks." The code recorded in the module will then be displayed as shown in Figure 2.36.

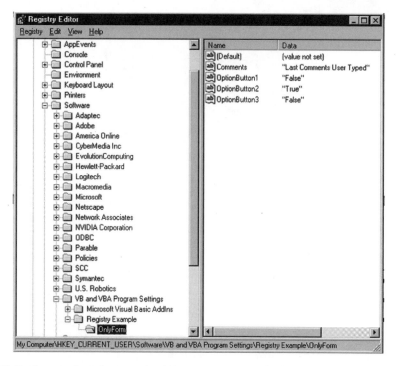

FIGURE 2.32 Registry entries when sample choices correspond to the GUI in the preceding figure.

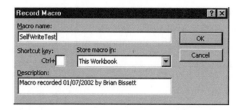

FIGURE 2.34 Terminating the recording
FIGURE 2.33 Recording a macro.   of a macro from the Visual Basic toolbar.

Notice that all macros recorded have at the very top of the subroutine as a comment "Macro recorded {date} by {author}." This is a handy comment because it immediately identifies the subroutine as one that has been software generated. The macro recording function need not only be utilized when user is unsure of the properties or methods needed to program a sequence by hand. Using the macro recorder to generate code for complex formatting applications on worksheets is a real timesaver as well.

One word of caution is in order. It may be necessary to tweak the code generated by the macro recorder in order to adopt it successfully for use in a macro. This is especially true if selections will vary within worksheets from one run to the next. In addition, the macro recorder does not always use the most compact, understandable, and efficient means of coding. The user may wish to edit the recorded code for clarity. Also, the macro recorder often tends to specify properties and methods that need not be coded (it specifies default properties and methods). It is often worthwhile to perform a coding check on recorded macros to determine how such subroutines can be streamlined. Even with its shortcomings, the macro recorder is a highly valuable tool for a programmer at any level.

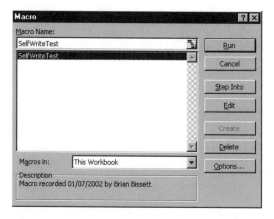

FIGURE 2.35 How to view recorded macros.

```
Book1 - Module1 (Code)
(General)                                    SelfWriteTest

    Sub SelfWriteTest()
    '
    ' SelfWriteTest Macro
    ' Macro recorded 01/07/2002 by Brian Bissett
    '

    '
        Range("C1").Select
        ActiveCell.FormulaR1C1 = "Title of this Report"
        Range("C1").Select
        Selection.Font.Bold = True
        Range("A1:F1").Select
        With Selection
            .HorizontalAlignment = xlCenter
            .VerticalAlignment = xlBottom
            .WrapText = False
            .Orientation = 0
            .ShrinkToFit = False
            .MergeCells = False
        End With
        Selection.Merge
        Range("A2").Select
        ActiveCell.FormulaR1C1 = "1"
        Range("A3").Select
        ActiveCell.FormulaR1C1 = "2"
        Range("A4").Select
        ActiveCell.FormulaR1C1 = "3"
        Range("A5").Select
        ActiveCell.FormulaR1C1 = "4"
        Range("A4:A5").Select
        Selection.AutoFill Destination:=Range("A4:A11"), Type:=xlFillDefault
        Range("A4:A11").Select
    End Sub
```

FIGURE 2.36 Example of module and subroutine generated by macro recorder.

3   Robotic Automation Using Visual Basic

3.1   COMMUNICATIONS OVERVIEW

The most reliable form of communication is synchronous communications. In a synchronous communication system, events happen at fixed intervals specified by the system clock. Synchronicity is desirable behavior because it makes event planning much easier by knowing exactly when something is coming or has the potential to be coming. Events are usually logged on the rising edge of the system clock. Unfortunately, communications between a computer and a peripheral are asynchronous. A predefined timing scheme, as to when characters are sent, does not exist. More importantly, the data stream can arrive at the device at any time. This poses special problems when trying to develop reliable systems to communicate between a computer and a peripheral.

Many elements are involved in communicating between a robot and a controller. Successful communication is much like a math problem — several steps are involved and each must be exactly right to obtain the desired result. Two major standards of communication via serial or parallel port, RS-232 and RS-485, will be discussed. While RS-232 remains extremely popular, RS-485 is fast becoming the workhorse of the industry.

RS-485 allows multiple nodes to communicate bi-directionally over a single twisted shielded pair. The maximum number of devices that can be connected to the bus is 32, which is usually more than adequate. RS-485 carries signals differentially, which means two signals are transmitted down two wires, and at the receiver the two signals are subtracted from each other. The resulting differential signal is the data stream sent from the transmitter. In theory, any noise acquired on the line will be present in both wires and will subtract out when the data reaches its final destination. In practice, this scheme works quite well.

The most popular interface standard in the world is RS-232, which was first introduced in 1962. It was developed for single-ended data transmission at relatively slow data rates (20 kbps) over short distances (<50 ft). Unlike its differential cousin RS-485, RS-232 is limited to communication between two nodes. The fact that RS-232 transmits data in a single-ended fashion makes this standard more prone to noise problems.

Obviously, when attempting communication between any two devices, the same standard must be used in order to communicate. If an option is available, always choose RS-485. Many devices support both standards and serial cards to transmit data in an RS-485 format can be obtained from several manufacturers.

To illustrate communication and control devices via a PC using Visual Basic, a step-by-step device driver will be written for a Kloehn syringe. In the pharmaceutical industry, the two major producers of syringes are Kloehn and Cavro. The protocols between the two syringes are almost identical. In learning to read and understand the protocol, the user will be able to change the parameters in this application to control any device.

3.2   THE MSCOMM CONTROL

The MSComm control takes the drudgery out of communication when using a serial port. One control is required for each serial port used for communication.

First, set the parameter for the serial port the control is to represent. This procedure is done from code by using the CommPort property:

```
Kloehn!MSComm.CommPort = 1
```

Once the serial port to be controlled is specified, the port must be opened before any communications can take place. Make sure the port is not already open before sending a command to open it or an error will result. The following code fragment will accomplish this:

```
Select Case Kloehn!MSComm.PortOpen

    Case Is = True

        Debug.Print "Port already is Open"

    Case Is = False

        Debug.Print "Opening Port"

        Kloehn!MSComm.PortOpen = True

End Select
```

The size of the input buffer can also be modified to reflect the requirements of the application. If characters were not removed from the buffer in a timely manner, it would be advantageous to have a larger input buffer. Conversely, if the input buffer is polled and emptied at regular short intervals, the application would be better suited to a smaller buffer. The code for changing the buffer input buffer size is:

```
Kloehn!MSComm.InBufferSize = 10200
```

Handshaking is a sequence of events, governed by hardware or software, requiring mutual agreement of the state of the operational modes prior to information exchange. If handshaking is necessary for the communication to take place it can be set with the Handshaking property:

```
Kloehn!MSComm.Handshaking = 0
```

From the Kloehn manual, the communication settings for a Kloehn syringe are: 9600 baud, 8 bits, no parity, and 1 stop bit. Set up from code by using the Settings property:

```
Kloehn!MSComm.Settings = "9600,N,8,1"
```

These settings can be defined as:

- Baud (Bd) — unit of modulation rate. One baud corresponds to a rate of one unit interval per second, where the modulation rate is expressed as the reciprocal of the duration in seconds of the shortest unit interval. In other words, the speed at which the data travels. The larger this number is the better.
- Bit — abbreviation for binary digit. It is a character used to represent one of the two digits in the numeration system with a base of two, and only two possible states of a physical entity or system. Those states are high or 1 and low or 0. A state is usually set to 8 because a byte has 8 bits.
- Byte (B) — a sequence of adjacent bits (usually 8) considered as a unit. This usage is predominant in computer and data transmission literature. When so used, the term is synonymous with "octet."
- Parity check — a test that determines whether the number of ones or zeros in an array of binary digits is odd or even. Essentially, a parity is a data integrity check. Parity is used in error-detecting and error-correcting mechanisms. Odd parity is standard for synchronous transmission and even parity is standard for asynchronous transmission.

Once the port has been opened, and the settings have been achieved, the user should be concerned with three major properties:

- The Input property removes characters from the receive buffer.

```
Buffer_Contents$ = MSComm.Input
```

- The Output property writes characters to the transmit buffer.

```
Kloehn!MSComm.Output = Cmd$
```

- The CommEvent property is set when an OnComm event is generated. An OnComm event is generated whenever a communication "event" or error occurs. For example, when data is received and stored in the receive buffer the CommEvent property is set to 2.

With the information shown in Table 3.1 and Table 3.2, data can now be sent and received, but the buffers must be sampled at the proper intervals in order to have reliable communications.

3.3 USING THE TIMER CONTROL

The Timer control will generate a timer event at evenly spaced intervals specified by the user. The Interval property dictates the length of the interval. The Interval property is measured in milliseconds, and the maximum value is 64,767. Therefore, the maximum interval is over a minute (≈ 64.7 seconds). Although the Interval property is measured in milliseconds, the computer's system clock generates only 18 clock ticks per second. The smallest accurately timed interval is then $^1/_{18}$ of a second. Some events, i.e., demands on system resources, affect the accuracy of the interval. A good rule of thumb — do not use the Timer when an interval precision of greater than 0.1 seconds is required. In fact, use C or C++, not Visual Basic, if intervals this precise are required.

The Timer event is simply a subroutine that repeatedly executes with a break between executions specified by the Interval property. To access the subroutine that will run when the Timer event occurs, simply double click on the Timer icon. Problems will occur if the interval set is shorter than the time it takes for the subroutine to execute. The time it takes for the subroutine to execute can be determined using the following code snippet:

```
'Place this statement at the top of any Module to Define the
Win API Sleep Function

Declare Sub Sleep Lib "kernel32.dll" (ByVal dwMilliseconds As
Long)

'Create a button and add this code when the button is clicked:

Private Sub Button_Click()

'Subroutine Timing Example

Dim t0 As Single, tf As Single

t0 = Timer

'Do nothing for 2000 milliseconds (2 Seconds)

Sleep 2000

tf = Timer

'In the event Midnight is crossed

If tf < t0 Then
```

```
    t0 = t0 - CLng(24) * CLng(60) * CLng(60)
End If

Debug.Print "Start: "; t0; " End: "; tf

Debug.Print "Time Elaspsed: "; (tf - t0)

End Sub
```

TABLE 3.1
MSComm Control Events

Constant	Value	Description
ComEvSend	1	Fewer than Sthreshold number of characters in transmit buffer
ComEvReceive	2	Received Rthreshold number of characters (This event is generated continuously until the input property is used to remove the data from the receive buffer.)
ComEvCTS	3	Change in clear-to-send (CTS) line
ComEvDSR	4	Change in data-set-ready (DSR) line (This event is only fired when DSR changes from 1 to 0.)
ComEvCD	5	Change in carrier detect line
ComEvRing	6	Ring detected (Some UARTs do not support this event.)

Source: From *Microsoft Visual Basic 6.0 Help File*, Microsoft Corporation, Redmond, WA, 2000. With permission.

TABLE 3.2
MSComm Control Constants

Handshake Constants

Constant Value Description
comNone 0 No handshaking
comXonXoff 1 XOn/XOff handshaking
comRTS 2 Request-to-send/clear-to-send handshaking
comRTSXOnXOff 3 Request-to-send and XOn/XOff handshaking

OnComm Constants

comEVSend 1 Send event
comEvReceive 2 Receive event
comEvCTS 3 Change in clear-to-send line
comEvDSR 4 Change in data-set-ready line
comEvCD 5 Change in carrier detect line
comEvRing 6 Ring detect
comEvEOF 7 End of file

Error Constants

comEventBreak 1001 Break signal received
comEventCTSTO 1002 Clear-to-send timeout
comEventDSRTO 1003 Data-set-ready timeout
comEventFrame 1004 Framing error
comEventOverrun 1006 Port overrun
comEventCDTO 1007 Carrier detect timeout
comEventRxOver 1008 Receive buffer overflow

Source: From *Microsoft Visual Basic 6.0 Help File*, Microsoft Corporation, Redmond, WA, 2000. With permission.

Here the value of the system Timer is stored in a single precision variable at the start of the subroutine (t0) and at the end of the subroutine (tf). In the literature, the Timer variable is always stored to an integer, because the Timer function is meant to be used to return values in 1 second increments. In practice, the Timer can return values in millisecond increments as shown. Be aware that the values returned are no more accurate than $1/18$ of a second.

Also, this subroutine uses the Win API (application programming interface) Sleep function. The Sleep function pauses the program execution for a certain amount of time specified in milliseconds. Some programmers will use a "Do Nothing" loop to pause a program. This is not a good programming practice because loops will execute faster or slower on different machines, depending upon processor speed, memory, background processes, etc. A loop that may provide a sufficiently long or short pause on one machine may not do so on another. Programmers should strive to write code that will be as portable as possible. Always use the Sleep function to pause program execution. One final word on the Sleep function — it only pauses the program execution, it does not allow background process to continue running and catch up with the application. A method of doing this will be covered later.

Lastly, pay attention to this piece of code in the subroutine:

```
'In the event Midnight is crossed

If tf < t0 Then

   t0 = t0 - CLng(24) * CLng(60) * CLng(60)

End If
```

This snippet of code takes care of a condition referred to as the "midnight" bug. The Timer function is reset to zero at midnight. If t0 is set at 11:59 p.m. and tf is set at 12:01 a.m., then t0 will always be less than tf. So, if the program is waiting for a specified period of time to elapse by checking (tf − t0), it will be waiting forever. Compounding this problem is that this bug will cause the program to lockup only intermittently. If the Timer does not cross midnight between the time t0 and tf are set, the program will work fine. This makes the problem very difficult to find.

A good rule of thumb is not to set the Timer's Interval property smaller than 5 times the length of time it takes the Timer event subroutine to execute. It is not a good situation to have the Timer event subroutine called as it is executing (or a so-called recursive Timer event subroutine). This is asking for trouble.

Setting the Timer's enabled property to false can halt the execution of the subroutine dictated by the Timer event. If the subroutine to be called by the Timer event is to function with the start of the program, make sure this property is set to true during design time.

While on the subject of timing, another Win API function deserves to be mentioned, GetTickCount. GetTickCount will return a value representing how much time has elapsed (in milliseconds) since Windows was started. The advantage of the GetTickCount function is that it does not reset to 0 at midnight, but resets to 0 every 49.7 days. Therefore, if an application is being constructed which will not run constantly for more than 49 days (Windows is hard pressed to continue running for 7 days without crashing), then the programmer need not worry about the GetTickCount value being reset. GetTickCount is declared using the following statements:

```
'Place this statement at the top of any Module

Declare Function GetTickCount Lib "kernel32.dll" () As Long

'This will print the Current Tick Count to the Debug Window

Debug.Print GetTickCount
```

Although Timers are good for waiting a specified period of time before resuming a task, sometimes it is desirable to pass control to the operating system and let it process other events in its queue before continuing. For example, sometimes a program will change focus from one form to another form. Everyone has seen an instance where focus has been changed from one form to another, yet the new form is not redrawn by the operating system. Often, parts of the last form of focus will be seen over the current form of focus. The DoEvents function can eliminate these problems, but its use must be judicious and well thought out.

The DoEvents function will return an integer representing the number of open forms in stand-alone versions of Visual Basic. Although it returns this value, it is seldom used for anything, and DoEvents will often be called without a use for the value it returns. In all other applications, DoEvents will return a zero.

The DoEvents function passes control to the operating system. Control is returned after the operating system has finished processing all the events in its queue, and all the keystrokes in the SendKeys queue have been sent. An inherent danger is that the operating system might not return control to the calling application for some reason. However, for all but the simplest of applications, calls to DoEvents are necessary.

DoEvents are most useful for simple things like allowing a form to redraw itself after a change. The DoEvents function allows other tasks in the operating system's queue to continue completely independent of your application, with the operating system handling cases of multitasking and time slicing.

Another word of caution, **do not** make recursive calls to the DoEvents function. Making recursive calls is an excellent way to lock up an application. Microsoft has another piece of advice on using the DoEvents function: "In addition, do not use DoEvents if other applications could possibly interact with your procedure in unforeseen ways during the time you have yielded control." If the interaction will be unforeseen, how can a programmer avoid it?

The following subroutine will execute the DoEvents function in a nearly bulletproof manner:

```
Public Sub Idle_Time(ByVal nSecond As Single)

Dim t0 As Single, tn As Single

Dim dummy As Integer, temp$

t0 = Timer

If DoEvents_Flag = True then Exit Sub

DoEvents_Flag = True

Do While Timer - t0 < nSecond

    dummy = DoEvents()

    'If we cross Midnight, back up one day

    If Timer < t0 Then

        t0 = t0 - CLng(24) * CLng(60) * CLng(60)

    End If

Loop

DoEvents_Flag = False

tn = Timer

End Sub
```

If the programmer wishes to pause the application for a specified period of time, and allow tasks to empty for the operating system's queue during this time, the Idle_Time subroutine should

TABLE 3.3
Kloehn Syringe Protocol

Byte #	Description	Hex
1	Line synchronization character <STX>	FF
2	Start Transmit character	02
3	Device address	31 - 5F
4	Sequence number	31–3F
5	Command(s) (n bytes)	(Varies)
5+n	End of command(s) <ETX>	03
5+n+1	Check sum	(Varies)

- Byte 1: The line synchronization character, FF hex, indicates a command package is coming.
- Byte 2: The start transmit character, 02 hex, signals the beginning of a new package.
- Byte 3: The device address is an address number for a device or for a group of devices. Up to 15 devices can be addressed.
- Byte 4: Sequence number indicates the package sequence. If an error occurs during the communication, the host sends the last package again to the device with a new sequence number. The sequence number starts with 31 hex (ASCII 1). When repeating a command, the host sets bit 3 of the sequence number byte to 1 and increases the sequence number by 1. The valid sequence numbers are hexadecimal 31 for the first package, hexadecimal 3A for the second package (the first repeated package), 3B for the third package, and etc. The maximum number of repeat is 7 with a sequence number of 3F.
- Byte 5: The command or a sequence of commands starts with byte 5. A command or a command sequence with length n bytes uses byte 5 to byte 5+n1.
- Byte5+n: The end-of-command(s) character, 03 hex, indicates the end of a command or command sequence.
- Byte5+n+1: The check sum is calculated by an exclusive-or operation on all bytes except line synchronization byte and check sum byte.

Source: From *50300 Hardware User's Manual*, Kloehn Ltd., Las Vegas, NV, 1996. With permission.

be called. The DoEvents_flag should be declared as a global variable at the module level. Whenever a DoEvents call is desired, call the Idle_Time subroutine. If the subroutine should be called recursively, the DoEvents_flag will not allow the DoEvents function to be called until the last DoEvents call has finished processing. In addition, notice that this subroutine has the "midnight bug" fix provided within it.

3.4 DECIPHERING A COMMUNICATION PROTOCOL

Once the developer knows how to send commands via the MSComm control, and has knowledge of system timing, the question becomes how to build a command that the device will "understand." Table 3.3 illustrates the communications protocol for a Kloehn syringe. It is nearly identical to the protocol for Cavro syringes. These two syringes are currently the most popularly used syringes in pharmaceutical labs. Table 3.3 describes the command package format of the OEM protocol.

To build a command, a knowledge of the ASCII character codes and their hexadecimal equivalents is vital. Table 3.4 contains this information.

Table 3.4 has three columns for each character represented: Dec, Hex, and Char. Char is the actual character the code represents. Dec is the code number in decimal format or ordinary numbers. Hex is the code number in hexadecimal format. The best way to convert Decimal numbers to Hexadecimal numbers is to buy a calculator that can convert them.

TABLE 3.4
ASCII Character Codes

Dec	Hex	Char	Dec	Hex	Char	Dec	Hex	Char	Dex	Hex	Char
0	00	NUL	32	20		64	40	@	96	60	`
1	01	SOH	33	21	~	65	41	A	97	61	a
2	02	STX	34	22	.	66	42	B	98	62	b
3	03	ETX	35	23	#	67	43	C	99	63	c
4	04	EOT	36	24	$	68	44	D	100	64	d
5	05	ENQ	37	25	%	69	45	E	101	65	e
6	06	ACK	38	26	&	70	46	F	102	66	f
7	07	BEL	39	27	'	71	47	G	103	67	g
8	08	BS	40	28	(	72	48	H	104	68	h
9	09	TAB	41	29	)	73	49	I	105	69	i
10	0A	LF	42	2A	*	74	4A	J	106	6A	j
11	0B	VT	43	2B	+	75	4B	K	107	6B	k
12	0C	FF	44	2C	'	76	4C	L	108	6C	l
13	0D	CR	45	2D	-	77	4D	M	109	6D	m
14	0E	SO	46	2E	.	78	4E	N	110	6E	n
15	0F	SI	47	2F	/	79	4F	O	111	6F	o
16	10	DLE	48	30	0	80	50	P	112	70	p
17	11	DC1	49	31	1	81	51	Q	113	71	q
18	12	DC2	50	32	2	82	52	R	114	72	r
19	13	DC3	51	33	3	83	53	S	115	73	s
20	14	DC4	52	34	4	84	54	T	116	74	t
21	15	NAK	53	35	5	85	55	U	117	75	u
22	16	SYN	54	36	6	86	56	V	118	76	v
23	17	ETB	55	37	7	87	57	W	119	77	w
24	18	CAN	56	38	8	88	58	X	120	78	x
25	19	EM	57	39	9	89	59	Y	121	79	y
26	1A	SUB	58	3A	'	90	5A	Z	122	7A	z
27	1B	ESC	59	3B	<	91	5B	[	123	7B	{
28	1C	FS	60	3C	=	92	5C	\	124	7C	|
29	1D	GS	61	3D	>	93	5D	]	125	7D	}
30	1E	RS	62	3E	>	94	5E	^	126	7E	~
31	1F	US	63	3F	?	95	5F	_	127	7F	DEL

Note: Dec 32/Hex 20 is ≈ {space}.

It is necessary to know these codes because some of the characters in a communication protocol *cannot be accessed via the keyboard*. The only way to build them into a command string is to use the codes. This is true of the control characters and special characters listed in Table 3.4. To access any character shown in the table use the Chr() command. For example:

```
Debug.Print Chr$(65), Chr$(44), Chr$(122)
```

Output:

```
A , z
```

Looking at the command format of the OEM protocol, the first thing to determine is which bytes in the command string are subject to change from command to command. In this case, bytes 1,2,(5+n) will never change. Bytes 3,4,5,(5+n+1) will vary with each command sent. The logical thing to do is to construct a function that will build the command string, hardcoding the bytes that will never change, and passing the bytes as parameters that are subject to change. Apply the following subroutine:

```
Public Function Build_OEM_String(XLSeqNum As Integer, XLAddr
As Integer, XLCmd$) As String

'Build Command String in OEM Kloehn Format

Dim Line_Sync As String

Dim STX As String

Dim ETX As String

Dim Check_Sum As Single

Dim build_string As String

'Dim build_checksum As String

'Define Fixed Bytes

Line_Sync = Chr$(&HFF) 'OR Chr$(255) 'Byte No. 1

STX = Chr$(2) 'Start of Text - Byte No. 2

ETX = Chr$(3) 'End of Text - Byte No. (5+n)

'ShowMsgData Messages!MsgField, "Current Command: "; XLCmd$

build_string = ""

build_string = build_string & STX

build_string = build_string & Chr$(Dev_No)

build_string = build_string & Trim$(Str(XLSeqNum))

build_string = build_string & XLCmd$

build_string = build_string & ETX

build_string = build_string & Chr$(ChkSum(build_string))

build_string = Line_Sync & build_string

Build_OEM_String = build_string

Hex_msg$ = ASCtoHex(Build_OEM_String)

Debug.Print "OEM Command String Sent: "; Hex_msg$

End Function
```

Two functions have not yet been defined: ASCtoHex and ChkSum. An explanation and code for them will follow shortly.

Looking at the subroutine, the fixed bytes are defined first with:

```
Line_Sync = Chr$(&HFF) 'OR Chr$(255) 'Byte No. 1

STX = Chr$(2) 'Start of Text - Byte No. 2

ETX = Chr$(3) 'End of Text - Byte No. (5+n)
```

Notice that all three of these characters are special in that they cannot be accessed via the keyboard. The Line Sync character is of particular interest. It is not listed in Table 3.4 but can be accessed using the decimal 255 or its hexadecimal equivalent &HFF. (This is where a good calculator comes in handy.)

The variable parameters that are passed to the function are in order: byte 3 (the device address) as integer, byte 4 (the sequence number) as integer, and byte 5 (the command to be executed) as string. Byte 5 is of particular interest as it can vary in length depending upon the command. Byte (5+n+1) (the last byte to be sent in the command string) is a checksum character and it must be calculated for each command via the function ChkSum.

Checksums are stored or transmitted with data and are intended to detect data integrity problems. A checksum is a numeric representation of a group of data items, whose sum is used for data checking purposes. The checksum is transmitted with the group of data items and is calculated by treating the data items as numeric values. In this particular case, the checksum is calculated by Exclusive Or'ing all the bytes in the command string except for the line sync (byte 1) and the checksum byte itself (the last byte). By passing the generated command string to the following function the checksum will be calculated and returned.

```
Function ChkSum(var$) As Integer

'Calculates vertical checksum needed for transmit string

'var$ is the string for which to calculate a checksum

Dim i, temp

temp = 0

For i = 1 To Len(var$)

    temp = temp Xor Asc(Mid$(var$, i, 1))

Next i

ChkSum = temp

End Function
```

In the Build_OEM_String subroutine, the command string is first built without the line sync character, then the checksum byte is calculated and added to the command string. Finally, the line sync is added to the beginning of command string with the calculated checksum. This is in accordance with the OEM protocol, and is the simplest way to construct the string.

Byte 3 (the device address) is the address to which the command is being sent. On some buses, notably RS-485, all devices connected to the bus will receive every message sent over the bus. Only the device, which has the same address as the address encoded in the sent message, will respond to the command.

Byte 4 (the sequence number) is an important concept. Whenever communication takes place over a bus some packets (command strings) are bound to be lost. This can occur because of a variety of reasons that include noise, improper termination, poorly designed firmware, etc. If a packet (or command string) is unreadable, the device will look to the next packet with an incremented sequence number to replace the unreadable packet. Two kinds of communication errors are possible. A corrupt packet occurs when one of the boundaries of the packet is recognizable but the data within the packet is unintelligible. A lost packet is one whose frame (or boundaries) cannot even be recognized by its intended recipient.

Byte5 (the command to be executed) is unique in that it is the starting position for the only command that may contain more than one character (or byte). Table 3.5 shows some of the more common commands a Kloehn syringe can execute.

For example to reset the syringe, the 5th, 6th, and 7th bytes should be set to the three-character string: "W4R". The "R" at the end of the command signals the syringe to execute the command immediately.

TABLE 3.5
Kloehn Syringe Commands

Machine Code Command List

- Commands which have a (p) can be embedded within a program or executed immediately if followed by the "R" command.
- Commands which have a (po) can be used only within a program.
- Commands which have an (e) are executed when they are recognized and are not stored within a program.
- Commands which have an (s) are normally setup commands, and the results of these commands are stored in non-volatile memory by the "!" command.
- After a Reset or power-up, the first command should be a "Q" and the reply, if any, should be ignored. A "W4", "Y4", or "Z4" must be sent before syringe move commands can be executed.
- Numerical ranges in parenthesis are for the 24K model. The notation [48,000] means to replace the listed value with the bracketed value for the 48K model.
- Commands which have a [@n] may use "@n" for the parameter value of "n", and the parameter value is determined as a normalized fraction of its full-scale range
- Commands which have a [@a] may use "@n" for the parameter value of "n", and the parameter value is the actual input number that is read.
- Commands which do not have any [@_] brackets cannot use "@n" for the parameter value of "n". See section 3.15.4 for a list of commands which can use the "@n" value.

Frequently Used Syringe Commands

- An (p) [@n] Absolute syringe move. Move the syringe to position "n" with the "busy bit" in the status byte set to "busy" ("0"). (n: 0...24,000 [48,000])
- An (p) [@n] Absolute syringe move. Move the syringe to position "n" with the "busy bit" in the status byte set to "not buy" ("1"). (n: 0...24,000 [48,000])
- cn (p,s) [@n] Set the stopping speed. The factory default is 743. (n: 40...10,000)
- Dn (p) [@n] Relative dispense. Move the syringe "n" steps up from the current position and set the "busy bit" in the status byte to "busy" ("0"). (n: 0...24,000 [48,000])
- dn (p) [@n] Relative dispense. Move the syringe "n" steps up from the current position and set the "busy bit" in the status byte to "not busy" ("1"). (n: 0...24,000 [48,000])
- Pn (p) [@n] Relative pickup. Move the syringe "n" steps down (aspirate or air gap) from the current position and set the "busy bit" in the status byte to "busy" ("0"). (n: 0...24,000 [48,000])
- pn (p) [@n] Relative pickup. Move the syringe "n" steps down (aspirate or air gap) from the current position and set the "busy bit" in the status byte to "ready" ("1"). (n: 0...24,000 [48,000])
- Sn (p,s) [@n] Set the top speed. Note: the preferred method of setting top speed is to use the "V" command. (n:0...34)
- Vn (p,s) [@n] Set the top speed to "n" steps per second. The factory default is 5000. (n: 40...10,000)
- vn (p,s) [@n] Set the Start Speed to "n" steps per second. The factory default is 743. (n: 40...1000)
- Wn (p) Initialize the syringe on systems with or without a valve. (n: 4 or 5) n = 4 Move the syringe to the "soft stop" position after moving the valve to port A (must be used after a Reset or a power-up, before syringe move commands can be recognized). This is the same as pressing the "Initialize" (lower) switch on the front panel. n = 5 Set the current syringe position as the "home" position.

Frequently Used Valve Commands

- B (p) Move a three way standard valve to the "bypass" position (port A-to-port B).
- I (p) Move a three way standard valve to the "input" position (port A-to-syringe).
- O (p) Move a three way standard valve to "output" position (port B-to-syringe).

Source: From *50300 Hardware User's Manual*, Kloehn Ltd., Las Vegas, NV, 1996. With permission.

The last function, ASCtoHex, takes each character of the string passed to it and returns a string of spaced hexadecimal numbers which represent the characters of the passed string in hexadecimal format. This is handy because if the string is passed in its "literal form" the control and special characters could print out as anything. When called in the Build_OEM_String function, the

ASCtoHex function returns the hexadecimal stream of characters that will be sent out of the serial port to the device.

```
Public Function ASCtoHex(convert$) As String

For Index = 1 To Len(convert$)

    ic$ = Mid$(convert$, Index, 1)

    If Index = Len(convert$) Then

        temp$ = temp$ & Hex(Asc(ic$))

        Else

        temp$ = temp$ & Hex(Asc(ic$)) & " "

    End If

Next Index

ASCtoHex = temp$

'Debug.Print "ASCtoHex: "; temp$

End Function
```

3.5 SENDING A COMMAND

Now that a means is available to build the proper command string for every different command and device, a means must be devised to actually send the command out the port to the device. This can be accomplished with the following subroutine:

```
Private Sub Execute_Button_Click()

Dim Repeated As Integer, SeqNo As Integer

ReplyOK = False: Repeated = 0

Repeat_Cmd:

Repeated = Repeated + 1

If Repeated > 10 Then

    MsgBox "Device Not Responding to Commands", vbExclamation,
    "System Error"

    Exit Sub

End If

'Enable UART to Transmit RS-485 Communication

Kloehn!MSComm.RTSEnable = True

Call Idle_Time(0.005)

Statts(Real_DevNo) = "Busy"

Kloehn!Status_Button.Value = True

Call Determine_Wait(Kloehn!Cmd_String.Text)
```

```
'Send Each Command Down the line 7 Times each with different
Seq No.

For SeqNo = 1 To 7

    Call SendCmd(Build_OEM_String(SeqNo, Dev_No,
    Kloehn!Cmd_String.Text))

Next SeqNo

'Enable UART to Recieve RS-485 Communication

Call Idle_Time(0.005)

Kloehn!MSComm.RTSEnable = False

'Wait 50ms For Reply

Call Idle_Time(0.05)

If ReplyOK = False Then GoTo Repeat_Cmd

End Sub
```

The SendCmd subroutine sends the command to the serial port's output buffer to be sent. Its code is very simple:

```
Sub SendCmd(Cmd$)

    Kloehn!MSComm.Output = Cmd$

End Sub
```

Some explanation is in order here. First, the subroutine is intended to function when a button is pressed. The button can always be pressed from code using its value property.

The subroutines variables are:

- Repeated — keeps track of how many times the command has been repeated. Set to 0 at beginning of subroutine.
- SeqNo — the sequence number of the command string currently being sent.
- ReplyOK — global variable set to false on initialization of subroutine. Set to true when and if a reply is received from the syringe that indicates the command was accepted and executed by the syringe.

Before the command is sent to the MSComm control (and out the serial port), the RTSEnable property of the MSComm control is first set to true. For RS-485 communications, the RTS line must be set high before any commands can be sent out to the bus. Conversely, to receive any commands from the bus, the RTS line must be set to low (or false).

The command following the setting of the RTS line high is where things get interesting. Visual Basic is an interpreted language rather than a compiled language. It runs slower relative to compiled languages such as C++. Because the RTS line was set to high does not mean it will be high before the next immediate command executes. In C++ the command would be executed almost instantaneously. However, I have witnessed delays in VB when setting this property that caused the messages sent to be unreadable because the RTS line was still low when the message was sent. A short call of a few milliseconds to the Idle_Time subroutine will eliminate this problem every time.

The function Determine_Wait sets an appropriate waiting period before allowing the syringe to be set again as ready. If the syringe is aspirating or dispensing large volumes of fluid this waiting time will become longer. More on this later when queuing is discussed.

Finally, the command is actually sent. Notice a loop is set up to increment the sequence number to repeat the command the maximum number of times in case the command string is lost. Once the command strings are sent, the RTS line is set to false (or low) to allow replies from the syringe to travel along the bus into the serial port's buffer. The program then idles for 50 milliseconds waiting for a valid reply. If a valid reply is received, then the Global ReplyOK variable is set to true and the program is free to execute the next command. Should a valid reply not be received, the program will attempt to resend the command up to 10 times after which an error message will appear.

For example, when a command is sent to initialize a syringe with a device number of 1, the code will send the following command string:

```
OEM Command String Sent: FF 2 31 31 57 34 52 3 30

OEM Command String Sent: FF 2 31 32 57 34 52 3 33

OEM Command String Sent: FF 2 31 33 57 34 52 3 32

OEM Command String Sent: FF 2 31 34 57 34 52 3 35

OEM Command String Sent: FF 2 31 35 57 34 52 3 34

OEM Command String Sent: FF 2 31 36 57 34 52 3 37

OEM Command String Sent: FF 2 31 37 57 34 52 3 36
```

Looking down the columns of hexadecimal characters:

```
<line sync> <STX> <Device#> <Seq#> <Cmd> <ETX> <Chksum>
```

Taking the first String Sent:

```
Characters: <Sync> <STX> 1 1 W 4 R <ETX> <Chksum>

OEM Command String Sent: FF 2 31 31 57 34 52 3 30
```

Notice that the fourth column (Seq#) is incremented from 1 to 7. Also notice that a unique checksum is obtained for each command string, since at a minimum the sequence numbers will always be different. The command initialize is W4R and is characters 5 to 7.

Sending the command is only one part of the process however. Once the command is sent the device will send a reply. The code must be able to interpret the reply and take appropriate action based on the information contained in the reply.

3.6 INTERPRETING A REPLY

For the Kloehn syringe a reply can be expected in the format shown in Table 3.6. This is a device response package format for OEM protocol.

Interpreting the reply is much like generating the command string, except in reverse. Here the reply is fetched from the serial port's buffer via the MSComm control, and the reply is taken apart byte by byte.

First, a mechanism of separating each distinct byte of the reply message is needed. The process is called tokenizing a string and, in this case, each byte will be a token. Visual Basic does not have a built in function for tokenizing a string but one has been created:

```
Public Function Token(symbol$, stoken$, token_no%, ret_token$)

Dim count As Integer, startpos As Integer, endpos As Integer,
start_flag As Integer

count = 0: start_flag = 0
```

TABLE 3.6
Kloehn Syringe Device Reply Protocol

Byte #	Description	Hex
1	Line synchronization character	FF
2	Starting character <STX>	02
3	Host address	30
4	Status and error byte	(varies)
5	Response, if any (n bytes)	(varies)
5+n	End-of-response mark <ETX>	03
5+n+1	Check sum	(varies)
5+n+2	Extra ending character <blank>	FF

- Byte 1: The line synchronization, FF hex, indicates a command package is coming.
- Byte 2: The starting character, 02 hex, signals the beginning of a new package.
- Byte 3: The host address, 30 hex, is the address number for the host computer.
- Byte 4: The status and error byte describes the device status. Please refer to Appendix C for the definitions of the status and errors.
- Byte 5: There may or may not be response byte(s) for a command. In general, all query commands, read input commands, and configuration query commands (~A, ~B, ~P, ~V, etc.) produce response bytes. Other commands do not produce a response.
- Byte 5+n: The end-of-response mark, 03 hex, indicates the end of the response byte(s).
- Byte 5+n+1: The check sum is calculated by an exclusive-or operation on all bytes except the line synchronization byte and the check sum byte.
- Byte 5+n+2: The extra ending character, FF hex, is an extra character to ensure the package is properly sent. This character might not be displayed the host terminal.

Source: From *50300 Hardware User's Manual*, Kloehn Ltd., Las Vegas, NV, 1996. With permission.

```
startpos = 1

For i = 1 To Len(stoken$)

   Select Case token_no%

   Case 1

      If Mid$(stoken$, i, 1) = symbol$ Then

         ret_token$ = Mid$(stoken$, 1, (i - 1))

         GoTo Function_End

      End If

   Case Else

      If Mid$(stoken$, i, 1) = symbol$ Then

         count = count + 1

            If count = (token_no% - 1) Then

               startpos = i + 1

               start_flag = 1
```

```
                              End If

                              If count = token_no% Then

                                  endpos = i - 1

                                  ret_token$ = Mid$(stoken$, startpos, ((endpos
                                  - startpos) + 1))

                              GoTo Function_End

                          End If

                     End If

                End Select

                'If i = Len(stoken$) Then

                    'ret_token$ = Mid$(stoken$, startpos, ((Len(stoken$) -
                    startpos) + 1))

                    'Exit For

                'End If

        Next i

        If start_flag = 1 Then

            ret_token$ = Mid$(stoken$, startpos, ((Len(stoken$) -
            startpos) + 1))

            Else

            ret_token$ = ""

        End If

        Function_End:

        End Function
```

Next, is an explanation of passed parameters to the token function:

symbol$ — the character (or delimiter) which separates each substring. Common delimiters include commas and spaces, but can be set to any character desired.

stoken$ — the string from which the tokens will be extracted.

token_no% — the position number of the token. Using 3 would provide the third token found in the string passed. If the token at this position does not exist, the function will return and empty string "".

ret_token$ — the extracted token from position specified with token_no%.

One last word on tokens, Visual Basic 6.0 has a new function called Split that essentially works in a similar fashion to the token function, except each individual token is placed as an item within an array. The methodology provided for handling device replies could be modified to work using this function.

Messages from the devices being communicated with are stored to a buffer in the serial port. The contents of this buffer can be extracted with the MSComm Control using the Input property. The key and critical question here is *when* should the contents of the buffer be extracted.

If the contents of the receive buffer are not extracted quickly enough, the receive buffer will overflow resulting in an error. If the contents of the receive buffer are extracted too often, the program will utilize resources unnecessarily, emptying a buffer that contains nothing (often referred to as buffer starvation).

One method is to wait for a Receive event to be logged in the MSComm control, and then empty the buffer. The advantage to using this event as a trigger to empty the buffer is that the buffer will always contain data when it is emptied. A disadvantage to using this method is that the buffer may contain several reply messages when it is emptied. During the latency between the time that the Receive event is generated, and the time that the event is acted upon by the program, the port may receive another reply message.

A better method is to poll the serial port using a timer at specified intervals. The contents of the buffer are then checked, and if the buffer contains anything, it is sent for analysis, otherwise the program goes on its merry way. For the Kloehn syringes, and many other devices running at 9600 baud, 28ms is an effective polling interval. The faster the device (or the higher the baud rate), the shorter the polling interval should be. Sometimes this has to be determined by trial and error, when replies are not getting lost then the polling interval is adequate.

The following subroutine uses an MSComm control named Kloehn_Comm to read the receive buffer of the serial port. A SerialPort_Timer is used to poll the receive buffer every 28ms. The contents of the buffer are retrieved using the Input property. If the buffer is not empty, then the reply is interpreted using the subroutine Interpret_Reply.

```
Private Sub SerialPort_Timer_Timer()

Asc_msg$ = Kloehn!MSComm.Input

If Asc_msg$ <> "" Then

'Debug.Print "ASCII Message: "; Asc_msg$

Hex_msg$ = ASCtoHex(Asc_msg$)

Debug.Print "Hex Message: "; Hex_msg$

'Kloehn!Response.Caption = Hex_msg$

Call Interpret_Reply(Hex_msg$)

End If

End Sub
```

Next is the subroutine that interprets the reply.

```
Sub Interpret_Reply(reply$)

Dim delim$, No_Tokens%, Track%

Dim msg$, i As Integer

delim$ = Chr$(32)

No_Tokens% = Int(Len(reply$)/3)

For i = 1 To No_Tokens% + 1

    Token delim$, reply$, i, msg$

    'Stop

    If msg$ = "FF" Then
```

```
      'Debug.Print " ** OEM Line Sync Detected **"
      Track% = 1
   End If
   If Track% = 1 And msg$ = "2" Then Track% = 2
   If Track% = 2 And msg$ = "30" Then
      Token delim$, reply$, i + 1, msg$
      'Stop
      Call CheckStatusByte(msg$)
   End If
Next i

End Sub
```

The preceding subroutine looks first for the line synchronization character "FF." This character is sent out in a long stream to let the device looking for a reply to know that it is coming. Once the line sync is detected, the subroutine looks for the <STX> character or "2." It then checks to make sure the next byte is "30." The next byte (byte 4) is the status byte and its value is passed to the subroutine CheckStatusByte. The CheckStatusByte subroutine determines what the reply message means by interpreting the value of the status byte against the data in Table 3.7. The status byte indicates the device status and any existing error conditions.

Bit #5 is the busy/ready bit which is logic "0" if busy executing commands and logic "1" if ready for new commands. The "busy" version of the syringe move commands should be used so that the true status of the device will be reported.

```
Public Sub CheckStatusByte(status_byte$)

'OEM Status and Error Code Checking

Select Case status_byte$
   Case "40", "60"
      GoSub Statts_OK
      'Diagnose!Mssg_Text.Text = "Reply: Standalone Kloehn"
   Case "48", "68", "4E", "6E"
      GoSub Statts_OK
      'Diagnose!Mssg_Text.Text = "Reply: Reserved Kloehn"
   Case "46", "66", "49", "69", "4A", "6A", "4B", "6B", "4C",
   "6C"
      GoSub Statts_Error
      'Diagnose!Mssg_Text.Text = "Reply: Kloehn Hardware Error"
   Case "42", "62", "43", "63", "44", "64", "45", "65", "4D",
   "6D"
      GoSub Statts_Error
```

TABLE 3.7
Status and Error Codes for Kloehn Syringes

ASCII	Binary	Hexadecimal	Decimal	Code	Description
bsy or rdy	7 6 5 4 3 2 1 0	bsy or rdy	bsy or rdy		
@ or `	0 1 X 0 0 0 0 0	40 or 60	64 or 96	0	No error
A or a	0 1 x 0 0 0 0 1	41 or 61	65 or 97	1	Syringe not initialized
B or b	0 1 X 0 0 0 1 0	42 or 62	66 or 98	2	Invalid command
C or c	0 1 x 0 0 0 1 1	43 or 63	67 or 99	3	Invalid operand
D or d	0 1 x 0 0 1 0 0	44 or 64	68 or 100	4	Communication error
E or e	0 1 x 0 0 1 0 1	45 or 65	69 or 101	5	Invalid R command
F or f	0 1 x 0 0 1 1 0	46 or 66	70 or 102	6	Low voltage
G or g	0 1 x 0 0 1 1 1	47 or 67	71 or 103	7	Device not initialized
H or h	0 1 x 0 1 0 0 0	48 or 68	72 or 104	8	Program in progress
I or i	0 1 x 0 1 0 0 1	49 or 69	73 or 105	9	Syringe overload
J or j	0 1 x 0 1 0 1 0	4A or 6A	74 or 106	10	Valve overload
K or k	0 1 x 0 1 0 1 1	4B or 6B	75 or 107	11	Syringe move not allowed in valve bypass position
L or l	0 1 x 0 1 1 0 0	4C or 6C	76 or 108	12	No move against limit
M or m	0 1 x 0 1 1 0 1	4D or 6D	77 or 109	13	NVM Memory failure
N or n	0 1 x 0 1 1 1 0	4E or 6E	78 or 110	14	Reserved
O or o	0 1 x 0 1 1 1 1	4F or 6F	79 or 111	15	Command buffer full
P or p	0 1 x 1 0 0 0 0	50 or 70	80 or 112	16	For 3-way valve only
Q or q	0 1 x 1 0 0 0 1	51 or 71	81 or 113	17	Loops nested too deep
R or r	0 1 x 1 0 0 1 0	52 or 72	82 or 114	18	Label not found
S or s	0 1 x 1 0 0 1 1	53 or 73	83 or 115	19	No end of program
T or t	0 1 x 1 0 1 0 0	54 or 74	84 or 116	20	Out of program space
U or u	0 1 x 1 0 1 0 1	55 or 75	85 or 117	21	Home limit not set
V or v	0 1 x 1 0 1 1 0	56 or 76	86 or 118	22	Call stack overflow
W or w	0 1 x 1 0 1 1 1	57 or 77	87 or 119	23	Program not present
X or x	0 1 x 1 1 0 0 0	58 or 78	88 or 120	24	Valve position error
Y or y	0 1 x 1 1 0 0 1	59 or 79	89 or 121	25	Syringe position error
Z or z	0 1 x 1 1 0 1 0	5A or 7A	90 or 122	26	Syringe may crash

Source: From *50300 Hardware User's Manual*, Kloehn Ltd., Las Vegas, NV, 1996. With permission.

```
    'Diagnose!Mssg_Text.Text = "Reply: Kloehn Communications
    Error"

  Case "41," "61," "47," "67"

    GoSub Statts_Error

    'Diagnose!Mssg_Text.Text = "Reply: Error Device Not
    Initialized"

End Select

Exit Sub

Statts_OK:

ReplyOK = True

Debug.Print "*** No Error ***"

Return
```

```
Statts_Error:

ReplyOK = False

'A Reply was recieved indicative of an Error

Debug.Print "** Communications Error **"

Return

End Sub
```

Remember, when sending a command using the Execute_Button_Click subroutine, the command is first sent and then the program pauses for 50 ms using the Idle_Time subroutine. After the 50 ms pause, the value of the ReplyOK variable is checked. If the ReplyOK Value is equal to true, then the subroutine is finished. If the ReplyOK Value is false, the subroutine will resend the command. The command will be resent up to 10 times until a reply is received indicating the transmission was successful.

The CheckStatusByte subroutine determines whether or not the machine understood and acted upon the command sent to it. It looks at the Status Byte of the reply message. The status byte provides a wealth of information that is explained in the Table 3.7.

First, the status byte will have an overall hexadecimal or decimal value. Different values mean different things. A description of what each value is indicative of is given in Table 3.7. The CheckStatusByte subroutine is rather simple in that it merely looks at the error codes and places them into one of two classifications: an okay response, or a response that indicates an error occurred and the command should be resent. Nothing stops the developer from having the code perform a different action for each value returned. In this manner, rather sophisticated error detection and correction algorithms can be written which take nearly every possible occurrence into account.

Second, notice that bit number 5 of the status byte is marked as an "X" in Table 3.7. The "X" indicates a "don't care condition". In other words the value of bit number 5 will have no bearing on what the error code returned will be. If bit number 5 is a "1", then 32 decimal (or 20 hexadecimal) will be added to the value of the status byte. In binary arithmetic, bit 5's decimal value is $2^5 = 32$. Notice that each condition contains two values in hexadecimal and decimal that refer to the same condition. They differ by 32 in decimal and 20 in hexadecimal. This takes into account the possibility that bit 5 can be either a "1" or a "0".

The reason bit 5 is made into a "don't care condition" is so that its value can be used to indicate something other than the value of the reply that is returned. In this particular case bit 5 is used to return the status of the syringe. If this bit is set to "1", then the device is ready to receive another command. If the bit is set to "0", then the device is busy. This bit only has significance if the busy version of the syringe commands is used, not the ready version.

The driver that is written to control this syringe does not make use of bit number 5's information. This is because a method exists of determining the syringes status that is less taxing on system resources and also less obvious to implement. The developer will know the syringe speed in steps per second and the number of steps the syringe is to be moved. The amount of time required for a syringe to execute a movement command can thus easily be calculated. The program then waits the appropriate amount of time and checks for a valid reply command once. This is much less taxing on system resources than constantly polling the status byte and waiting for bit 5 to change from "0" to "1".

One final thought for consideration, notice that many of the hexadecimal responses are described as reserved. These values have not been assigned by the system but are available for use should future expansion of the protocol require more error/response codes. However, in monitoring the responses generated by Kloehn syringes, reserved codes have shown up in replies. Whether this is from an early firmware upgrade not yet documented in the literature, or a firmware error that is sending reserved codes for some unknown reason is anyone's guess. The important thing is to realize that things may not happen the way they are supposed to.

When developing a device driver, it is a prudent idea to spy on the serial port and make sure it is not sending back replies that are not defined or do not make any sense. The developer cannot control the behavior of the firmware within the hardware, or the accuracy in following the communication protocol. The developer can compensate for hardware deficiencies or irregularities through clever design and development of software, but only if he is aware the deficiency exists.

3.7 SETTING UP A QUEUE USING THE TIMER CONTROL

Using the subroutines that were discussed previously, the byte values can be calculated to send a single command, the individual bytes can be assembled into a command string, and the command string can be sent out the serial port using the MSComm Control to be read by the intended device. Once the command has been sent, the software is able to interpret the reply sent from the hardware device, and to determine if the command has been successfully executed, or if an error has occurred. There is really nothing inherently difficult about doing this.

However, it is difficult to manage a stream of command strings sent to multiple devices over a single bus (serial port). Fifteen different Kloehn syringes can be hooked up to a single bus. Now, the software must keep track of a multitude of parameters to include: (1) which command in the stream (sent to a particular device) is next to be executed; (2) the status of each device; (3) the errors that have occurred on each device; (4) the current position of every syringe; (5) the current speed setting of each syringe; and (6) the time each syringe should idle (transition time to complete a move) before proceeding to the next command. Depending upon the device to be controlled, the list of parameters to keep track of may be larger or smaller.

To accomplish this, four parameters are set up as global variable arrays with 15 elements each (one element for every possible device number).

```
Global SSpeed(15) As Integer              'Current Syringe Speed
Global Statts(15) As String               'Current Syringe Status
Global Wait_Time(15) As Single     'Time Till Syringe Task Finished
Global Curr_Pos(15) As Single   'The Current Position the Syringe is in
```

When the program first begins all of these parameters are set to default values using the following subroutine that executes when the form loads:

```
Public Sub Set_Params()
For i = 1 To 15
   Wait_Time(i) = 0: Curr_Pos(i) = 0
   SSpeed(i) = 5000
   Statts(i) = "Unknown"
Next i
End Sub
```

Executing a single command at a time in a manual fashion is simple. The user simply types the command in the testbox, chooses the option button for the proper device number and serial port number, and then presses the execute button.

However, executing multiple commands requires that the commands be queued up. This is accomplished by using a series of subroutines. To send a command first call the following subroutine that sets the proper serial port number and device number:

```
Public Sub SetSyrCmd(SerialPort%, DeviceNo%) ', Protocol$)

If Kloehn_Port <> SerialPort% Then

    Call Change_Kloehn_Port(SerialPort%)

End If

Kloehn!DevNo(DeviceNo%).Value = True

Call Idle_Time(0.05)

End Sub
```

Next the actual command is sent using the following subroutine:

```
Public Sub SyrCmd(Cmd$)

Dim stop_value As Single

'If Device Not Ready to Receive Cmd then Wait

stop_value = Timer + 120

While Statts(Real_DevNo) = "Busy" And Timer < stop_value

'Debug.Print Statts(Real_DevNo), Timer, stop_value

DoEvents

Wend 'End While loop when Device is Ready or 2 min

'Write Cmd to TextBox

Kloehn!Cmd_String.Text = Trim$(Cmd$)

'Press Execute Button

Call Idle_Time(1)

Kloehn!Execute_Button.Value = True

End Sub
```

Notice that this subroutine does not automatically send the command immediately and unconditionally. The command will only be sent if the device for which it is intended is not busy. If the device is busy, then the program will wait up to two minutes for the device to become ready, and then send the command. In the mean time, while the program is waiting for device n to become ready, another command can be sent to a different device (provided that device is in a state of readiness). The global array variable Statts acts as a traffic director, permitting commands to proceed only if the device will be receptive to the command next in line to be sent. Using this type of methodology, different types of subtasks or threads are spawned to execute the tasks required in a specific order. For example, to execute an initialize command for device number one attached to serial port 2 (B) use the following code:

```
Call SetSyrCmd(2, 1)

Call SyrCmd("W4R")
```

The main concern now is determining if the device is busy or ready. This can be accomplished by two methods. Depending upon the type of command sent, only one or the other may work. The most reliable way to determine the status of any device is to query its status byte at regular intervals to see when it is ready. However, a factor does complicate this process.

This protocol has two methods for sending syringe movement commands. If an upper case letter is used, then the syringe status byte is set to busy until the syringe has completed its movement to the specified position. For this case, the status byte can be polled (or queried) using a timer to determine when the movement is completed.

However, if a lower case letter is used, then the syringes status is returned as ready immediately after the command is given, regardless of whether the move has been completed or not. If the syringe has been issued a command using a lower case letter then two things have to be determined before setting the device status to ready: (1) was the command received ok (an ACK was received) and (2) has enough time elapsed that the move has been completed. Item (1) is simple enough to determine. Item (2) is slightly more complicated.

To calculate the time required to idle between commands requires that the program keep track of two parameters whenever they are changed. Those parameters are the current position of the syringe and the current speed setting of the syringe. Remember that these values are stored in the global array variables SSpeed() and Curr_Pos(). It is important to know the current position because it is necessary to calculate the number of steps that the syringe will be moving = [abs(starting position − ending position)]. The speed is given in steps per second. Therefore, it is simple to construct a function that calculates the time a particular device should spend idle:

```
Public Function Kloehn_TDelay(Steps, speed) As Single

'Calculates Wait Time Required After Issuing Syringe Command

'The 0.5 is an additional safety factor

Kloehn_TDelay = Int((Steps/speed)) + 0.5

End Function
```

This type of timing scheme is used with a variety of instruments, robots, syringes, moving conveyers, etc. All will give some type of speed that is a function of how many units (steps, inches, cm) it can move in a unit of time (seconds, milliseconds, minutes).

The prior function is called when the Determine_Wait subroutine is called, which occurs every time a command is sent. Its code follows:

```
Public Sub Determine_Wait(Cmd$)

Dim Steps As Single, Position As Single

Select Case Left$(Cmd$, 1)

   Case "V"

      SSpeed(Real_DevNo) = Val(Mid$(Cmd$, 2, Len(Cmd$) - 2))

   Case "p", "P"

      Steps = Val(Mid$(Cmd$, 2, (Len(Cmd$) - 2)))

      Wait_Time(Real_DevNo) = Timer + Kloehn_TDelay(Steps,
      SSpeed(Real_DevNo))

      Curr_Pos(Real_DevNo) = Curr_Pos(Real_DevNo) + Steps

   Case "d", "D"
```

```
        Steps = Val(Mid$(Cmd$, 2, Len(Cmd$) - 2))

        Wait_Time(Real_DevNo) = Timer + Kloehn_TDelay(Steps,
        SSpeed(Real_DevNo))

        Curr_Pos(Real_DevNo) = Curr_Pos(Real_DevNo) - Steps

    Case "a", "A"

        Position = Val(Mid$(Cmd$, 2, Len(Cmd$) - 2))

        Steps = Abs(Curr_Pos(Real_DevNo) - Position)

        Curr_Pos(Real_DevNo) = Position

        If Steps > 0 Then

            Wait_Time(Real_DevNo) = Timer + Kloehn_TDelay(Steps,
            SSpeed(Real_DevNo))

        End If

    Case "W"

        Wait_Time(Real_DevNo) = Timer + 5

    Case Else

        Wait_Time(Real_DevNo) = Timer + 3

End Select

End Sub
```

Notice that this subroutine makes no distinction between an upper case command or a lower case command. It does not care whether the device status was set to busy or ready. It merely calculates the appropriate amount of time to wait for the movement to be completed. Once this has been determined, it is easy to determine the status of the device. The device changes status from ready to busy as soon as a command is sent to it. The status changes back from busy to ready as soon as two criteria have been satisfied: (1) a valid response has been received to the command and (2) enough time has passed for the action to have taken place (i.e., the syringe moves, the valve changes position, etc.).

The first condition is satisfied as soon as the command is issued, if a reply has not been received to the command sent, the command will be resent up to 10 times. The second condition is checked by means of another Timer control named Status_Timer whose interval is set to 100 (100 ms). Every time the interval passes the following subroutine is executed:

```
Private Sub Status_Timer_Timer()

For i = 1 To 15

    If Statts(i) = "Busy" Then

        If Timer > Wait_Time(i) Then

            Statts(i) = "Ready"

            Kloehn!Status_Button.Value = True

        End If

    End If
```

```
Next i

End Sub
```

By means of a loop, the status of all 15 possibly connected devices is checked and updated every 100 ms. This system is much less taxing on system resources because for every command sent only 1 reply must be interpreted from the bus — as opposed to continuously polling the bus and checking the replies to determine the status of each and every device.

Occasions will probably arise when it would be desirable to send out commands to several different devices, and then wait for every device to finish the tasks assigned to them before allowing any of the syringes to accept new commands. The following subroutine accomplishes this nicely. The subroutine will wait up to 2 minutes for all pending commands on all devices to finish.

```
Public Sub IdleTillDone()

'This Sub will Idle the Program Until all Pending Cmds are
Finished

Dim Finished As Boolean, stop_value As Single, count As Integer

Dim Total_Devices As Integer

'Find Total No of Devices in Known State

For i = 1 To 15

   If Statts(i) = "Ready" Or Statts(i) = "Busy" Then

      Total_Devices = Total_Devices + 1

   End If

Next i

Finished = False

stop_value = Timer + 120

While Finished = False And Timer < stop_value

count = 0

   For i = 1 To 15

      If Statts(i) = "Ready" Then

         count = count + 1

      End If

   Next i

   If count = Total_Devices Then Finished = True

   DoEvents

   'Debug.Print "Waiting.....", Count, Total_Devices

Wend 'End While loop when all Pending Cmds are Finished or 2 min

End Sub
```

Remember, from looking at the status byte codes that the 5th bit of the status byte indicates whether the device is ready or busy. A crucial part of communication with any device is the ability to transform bytes into their binary equivalents or bits. Often, one byte of information can convey information on many different parameters, with each bit, (or sometimes a group of bits) representing the state of different parameters.

In the subroutine CheckStatusByte, the status byte was utilized only to determine if an error occurred when the last command was sent. The status byte does however encode other information within its value, notably the busy/ready flag for the device.

Recall from the device response format that a response may be returned starting at byte 5. Currently, the device driver makes no provision for decoding the busy/ready flag, or extracting a response should one be returned. By modifying the subroutine Interpret_Reply and creating one new subroutine this information can be extracted, and if necessary utilized.

```
Sub Interpret_Reply(reply$)
Dim delim$, No_Tokens%, Track%, Resp$
Dim msg$, i As Integer
delim$ = Chr$(32)
No_Tokens% = Int(Len(reply$)/3)
For i = 1 To No_Tokens% + 1
    Token delim$, reply$, i, msg$
    'Stop
    If msg$ = "FF" Then
        'Debug.Print " ** OEM Line Sync Detected **"
        Track% = 1
    End If
    If Track% = 1 And msg$ = "2" Then Track% = 2
    'Do Not include Status Byte in Response
    If Track% = 3 Then
        Token delim$, reply$, i + 1, msg$
        If msg <> "3" Then
            Resp$ = Resp$ & " " & msg$
        Else
            Track% = Track% + 1
        End If
    End If
    If Track% = 2 And msg$ = "30" Then
        Token delim$, reply$, i + 1, msg$
        'Stop
        Call CheckStatusByte(msg$)
        Call CheckBitFive(msg$)
        Track% = Track% + 1
```

```
        End If

    Next i

    If Resp$ <> "" Then Debug.Print "Response:  "; Resp$

End Sub
```

Notice that the preceding revised subroutine passes the status byte to a new subroutine Check-BitFive which will be defined later. In addition, it extracts any response that may have been sent in the reply to the variable Resp$. From the protocol, all responses fall between the status byte and the end of response mark (03h). If hexadecimal 03 immediately follows the status byte, then no response has been sent in the reply.

```
Sub CheckBitFive(status_byte$)

'This subroutine determines value of all Bits in the Status Byte

'Bit No. 5 is of particular interest as it determines:

'1 = "Ready" Or 0 = "Busy"

Dim BitNo(7) As Integer, DecVal As Integer

Dim Check_Byte As Integer, Display$

'Deconstruct Hex Value into Binary Form

If status_byte$ = "" Then Exit Sub

status_byte$ = "&H" & status_byte$

DecVal = CDec(status_byte$)

Debug.Print "Decimal Value: "; DecVal

Check_Byte = 255 - DecVal

For Bit = 7 To 0 Step -1

    If Check_Byte > = 2 ^ Bit Then

        Check_Byte = Check_Byte - 2 ^ Bit

        BitNo(Bit) = 0

        Else

        BitNo(Bit) = 1

    End If

    Display$ = Display$ & Trim$(Str(BitNo(Bit)))

Next Bit

Debug.Print "Bit Number: "; "76543210"

Debug.Print "Binary Form: "; Display$

End Sub
```

This subroutine is useful in that it can take any byte and extract its binary form. First, the decimal value of the passed hexadecimal byte is extracted using the CDec() function (the CByte() function could also be utilized). Next a value (Check_Byte) is obtained by subtracting the Decimal value from 255 (the maximum binary value for an 8 bit byte). This value is then used in the loop to extract the value for each bit which is stored in the array BitNo(). The result of the busy/ready flag is stored in BitNo(5). When running this code the following will be printed to the immediate window if no response is present in the reply:

```
Hex Message: FF 2 30 48 3 79 FF

*** No Error ***

Decimal Value: 72

Bit Number: 76543210

Binary Form: 01001000
```

(Notice 48 h ⇔ 72 d)
If a reply is present something similar to the following will be printed to the immediate window.

```
Hex Message: FF 2 30 60 32 33 31 35 20 3 74 FF

*** No Error ***

Decimal Value: 96

Bit Number: 76543210

Binary Form: 01100000

Response: 32 33 31 35 20
```

(Notice 60 h ⇔ 96 d, and that the reply is between the status byte and 3 h.)

A complete device driver for a Kloehn Syringe using the subroutines covered in this section is provided at http://www.pharmalabauto.com and is shown in Figure 3.1. The methodologies employed previously should allow the reader to construct a device driver for any device with a known protocol.

3.8 CONTROLLING MULTIPLE INSTRUMENTS FROM DIFFERING MANUFACTURERS

While the prior application is robust at controlling a set of Kloehn syringes, an assay will not usually be composed of like components from a single manufacturer. As a matter of course, an

FIGURE 3.1 The completed Kloehn syringe project.

assay will be constructed from several instruments, many of which are from different manufacturers. Many different schemes can be employed to create a single unified program to control all of these different instruments, but the method presented in this chapter is superior to many for the following reasons. First, it allows the designer to modularize the system. When a task is broken into modules, each module can be debugged separately, which is infinitely preferable to having to debug the entire method each time there is a problem. Second, this scheme calls for the designer to implement a rules based system which makes change to the overall method possible by simply changing the rules by which the method operates.

As an example, take an assay to be developed that consists of two syringes, a robotic arm to manipulate a needle, and a microtiter platereader. Obviously designing and writing the code for such a system is beyond the scope of this text. However, once you know how to write a device driver (as was demonstrated earlier in this chapter), you can write a device driver for every instrument to be used in your system. At that point, to create a viable working system, all that remains is to have a method of integrating all the various instrument's drivers.

As a starting point for creating an assay, create three device drivers, one for the syringes, one for the robotic arm, and one for the platereader. Each device driver should be created as an independent module. In other words, no matter what language the device driver is written in, the source code could be loaded into the development environment and be used to control the instrument independently. For example, the device driver to control the platereader could be used to load and unload microplates, and trigger the platereader to take readings of the microplates after they have been loaded. It would probably be desirable for the developer to take each driver and create an OCX, OLE, or VBX control out of the device driver code. Then, each driver could be imported into whatever development environment the designer wished to utilize by simply loading the control into their project where it would appear in the custom controls toolbar (full of both standard and custom OCX, OLE, and VBX tools).

Now it is not enough to create three device drivers to control the various instruments to be utilized in the assay. The designer must think three levels of abstraction ahead of where they are going with the project. The user developing the method must be able to have control over the parameters of each device. Also, the developer will need to extract specific information from the instrument. Ideally, every device driver should be able to set and extract every parameter in the instrument. However, the reality of the situation is that every organization operates with time constraints and it may not be necessary to incorporate total functionality into every device driver that an assay will require to function.

As an example, a device driver to control a platereader should have at a minimum the ability to: (1) send or extract a plate into the reader; (2) start a scan on the reader; and (3) set the scanning parameters, i.e., gain, source intensity, etc. The driver should similarly be able to extract from the reader (1) the location of the plate (in/out), (2) the values of the last microplate scan (results), and (3) the current scanning parameters.

Although the parameters required for a device driver will vary from instrument to instrument, two parameters must be available if the user wishes to integrate the device driver as part of an overall system or method. Those parameters are Status and Error.

The Status parameter can have one of three possible states: busy, ready, or unknown. Typically, unknown is only encountered in the startup of the software or when power is first applied to the device. After a powered-up device has been reset by the software, the Status should always be known (or be able to be requested and become known). If the status of an instrument becomes unknown in the middle of executing an assay, some sort of corrective action must be taken.

The second parameter, Error, is Boolean in nature. Either no errors have occurred or an error has occurred. That being said, if an error has occurred it can fall into one of three categories. It can be a fatal error from which recovery is not possible, in which case the method or assay must be terminated (preferably saving any previously acquired data first). It can be a nonfatal error that prohibits further commands from being executed until some sort of action is taken, i.e., the device is reset, etc. Finally, it can be a nonfatal error that does not inhibit the device from executing further commands, but is merely logged to the system. Obviously the controlling software must be able to handle all three kinds of errors.

When the issue of controlling the various devices is solved, the next logical concern is making the instruments function as a cohesive group, as opposed to a group of individual devices that execute commands without regard to the needs of the whole group. This is really a two step process. First, a determination must be made as to whether a device is capable of executing a command. Notice that here a determination is only made if the device can execute a given command, not what effect the execution of that command will have upon group members.

Recall when the Kloehn syringe device driver was developed earlier in this chapter that a queue had to be developed to allow control of multiple syringes. A similar methodology must be used when controlling multiple devices of differing manufacturers. In this type of situation it is often useful, for reasons of clarity, to use a public (or global) variable with multiple dimensions. For example:

```
Public Instrument(Device, IDNumber, Parameter) As Integer
```

This example, is the most compact of the options, using a three-dimensional array to hold all the information on any particular device in the system. In this example the information held for each device is coded and stored in the array in the form of an integer.

The first element of the array is the actual device that is being referenced. In the example outlined earlier, the device could be a platereader, a robotic arm, or one of two syringes. The designer of the system might choose to code the possible devices as follows:

```
Platereader = 1

Robotic Arm = 2

Syringe = 3
```

The second element of the array is referred to as the IDNumber. Recall that each device on the bus must have its own unique address (or device number). Remember commands are sent to all the instruments connected on the bus, but only those whose address is encoded in the command string will respond to the sent command. Now, it is also possible to have multiple buses (i.e., using multiple serial ports) in which case a fourth dimension should be added to the public variable instrument. For example:

```
Public Instrument(Device, SerialPortNumber, IDNumber,
Parameter) As Integer
```

For the sake of simplicity in this discussion, use the case where all instruments reside upon the same bus (or are connected to the same serial port). The designer might then choose to code the IDNumbers of the devices in the system as follows:

```
Syringe 1 = 1

Syringe 2 = 2

Platereader = 3

Robotic Arm = 4
```

The last dimension in the example array is Parameter. The parameter to be set or referenced is coded in this dimension. Taking the simplest case, only code the status and error parameters. The designer might then choose to code the two parameters of the devices in the following way:

```
Status = 1

Error = 2
```

Some examples from this coding scheme follow (where X = don't care condition):

- Instrument(1,3,X) -> references the platereader
- Instrument(2,4,X) -> references the robotic arm
- Instrument(3,1,X) -> references syringe 1
- Instrument(3,2,X) -> references syringe 2

Getting more specific, information must be returned about each parameter (in this case status and error). The designer may choose to code the parameters to be returned in the following way:

```
Status = Busy = 1

Status = Ready = 2

Status = Unknown = 0

Error has occurred = 1

Error has not occurred = 0
```

Using this enhanced coding scheme, if the designer looks to the public variable instrument with the following arguments for dimensions, they can extract the following information:

- Example 1
  - Instrument(1,3,1) refers to platereader status
  - Instrument(1,3,1) = 0 (platereader status is unknown)
  - Instrument(1,3,1) = 1 (platereader status is busy)
  - Instrument(1,3,1) = 2 (platereader status is ready)
- Example 2
  - Instrument(3,2,2) references errors on syringe 2
  - Instrument(3,2,2) = 0 (an error has not occurred on syringe 2)
  - Instrument(3,2,2) = 1 (an error has occurred on syringe 2)

While this coding mechanism is the most compact, it is not the most intuitive to understand. This is especially true for an individual who is called upon to rework the code and is not the author of the original code. It is also practical to utilize a coding scheme where the instrument parameters can be accessed via a number of public variables. The following is an example of a scheme:

```
Public Status(Device) as Integer

Public Error(Device) as Integer

Public IDNumber(Device) as Integer

Public Bus(Device) as Integer
```

This scheme takes a different tact in that every parameter has been assigned a public variable. In addition, where the bus device resides and the address (IDNumber) it uses on the bus are also referenced through unique public variables. This scheme is much more understandable upon inspection but should be used when only a few parameters are to be utilized. Each parameter in turn spawns another public variable. Of course a hybrid that utilizes aspects of both coding schemes could be adopted. Ultimately, it is the designer of the system who chooses the coding scheme to control a method. Notice that in this scheme every public variable is a one-dimensional array, and the only parameter required to access information within the array is the device number.

Once the process of determining if a device is capable of executing a command has been accomplished, the more complicated problem arises of determining if the execution of that particular command, at that particular moment, is harmonious within the overall system. This determines if the execution of a given command will lead to the successful completion of a method or if it will inadvertently destroy the process to be automated. Why give a command that might harm the process to be accomplished? As a matter of course, any command coded into the process under ideal or normal conditions would serve to automate the process successfully. However, under real world conditions things can fail, which could lead to an otherwise innocuous command wreaking havoc upon the system.

In the system described, suppose that syringe 1 will be used to fill the microplate with buffer, and syringe 2 will be used to add compound to the microplate after syringe 1 fills the microplate with buffer. Under normal circumstances this process would operate fine. However, suppose for some reason syringe pump 1 failed to deliver the buffer to the microplate (maybe, the motor burns out and the syringe repeatedly sends an error response to the commands). If syringe 2 does not check to see if syringe 1 did its job, it will simply deliver the compounds to a microplate devoid of buffer. Further, if the system is totally automated, the platereader will take a measurement on a plate prepared improperly and possibly upload incorrect results into a database. The situation would be even more catastrophic if the compounds to be assayed were available in limited quantities, were proprietary, and took several weeks for the chemists to synthesize in house.

To prevent this, step number two in the process makes a set of instruments function as a cohesive group. Step two is simply implementing a "rules based mechanism" to oversee the operation of all instrumentation within the system (Figure 3.2). From a system perspective, the instruments themselves would reside at the lowest levels. The next level would comprise the various device drivers that control the instruments. The highest level would be the controlling mechanism that would control execution of method (probably, a computer program written in VB, C++, or Java).

The sample application outlined in Figure 3.2 consists of two syringes, a robotic arm, and a microtiter platereader. The following rules may be implemented:

- No command will be sent to a device unless that device's status is ready
- The platereader will not scan a plate unless it is full of buffer and sample
- Syringe 2 will not dispense the compound into the microplate until the robotic arm reaches the location it was last directed to over the microplate well

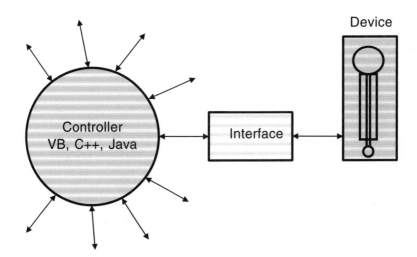

FIGURE 3.2 A system level perspective of instrumentation control.

- If a syringe logs an error or its status becomes unknown, the syringe will be initialized in an attempt to put the syringe back into a ready state
- If the robotic arm is unable to verify its location within the x,y,z coordinate system, the assay will be halted
- Syringe 1 will not aspirate buffer if the liquid level sensor determines the buffer reservoir is empty

The best way to implement a rules based system is to be as generic as possible. Each rule should be verified by means of a function that returns true if the application is free to proceed, and false if the application needs to wait or take corrective action. For example, a generic function that is capable of checking the status of every instrument is very easy to construct. Prior to sending a command, call this function to determine the status of the instrument where the command is being sent. If the instrument is ready, send the command. If the instrument is busy, the program will wait until the instrument is ready to receive the command. If the status of the instrument is unknown, then some sort of corrective action must be taken to determine its true state or reset the state of the instrument.

The application of well thought out system design rules and considerations should allow anyone with sufficient technical expertise to construct an automated method utilizing many different types of instrumentation.

3.9 HARDWARE CONSIDERATIONS IN ROBOTIC AUTOMATION

Sometimes a software control application that seems to function perfectly can be constructed. At other times, for one reason or another, machines will fail to function properly or the system may freeze altogether. Robotic automation — not encompassed by software alone — is the process of software and hardware acting together in unison. Most people take the hardware for granted and assume it will always function properly. When a single device on a bus (or more often, a single device connected to a serial port) is used, the device will in fact nearly always function flawlessly. However, when the process begins of integrating many different devices from many different vendors on a bus, hardware or firmware incompatibilities may exist.

In extreme cases, the only solution is to place differing hardware on separate buses to avoid the possibility of the hardware devices conflicting with one another. Conflicts are often not the fault of the manufacturer as it would be impossible for them to test the compatibility of their products with every potential piece of hardware that might be used.

One of the most troubling aspects of hardware problems is that they often occur intermittently, which makes troubleshooting them all the more difficult. Like, when a car makes a funny noise, but it remains unexpectedly quiet when brought to the mechanic. When the desired results are not received from a software command, and the fault cannot be found within the software itself, follow rule number one:

- Rule number 1 — place instrumentation from differing manufacturers on their own bus. Or, dedicate a separate serial port for each different piece of equipment. Pay close attention to the word difference. If one has 4 Kloehn syringes in a system, the four syringes can be chained off one serial port. But, a robotic arm or switching valve that claims to be compatible with Kloehn syringes should be run off a separate bus.

If problems persist after following rule number one, see rule number two:

- Rule number 2 — each and every communication line in the system must be terminated properly.

FIGURE 3.3 The cause of reflections in transmission lines.

All wires have an inherent resistance, inductance, and capacitance. These parameters form the characteristic impedance (Z_0) of the transmission line. If a line is short enough, the inherent resistance, inductance, and capacitance is small enough to be ignored. However, at some length the inherent parameters within the transmission line become too significant to ignore. This length will vary depending upon the speed of the signals that will propagate down the line, the environment in which the line is run, the shielding and grounding scheme of the line, and the physical characteristics of the line.

When the length of a transmission line between a computer and an instrument becomes too long, reflections will occur back down the line toward the source. Reflections will occur because the load attached to the end of the line is not equal to the characteristic impedance (Z_0) of the transmission line. Because of this inequality, all the power transmitted at the source (the computer) does not reach the load (the devices on the bus). The power that is not absorbed by the loads must go somewhere, so it reflects back down the line toward the source.

A good analogy would be to think of two balls thrown against a wall (Figure 3.3). The first ball is made of bread dough, and the second is a regular tennis ball. When the ball made of dough is thrown against the wall it does not bounce back. All the energy it contains is imparted to the wall that it hits. The energy contained within the dough is used to deform its shape and in the process the dough gives off a small amount of heat, etc. When the tennis ball is thrown against a wall it too deforms slightly but then it bounces back as it retains most of the energy imparted to it by the source (the person who threw it).

Ideally, the load attached to the end of the transmission line must match the characteristic impedance (Z_0) of the transmission line, in which case the line is matched. When a line is matched to its load, a maximum amount of power will be transferred from the source to the load. Even on a matched line reflections will still occur, but they will be so small that they do not cause a problem. This occurs when the transmission line is "short" enough. Matching is accomplished through termination, which is adding a load that may be composed of inductors, capacitors, and resistors.

What does this have to do with laboratory automation? Well, if a number of instruments are connected to a bus (line), and that line is improperly terminated, reflections will occur down that line back toward the source. Those reflections will arrive at every instrument connected to the bus on the way back toward the source (computer). If one of those reflections should imitate a legitimate control signal, that command will be executed on that instrument. Worse still is this effect will occur randomly, as the reflections back toward the source will be somewhat chaotic in nature.

This phenomenon is explained in Figure 3.4. The arrows on the clock line represent the rising edge of the clock where the control signals are sampled. The noise line reflects toward the source. The bold line represents a legitimate control signal.

The straight horizontal lines represent the threshold values for CMOS and TTL logic level detection. For TTL, the unknown region is between 0.8 – 3.3 volts. A voltage below 0.8 will be treated as a logic low and a voltage above 3.3 will be treated as a high. A voltage in between could be read as either state. For CMOS, the unknown region is between 1.3 – 3.7 volts. A voltage below 1.3 will be treated as a logic low and a voltage above 3.7 will be treated as a high, and a voltage in between could be read as either state.

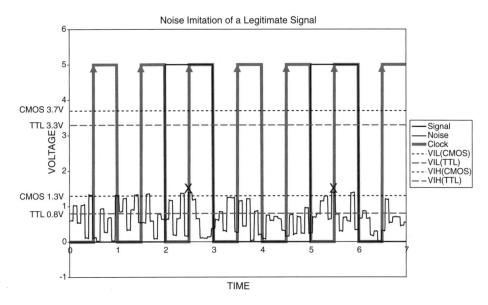

FIGURE 3.4 Reflection imitating a legitimate control signal.

Figure 3.4 shows a control signal made up of 7 bits. A control signal typically comprises 8 bits or one byte but having only 7 gives the figure more clarity. Looking at the control signal, it is clear that the bit values in order from lowest to highest (left to right on chart) are:

Bit No: 1,2,3,4,5,6,7
Value: L,L,H,L,L,H,L (L = logic low, H = logic high)

Now, look at the reflected noise signal and read the values where the noise will be sampled (the rising edge of the clock).

Bit No: 1,2,3,4,5,6,7
Value: L,L,?,L,L,?,L (L = logic low, ? = unknown state)

Notice that bits 3 and 6 are sampled at a point where the reflected noise has a high enough value to be clocked in the unknown region for both TTL and CMOS. The consequence of this is that both bits 3 and 6 could be clocked in as high. The ramifications of this are that the noise could imitate the legitimate control signal: L,L,H,L,L,H,L. If this control signal caused the instrument to reset itself, then reflected noise could cause the system to intermittently reset the machine. If this control signal caused the instrument to take a reading, then a reading could be taken at the wrong time.

Fortunately, tried and true methods of matching transmission lines exist. In order of complexity they are: (1) no termination, (2) series termination, (3) parallel termination, (4) RC termination, and (5) failsafe termination. Additional reference material on this subject can be found in Appendix D.

The no termination option is feasible if the cable is short and if the data rate is low. Reflections occur, but they settle after about three roundtrip delays. For a short cable, the round trip delay is short, and if the data rate is low, the unit interval is long. Under these conditions, the reflections settle out before sampling, which occurs at the middle of bit interval.

A low data rate ensures that reflections have sufficient time to settle before the next signal transition. At the same time, a short cable length ensures that the time required for the reflections

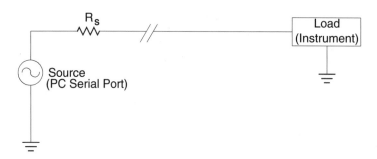

FIGURE 3.5 Series termination.

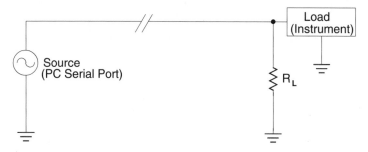

FIGURE 3.6 Parallel termination.

to settle is kept to a minimum. A low data rate and a short cable length are the most significant shortcoming of the unterminated option.

Low speed is generally defined as one of the following: (1) below 200 kbits/sec, (2) the cable delay (the time required for an electrical signal to transverse the cable) is substantially shorter than the bit width (unit interval), and (3) the signal rise time is more than four times the one way propagation delay of the cable (i.e., not a transmission line). As a general rule, if the signal rise time is greater than four times the propagation delay of the cable, the cable is no longer considered a transmission line.

Another termination option is popularly known as either series or backmatch termination (Figure 3.5). The termination resistors, Rs, are chosen because their value plus the impedance of the driver's output equals the characteristic impedance of the cable. Now, as the driven signal propagates down the transmission line, an impedance mismatch is still encountered at the far end of the cable (receiver inputs).

However, when that signal propagates back to the driver, the reflection is terminated at the driver output. Only one reflection occurs before the driven signal reaches a steady state condition. The time it takes for the driven signal to reach steady state is still dependent upon the length of cable the signal must traverse.

The most popular termination option (parallel termination) is to connect a single resistor across the conductor pair at each end (Figure 3.6). The resistor value matches the cable's differential mode characteristic impedance. If the bus is terminated in this way, no reflections occur, and the signal fidelity is excellent. The problem with this termination option is the amount of power dissipated in the termination resistors.

Parallel termination is arguably one of the most prevalent termination schemes today. In contrast to the series termination option, parallel termination employs a resistor across the differential lines at the far (receiver) end of the transmission line to eliminate all reflections. Eliminating all reflections requires that R_T be selected to match the characteristic impedance (Z_O) of the transmission line. As a general rule, it is usually better to select R_T to be slightly greater than Z_O. Overtermination

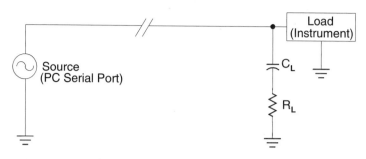

FIGURE 3.7 RC termination.

tends to be more desirable than undertermination, since overtermination has been observed to improve signal quality. R_T is typically chosen to be equal to Z_O. When overtermination is used, R_T is typically chosen to be up to 10% larger than Z_O. The elimination of reflections permits higher data rates over longer cable lengths. However, an inverse relationship occurs between data rate and cable length. That is, the higher the data rate the shorter the cable and conversely the lower the data rate the longer the cable. Higher data rates and longer cable lengths translate simply into smaller time unit intervals (TUIs) and longer cable delays. Unlike series termination, where high data rates and long cable lengths can negatively impact data integrity, parallel termination can effectively remove all reflections; thereby removing all concerns about reflections interfering with data transitions.

If power dissipation must be minimized, an RC termination (Figure 3.7) may be the solution. In place of the single resistor, a resistor in series with a capacitor is used. The capacitor appears as a short circuit during transitions, and the resistor terminates the line. Once the capacitor charges, it blocks the DC loop current and presents a light load to the driver. However, low pass effects limit use of the RC termination to lower data rate applications.

The value of R_T generally ranges from 100Ω to 150Ω (dependent upon cable Z_O) and is selected to match the characteristic impedance (Z_O) of the cable. C_T, on the other hand, is selected to be equal to the round trip delay of the cable divided by the cable's characteristic impedance (Z_O). The capacitor (C_T) should ideally be a high-speed polypropylene capacitor. Ceramic capacitors, although less robust, may also be used. Under no circumstances should a tantalum capacitor be used.

$$EQ1: C_T = (\text{Cable round trip delay})/Z_O$$

For example:

```
Cable Length = 100 feet
Velocity = 1.7 ns/foot
```

$$\text{Char. Impedance } (Z_O) = 100\Omega$$

Therefore,

$$C_T = (100 \text{ ft} \times 2 \times 1.7 \text{ ns/ft})/100\Omega \text{ or } = 3,400 \text{ pF}$$

The resulting RC time constant should be less than or equal to 10% of the unit interval (TUI). In the example provided, the maximum switching rate therefore should be less than 300 kHz. This termination should now behave like a parallel termination during transitions, but yield the expanded noise margins during steady state conditions.

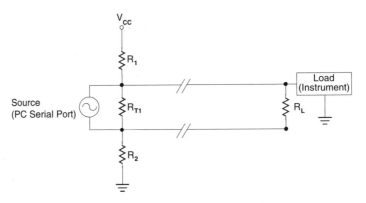

FIGURE 3.8 Failsafe termination.

When the driven signal transitions from one logic state to another, the capacitor C_T behaves as a short circuit and consequently, the load presented to the driver is essentially R_T. However, once the driven signal reaches its intended levels, either a logic high or logic low, C_T will behave as an open circuit. DC loop current is now blocked. The driver power dissipation will then decrease. The load presented to the driver also decreases. This occurs because the driver is now loaded with a large receiver input resistance typically greater than 4 kΩ instead of the typical R_T of 100Ω to 120Ω. This reduced loading condition increases the signal swing of the driver and results in increased noise margin.

As with all the previously discussed termination options, using RC termination has disadvantages. RC termination introduces a low pass filtering effect on the driven signal that tends to limit the maximum data rate of the application. The 3 dB cutoff point for a single pole low pass filter can be calculated by:

$$f_c = \frac{1}{2\pi R_T C_T}$$

This data rate limitation is the result of the impact that R_T and C_T, together, have upon the driven signal's rise time. The data rate limit is dependent upon the selection of R_T and C_T. Long RC time constants will have a greater impact upon the driven signal's maximum data rate, and vice versa. Because of these data rate limitations, the transmission lines best suited for RC termination are typically low speed control lines where level sensitivity is desired over edge sensitivity. Finally, the cost of parts required by RC termination can put it at a disadvantage in cost conscious applications.

Another popular option is a modified parallel termination that also provides a failsafe bias (Figure 3.8). The need for failsafe operation is both the principal application issue and most frequently encountered problem with RS-485. Failsafe biasing provides a known state in which no active drivers are on the bus. Other standards do not have to deal with this issue, because they typically define a point to point or multidrop bus with only one driver. The one driver either drives the line or is off. Because only one source is on the bus, the bus is off when the driver is off. RS-485, on the other hand, allows for connection of multiple drivers to the bus. The bus is either active or idle. When it is idle with no drivers on, the state of the bus can be high, low, or in the state last driven.

With no active drivers and low impedance termination resistors, the resulting differential voltage across the conductor pair is close to zero, which is in the middle of the receivers' thresholds. Thus, the state of the bus is truly undetermined and cannot be guaranteed. Some of the functional protocols that many applications use aggravate this problem. In an asynchronous bus, the first transition

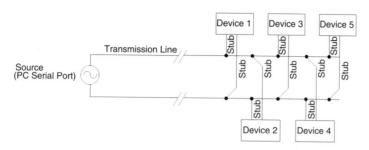

FIGURE 3.9 Stubs in transmission lines.

indicates the start of a character. It is important for the bus to change states on this leading edge. Otherwise, the clocking inside the UART is out of sync with the character and creates a framing error. The idle bus can also randomly switch because of noise. In this case, the noise emulates a valid start bit, which the UART latches. The result is a framing error, or worse, an interrupt that distracts the CPU from real work.

The way to provide failsafe operation requires only two additional resistors. At one end of the bus (the master node, for example), connect a pullup and pulldown resistor. This arrangement provides a simple voltage divider on the bus when there are no active drivers. Select the resistors so that at least 200 mV appears across the conductor pair. This voltage puts the receivers into a known state. Values that can provide this bias are 750Ω for the pullup and pulldown resistors, 130Ω across the conductor pair at the failsafe point, and a 120Ω termination at the other end of the cable. For balance, use the same value for the pullup and pulldown resistors.

The lack of R_1 and R_2, when the bus is idle, almost assures that the receiver output will not be in a known state. This is due to the insufficient voltage across R_T (on the order of 1 to 5 mV), as caused by the receiver's internal high value pullup and pulldown resistors. The presence of these internal pullup and pulldown resistors will guarantee receiver failsafe only for the open input condition. In order to switch the receiver into the logic high state, regardless of whether the bus is open or idle, a minimum of +200 mV (with respect to the inverted receiver input) must be developed across R_T. The sole purpose of R_1 and R_2 is to establish a voltage divider whereby at least +200 mV will be dropped across R_T.

However, the addition of external receiver failsafe biasing resistors does pose some concerns. The primary drawback relates to the increased driver loading with the addition of R_1 and R_2. The increased driver loading decreases the driver's output swing and, in turn, reduces the noise margin. Higher driver power dissipation is also symptomatic of the increased driver loading since the driver must source the additional current required by the external failsafe network. One last concern is that the extra cost and subsequent handling of two additional resistors (excluding R_T) might outweigh power termination's advantages in some applications. The advantages of failsafe termination point directly to its increased ruggedness. A transmission line terminated using the failsafe option will be able to withstand larger common mode voltages.

The last topic for consideration in this area is stubs (Figure 3.9). Connecting an instrument to a cable creates a stub, and that connection point is known as a node. By definition every node has a stub. Minimizing the stub length minimizes transmission line problems. In essence, stubs themselves are minature transmission lines. Stubs appear at two points, between the termination and the device behind it, and between the main cable and the termination. The stub's length should be kept as short as possible. Keeping a stub's electrical length below one fourth of the signal's transition time ensures that the stub behaves as a lumped load (as part of the termination) and not as a separate transmission line.

If the stub is too long, a signal that travels down the stub reflects to the main line after hitting the input impedance of the device at the end of the stub. This impedance is high compared with that of the cable. The net effect is degradation of signal quality on the bus. Keeping the stubs as short as possible avoids this problem. Instead of adding a long branch stub, the main cable should be looped to the device to be connected. If a long stub must be used, drive it with a special transceiver designed for that purpose.

3.10 SUMMARY

Successful automation and control of instrumentation relies upon a solid foundation of both hardware and software design. Well designed software instrument drivers are critical as they provide the mechanism for the controlling software application to interface with the various hardware devices encapsulated in the system. Equally as crucial is the transmission scheme implemented to deliver the control signals to their destinations with a minimum amount of corruption. Both hardware and software design considerations must be well thought out for trouble free operation.

4   Low Level Hardware Interfacing Using VB

Computer programs are great at crunching numbers and manipulating data, and clearly this is the reason most people use a computer. Thus far, this book has centered upon having a program perform a specific task, or series of tasks, on a batch of data produced from some process. This process would typically be an assay purchased from a vendor to measure some physical, chemical, or biological property. At some point however, most users would like to utilize a computer program to control a process, not to merely manipulate data.

In any organization, a machine is purchased to perform process X. As time goes on, it becomes clear to the user performing process X, that the method utilized to achieve process X is suboptimal in some way. The user then wishes to alter the method in which process X is done, or perhaps even more radically, build a new machine from scratch to perform process X.

Now, it is no longer a matter of merely processing data, but having the ability to control a series of devices that will create the data. Chances are a process of any degree of sophistication will require actions such as: solenoids turning on and off, liquids pumped from one vessel to another, analog voltages read from sensors, digital voltages read from boards, and measurements remotely triggered, etc.

Nearly everyone has seen large-scale factory operations that are computer controlled for applications such as assembling cars, steel production, and product packaging. Implementation of such automation on a smaller scale using a personal computer within a laboratory requires hardware that interfaces with a computer via a controller card. All hardware interfaces use similar methodology for setup and control. Example programs in this chapter apply to using the Keithley Metrabyte Metrabus system, but the underlying principles can be applied to any hardware I/O system. The sample programs given will actually drive and operate cards on the Keithley system, and can be altered to suit an individual's own application.

4.1 CONTROLLER CARDS AND ADDRESSING

The controller card is the director of communications between the computer and all the other cards in the hardware system. In the Keithley Metrabyte system, the controller card plugs into an ISA slot on the user's computer. The hardware cards to be controlled are connected to the controller card by a 50 conductor ribbon cable (referred to as a bus). Devices may be chained off the bus, that is more than one device may be connected to the cable. Each individual card connected to the bus must have a specific address. As the mailman needs an address to deliver a letter, the controller card needs an address to know where to send I/O commands. More on addressing later.

The Keithley Metrabyte system has two types of controller cards. The MDB-64 card has a direct connection to the bus of the host computer. It also draws power from the host computer to power cards connected to the bus. While this is convenient, the number of cards that can be powered is limited. If this limit is exceeded then a fuse will blow on the MDB-64 card. When this limit is reached, an external power supply must be connected to power the cards, and the fuse that allows power to be drawn from the computer must be removed, or it will blow again.

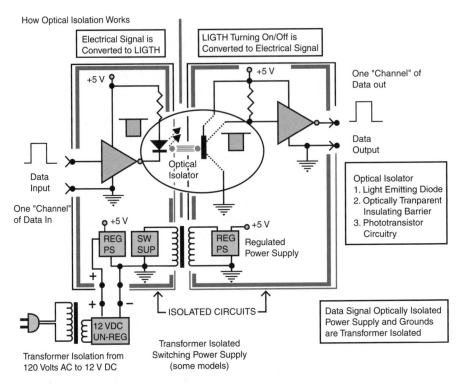

FIGURE 4.1 Optical isolation process. (From *Data Line Isolation Theory*, Technical Article #10, B&B Electronics Mfg. Co., Ottawa, IL, http://www.bb-elec.com, November 1998. With permission.)

The second type of controller card is the MID-64, which has optical isolation from the bus of the host computer. No direct connection exists between the card and the bus of the host computer. When a state changes on the Metrabyte bus, a light is projected on the substrate of a light sensitive semiconductor causing a voltage change from logic low to logic high. Because a physical connection to the computer's bus does not exist, a high voltage spike on the Metrabus system will not damage the host computer. Power must be supplied to the cards on the bus by means of an external supply. Because the host computer is protected from damage, this system is superior. The MID-64 card costs twice as much as the MDB-64 card. However, the money is well spent since the host computer is saved from damage.

The functionality of the MID-64 or the MDB-64 cards is illustrated in Figure 4.1. The cards plug into a PC's ISA bus. This long black connector inside the computer is typically located next to short white connectors that are PCI buses. ISA buses are rapidly falling into obsolescence, in fact some computers no longer have them. At this time Keithley Metrabyte does not produce a PCI controller card for the Metrabyte system, however an aftermarket vendor of Keithley components does. The installation and use of a PCI controller card will also be covered later.

To use an ISA card, its address must be specified and set on the card prior to installation. This is done through a dip switch on the card. Each switch represents one bit of the address. The switch can only be in two states, on or off. When a switch is in the off position, the total address is formed when the component of the binary address and its value are included. When a switch is in the on position, its value is not included in the total forming the address. This is somewhat counter intuitive.

In Figure 4.2, the address set is 768 decimal ($2^8 + 2^9 = 256 + 512 = 768$) The switches set to on do not contribute toward the address of each turned on switch. The contribution toward the total address is 2^n, where n is the position of the bit. In this case the bits start with position 2 (switch #8), and progress up to position 9 (switch #1) — again very counter intuitive. If switch 5

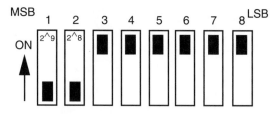

FIGURE 4.2 Setting the address via the dip switch. (From *Metrabus User's Guide*, Revision F, Keithley Metrabyte Division, Keithley Instruments, Inc., Cleveland, OH, July, 1994. With permission.)

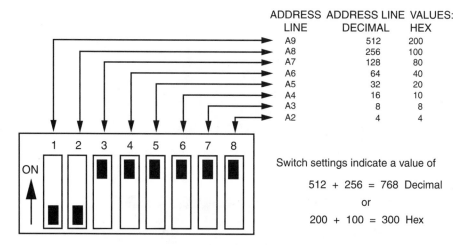

FIGURE 4.3 Detailed illustration of dip switch address settings. (From *Metrabus User's Guide*, Revision F, Keithley Metrabyte Division, Keithley Instruments, Inc., Cleveland, OH, July, 1994. With permission.)

were also off the address would be 2^5 + 2^8 + 2^9 = 32 + 256 + 512 = 800. Thus any combination of switch settings will result in a unique number. Figure 4.3 illustrates this in more detail.

In Figure 4.3, the address values are given both in decimal and hexadecimal format. Depending upon the function within the program to be used to control I/O, either one could be required. The easiest way to convert decimal numbers to hexadecimal, and to add hexadecimal numbers, is using a calculator.

Next, set the card to a free address, which is an address not utilized by another devise. Use the Win NT Diagnostics program (Start Menu -> Administrative Tools -> Windows NT Diagnostics) to find a free address. The resources tab will display open memory locations. Be advised however that cards often have unexpected names (Figure 4.4).

Installing a PCI card on the other hand is a whole different ballgame. The PCI card does not have a dip switch because the operating system chooses and assigns the address to be used by the card. This is called plug and play. In order for the user to know which address is assigned, some companies, such as ComputerBoards, provide a software utility that will list cards installed and their assigned addresses (Figure 4.5).

Even if the supplier, who makes the cards that will be used, does not supply software to determine the assigned base address, the resources tab in the Windows NT Diagnostics program can be used to determine the board's base address (Figure 4.6).

The only pitfall to this method is that the device name may not obviously identify the card whose address is sought. In Figure 4.7, the ComputerBoard PCI -MDB64 card is listed as CBUL32. Why "CBUL32" was imaged onto the hardware as a name is anyone's guess. However, it can be readily seen that the PCI-MDB64 card occupies two continuous address spaces from 2470 to 247F

FIGURE 4.4 The resources tab of the Windows NT diagnostic tools shows memory allocated to installed devices (cards, drives, etc.).

FIGURE 4.5 The Instacal program from ComputerBoards shows which cards are installed.

and from 2480 to 24FF. The base address of the card is always the lowest address, in this case 2470 hex. Converting 2470 hex yields 9328 decimal.

Always look at the devices present before a new card is installed. The new device listed after installation is the card that was just installed — regardless of its name. The instllation of PCI cards has one unique peculiarity. Sometimes a card will work in one slot but not in another. This sounds ridiculous but it is true. It does not happen often, but if a PC will not boot up properly after installing a card in one PCI slot, move it to another slot and see if it will work.

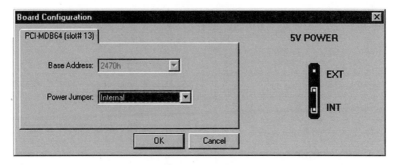

FIGURE 4.6 The Instacal program also shows the base address assigned to each card in the system in hexidecimal format.

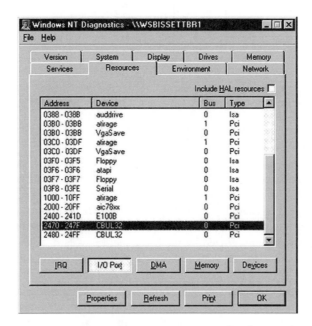

FIGURE 4.7 The PCI-MDB64 card occupies addresses 2470 to 24FF hexadecimal.

The address assigned to the card by the operating system should remain the same as long as the card is not moved from the slot where it was installed. Installation of other cards will not usually alter the addresses of preinstalled cards (but, under Windows anything is possible).

One last point worth mentioning, the controller card need not be used just with other cards from the Metrabus family. A custom circuit card, which is to be driven from the Metrabus controller cards, can be constructed. For reference, the pinout of the 50 pin Metrabus cable is illustrated in Figure 4.8.

4.2 INSTALLING AND USING THE PORTIO DLL

With a controller card properly installed, the next step in the process is to actually control the cards connected to the controller card on the bus. First, each card must have its own unique address, which is typically set by means of a dip switch on the card. Each card is then controlled by a single or series of bytes. Each byte contains 8 bits, which as stated earlier can be either in the on ("1") or off ("0") state. For example, an 8 channel relay controller board will utilize one byte to control

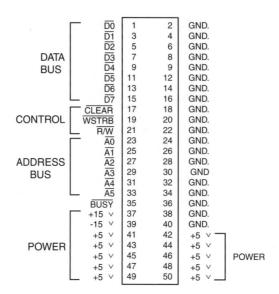

FIGURE 4.8 The Metrabus cable pinout assignments. (From *Metrabus User's Guide*, Revision F, Keithley Metrabyte Division, Keithley Instruments, Inc., Cleveland, OH, July, 1994. With permission.)

each of the relays on the board. Typically each bit will control one relay, if the bit is set to "1" the relay will be on, and if the bit is set to "0" the relay will be off.

The user needs to track the number of addressable locations each card is utilizing. In the example of the 8 channel relay controller, if a card utilizes 1 byte for its base address and 1 byte to control the 8 relays contained on the card, then the address space occupied by this card would be the base address + 1. For example, if the base address is 768d, do not set the addresses of any other cards on the bus to 768d or 769d. The next available address would be 770d because two bytes of the address space were utilized by the 8 channel relay card. If any other card on the bus is set to an address of 768d or 769d, an address conflict would result. When two or more cards share the same address space, a conflict results that makes it impossible to predict which card will carry out the commands issued. This is analogous to two people with the same name and address deciding which one will open the mail — impossible to predict. Obviously, it is very important when setting up cards on the bus to have nonoverlapping addresses.

Once the cards are connected to the bus and are addressed without conflicts, the individual bits and bytes are sent to a specific card for control. This step can be accomplished in a variety of ways. Some manufacturers provide ActiveX components that can be imported right into a programming environment, such as Visual Basic or Visual C++. Also, entire programs are dedicated to the control of such boards or cards, i.e., Labview or Testpoint. However, the most flexible method is the lowest level method, which is to actually write and read the addresses in and out of the port programmatically.

A very simple way to accomplish this is by using a free dynamic link library (DLL) available from Scientific Software Tools, Inc. (http://www.sstnet.com/) called the PortIO DLL. The PortIO DLL is a simple user mode DLL and NT kernel driver to allow direct hardware I/O access. This software is available for download both at Scientific Software Tools' web site and at Keithley Metrabyte's web site. The software includes examples for using the DLL both with Visual Basic and Visual C++, and the product can be adapted to any programming environment as long as the parameters are declared properly.

All of the examples in this section that relate to controlling cards utilize this DLL to access the hardware connected to the computer. When super fast speed is not required (the application is to accomplish a task within a quarter second paradigm as opposed to a microsecond paradigm), this driver works quite well.

4.3 CONTROLLING RELAYS

Relays are one of the most vital components in the automation of any system. A relay can be used not only to power on a specific instrument, but a relay can also be used to trigger events on instruments. For example, a relay connected to the terminals of a touch key switch can be used to simulate a keypress by closing the relay for a fraction of a second.

Relay cards fall into two basic categories. Heavy-duty relay cards can source several amps of current and are used primarily for providing power to instrumentation and ancillary equipment, such as pumps and vacuum degassers. Low current relay cards are used primarily at the circuit card level. Their use is relegated to tasks such as the automation of pushing buttons and grounding or providing voltage to traces on instrument circuit boards for the purpose of controlling a particular function on a device.

This section will cover the use of both a high current relay card and a low current relay card. Examples of software to control each card will be given along with a detailed explanation as to how the software works. The software examples are done in Visual Basic 6.0 and have been tested and debugged for each card. While many different brands of relay cards are on the market, they all operate using the same basic premises. The methodologies presented here are adaptable for use with other types of cards, should the user choose not to utilize Keithley Metrabus type cards. In fact, many clones that are compatible with the Keithley Metrabus system are available from manufacturers such as ComputerBoards, Inc.

4.3.1 THE MEM-08 RELAY CARD

The MEM-08 is a high current, double pole, double throw (DPDT) electromechanical relay board. While the board contains only 8 relays, each is capable of handling 5 amps, which is more than enough current to kill a human. This card is used to turn power on and off to various laboratory instruments.

In Figure 4.9 each relay has its own LED which indicates when the relay is in an energized state. Also notice the dip switch on the far right next to the ribbon connector cable. The user will use this switch to set the board address. Figure 4.10 illustrates how to set the board address. Note that the value is added to create the address for cards attached to the bus when the switch is in the on position. This is the exact opposite of how the address was created for the controller cards which plug into the PC's PCI or ISA slots.

The modularization process is very important to the beginning of an automation process. A project of any degree of complexity needs to perform a variety of tasks. Trying to implement all these tasks at the same time usually proves to be disastrous. Too many variables must be tracked down and too many potential conflicts between tasks must be resolved simultaneously. This is especially true when performing hardware and software integration, where it often is difficult to determine if the problem is software or hardware related.

FIGURE 4.9 The Keithley Metrabyte MEM-08 high current relay board.

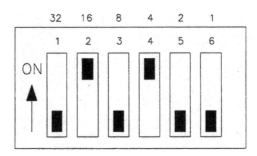

FIGURE 4.10 Setting the address on the dip switch (in this instance = 20d) (From *Metrabus User's Guide*, Revision F, Keithley Metrabyte Division, Keithley Instruments, Inc., Cleveland, OH, July, 1994. With permission.)

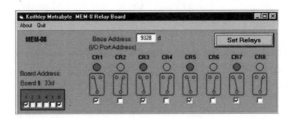

FIGURE 4.11 The MEM-08 sample application GUI.

A tactically smart approach to any large problem is to break the process down into several subtasks (or modularize it). Each subtask is then debugged and known to work before an attempt is made to integrate the subtask into the main process. In this section, each card on the bus is treated as a subtask. Every card has its own Visual Basic Project that contains forms and modules that can operate the card alone. Then, the software and the hardware can be tested separately to determine if they are working outside of the system. If the software and hardware works outside the system, but fails to work inside a process, then the fault is related to something occurring in the overall process – such as a device conflict. Operating in a methodical manner such as this greatly reduces the number of leads that must be checked when an error is encountered in a process.

When constructing software to control a piece of hardware, the first thing to construct is a graphical user interface (GUI) to represent the device to be controlled. Figure 4.11 shows the GUI developed to control the MEM-08 relay card. Note that the GUI is physically similar to the actual card. Also, the GUI is constructed using only the standard controls that come with Visual Basic. As a result, both upgrading from version to version and deployment are much easier because concern is eliminated about specialty controls being installed and licensed on machines in which this application will be deployed and developed.

In order to control the card by means of the PortIO.dll, the subroutines and functions built into the PortIO.dll must be declared and made available to the environment that is used for development. This function is accomplished easily by using the example module distributed with PortIO.dll. Chances are most applications will not use all the functions and subroutines declared in this module, but at least this way all are available for use in the project without worrying about having to add them later. The following code declares the subroutines and functions available in the PortIO.dll. (From PortI/O DLL, Shareware, Scientific Software Tools, Inc., Media, PA, 1998–2002.)

```
'* @doc INTERNAL

'* @module dlportio.bas |

'*

'* DriverLINX Port I/O Driver Interface
```

```
'* <cp> Copyright 1996 Scientific Software Tools, Inc.<nl>

'* All Rights Reserved.<nl>

'* DriverLINX is a registered trademark of Scientific Software
Tools, Inc.

'*

'* Win32 Prototypes for DriverLINX Port I/O

'*

'* Please report bugs to:

'* Scientific Software Tools, Inc.

'* 19 East Central Avenue

'* Paoli, PA 19301

'* USA

'* E-mail: support@sstnet.com

'* Web: www.sstnet.com

'*

'* @comm

'* Author: RoyF<nl>

'* Date: 09/26/96 14:08:58

'*

'* @group Revision History

'* @comm

'* $Revision: 1 $

'* <nl>

'* $Log:/DLPortIO/API/DLPORTIO.BAS $

'

' 1 9/27/96 2:03p Royf

' Initial revision.

'*

Public Declare Function DlPortReadPortUchar Lib "dlportio.dll"
(ByVal port As Long) As Byte

Public Declare Function DlPortReadPortUshort Lib "dlportio.dll"
(ByVal port As Long) As Integer

Public Declare Function DlPortReadPortUlong Lib "dlportio.dll"
(ByVal port As Long) As Long

Public Declare Sub DlPortReadPortBufferUchar Lib "dlportio.dll"
(ByVal port As Long, Buffer As Any, ByVal Count As Long)
```

```
Public Declare Sub DlPortReadPortBufferUshort Lib
"dlportio.dll" (ByVal port As Long, Buffer As Any, ByVal Count
As Long)

Public Declare Sub DlPortReadPortBufferUlong Lib "dlportio.dll"
(ByVal port As Long, Buffer As Any, ByVal Count As Long)

Public Declare Sub DlPortWritePortUchar Lib "dlportio.dll"
(ByVal port As Long, ByVal Value As Byte)

Public Declare Sub DlPortWritePortUshort Lib "dlportio.dll"
(ByVal port As Long, ByVal Value As Integer)

Public Declare Sub DlPortWritePortUlong Lib "dlportio.dll"
(ByVal port As Long, ByVal Value As Long)

Public Declare Sub DlPortWritePortBufferUchar Lib
"dlportio.dll" (ByVal port As Long, Buffer As Any, ByVal Count
As Long)

Public Declare Sub DlPortWritePortBufferUshort Lib
"dlportio.dll" (ByVal port As Long, Buffer As Any, ByVal Count
As Long)

Public Declare Sub DlPortWritePortBufferUlong Lib
"dlportio.dll" (ByVal port As Long, Buffer As Any, ByVal Count
As Long)
```

The parameter of utmost importance here is the physical address of the board. Without knowing the board's address, no information can be written to the board. While the programmer will know how to calculate the address set on the dip switch, the end user may not. By setting the dip switch on the GUI to look identical to the dip switch settings on the board (where a √ = on), the software will automatically calculate the board address in decimal form. This task is accomplished by the following code:

```
Sub BAddr_Change()

Curr_BAdr = 0

For Switch_No = 1 To 6

If Dip_Switch(Switch_No).Value = 1 Then

Curr_BAdr = Curr_BAdr + 2 ^ (6 - Switch_No)

End If

Next Switch_No

Board_Address.Caption = "Board $: " & Str(Curr_BAdr) & "d"

End Sub
```

When the project is run and the form (GUI) is loaded, the following code is executed:

```
Private Sub Form_Load()

'Specify Address to Board

Call BAddr_Change

'Port_Addr obtained from GUI Textbox labeled "Base Address"
```

```
DlPortWritePortUchar (Val(Port_Addr) + 1), Curr_BAdr

End Sub
```

First, the code obtains the address of the MEM-08 board. Then, the software points to where the information will be written. Remember, that all the cards must have their own unique nonoverlapping addresses. This command (`DlPortWritePortUchar (Val(Port_Addr) + 1), Curr_BAdr`) determines to what address the code will be sending the information. The code is writing to the actual board address + 1 because the byte that controls the relays is one higher than the physical address. In this particular case, the MEM-08 board only utilizes two bytes at two addresses. However, cards with more devices will utilize more bytes and addresses.

The final step actually actuates a particular relay. The following section of code shows how this can be accomplished:

```
Private Sub Trip_Relays_Click()

'Specify Address of Board

Call BAddr_Change

DlPortWritePortUchar (Val(Port_Addr) + 1), Curr_BAdr

data_byte = 0

For Relay_no = 1 To 8

If CR(Relay_no).Value = 0 Then

Activate(Relay_no).FillColor = &H8080FF

Line2(Relay_no).Visible = True

Line1(Relay_no).Visible = False

Else

Line2(Relay_no).Visible = False

Line1(Relay_no).Visible = True

Activate(Relay_no).FillColor = &HFF&

data_byte = data_byte + 2 ^ (8 - Relay_no)

End If

Next Relay_no

'Call DLL and Trip Relays

DlPortWritePortUchar Val(Port_Addr), data_byte

End Sub
```

Notice the sequence of events that takes place here. First, the software again determines the address of the board and points to where the next stream of commands (bytes) will be sent. This is good practice as the user may at any time update the GUI, so it is wise to recheck the address immediately prior to sending any commands.

The subroutine now enters a loop, and during each pass it checks the status of each individual checkbox that represents the desired state of each relay on the board. If a relay is checked (`CR(Relay_no).Value = 1`), then the contribution that bit contributes toward the total byte value is added to the running total of the databyte (`data_byte`). Upon finishing the loop, the value of the databyte (`data_byte`) is known and sent to the card using the command

FIGURE 4.12 The Keithley Metrabyte MEM-32 relay card.

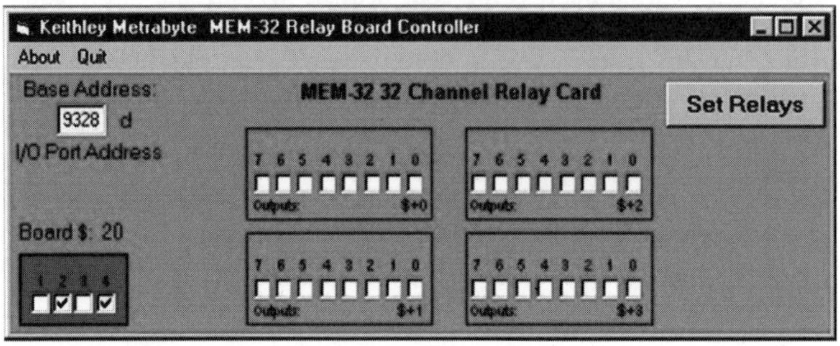

FIGURE 4.13 The MEM-32 sample application GUI.

(DlPortWritePortUchar Val(Port_Addr), data_byte). Once this value is received and interpreted by the card (a matter of milliseconds), the chosen relays are turned on. Notice also that this command is sent to the actual board address, not the actual board address + 1. This is because the software has previously pointed to where the next databyte should go (board address + 1) using the command (DlPortWritePortUchar (Val(Port_Addr) + 1), Curr_BAdr).

Some additional processes happen during this subroutine as well. For example, the graphic illustration of the relay is shown to change connections from one side to the other. Also, the LEDs that show the states of the individual relays are updated. One minor thing to point out is that the graphic shown is not an accurate depiction of a DPDT relay, it only serves as a visual indicator of a change in states.

4.3.2 THE MEM-32 RELAY CARD

While the MEM-08 relay board is a very useful tool, a relay capable of sourcing 5 amps is not required in many applications. The MEM-32 (Figure 4.12) has the advantage of having 32 relays per card, but each relay is only rated to handle $1/_2$ amp. The MEM-32 card requires four consecutive addresses on the bus as four 8-bit bytes are needed to control all 32 relays [4 bytes × 8 bits/byte = 32 bits (1 bit to control each relay)]. Using four bits as opposed to 1 bit on the MEM-08 card slightly increases the level of complexity in the programming of this card, but not to a great degree.

Unlike the MEM-08, the MEM-32 relay board does not have individual LEDs to indicate when a particular relay has been actuated. It has the relays and their connections, along with the dip switch required to set the card's address on the bus. As in Figure 4.12, a GUI has been created in Figure 4.13 to represent the actual card.

The dip switch in Figure 4.13 only has 4 switches, unlike the MEM-08 that has 6 switches. Since the MEM-32 card requires four consecutive addresses on the bus (Figure 4.14), the lowest interval that the user may alter the address on this card is $4 ($ = address). Of course this safety feature will not help if a card with higher precision address spacing such as the MEM-08 is set to overwrite its address space.

The GUI has four groups of eight checkboxes, each surrounded by a rectangle. This represents the four groups of eight relays contained on the card. For each group of eight relays, the card contains a notation such as Board Address + n (where n = 0 to 3). Each rectangle contains a notation of $+n to indicate which set of checkboxes (byte) controls which relays represented on the card.

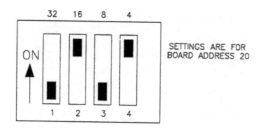

FIGURE 4.14 The MEM-32 board address dip switch has a minimum address spacing of 4. (From *Metrabus User's Guide*, Revision F, Keithley Metrabyte Division, Keithley Instruments, Inc., Cleveland, OH, July, 1994. With permission.)

Like the MEM-08 controller software, the first thing the code obtains is the address of the MEM-32 board. Then the software points to where the software will write the information. However, setting the states of the relays is done somewhat differently. The following section of code illustrates how this is accomplished:

```
Public Sub Calc_OutBytes()

Dim ByteNo As Integer

Dim BitNo As Integer

For ByteNo = 0 To 3

'Set Address Pointer

Debug.Print (Val(Port_Addr + 1)), (Curr_BAdr + ByteNo)

DlPortWritePortUchar (Val(Port_Addr + 1)), (Curr_BAdr + ByteNo)

Output_DataByte(ByteNo) = 0

For BitNo = 0 To 7

If OBit(BitNo + (ByteNo * 8)).Value = 1 Then

Output_DataByte(ByteNo) = Output_DataByte(ByteNo) + 2 ^ BitNo

End If

Next BitNo

Debug.Print "Byte No."; ByteNo; " = "; Output_DataByte(ByteNo)

DlPortWritePortUchar Val(Port_Addr), Output_DataByte(ByteNo)

Next ByteNo

End Sub
```

In this code, a loop is set up to cycle through the four bytes that each control a set of 8 relays. A second loop is set up to add the contribution of each individual bit, should that particular relay be chosen to be turned on. The value of each corresponding byte to be written is stored in the array Output_DataByte. At the beginning of the calculation process for each byte, the pointer is set to point at the address of each byte that is to be calculated with the command DlPortWritePortUchar (Val(Port_Addr + 1)), (Curr_BAdr + ByteNo). When the value for each byte has been determined (at the termination of the "BitNo" loop), the byte value is written out to the card using the command DlPortWritePortUchar Val(Port_Addr), Output_DataByte(ByteNo).

FIGURE 4.15 The MBB-32 digital I/O card.

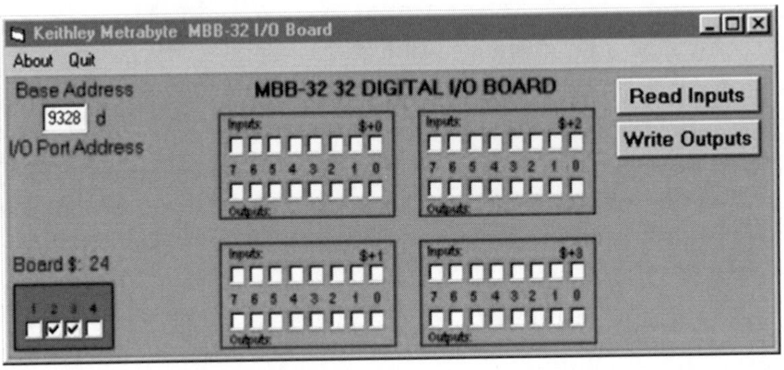

FIGURE 4.16 The MBB-32 sample application GUI.

4.4 THE MBB-32 DIGITAL I/O CARD

Up until now, the focus has been on writing information *to* a particular card, as opposed to receiving information *from* the card. The MBB-32 card is capable of both writing digital outputs (0 - 5.0 V) and receiving digital inputs. In fact, the card has 32 separate digital inputs and outputs. Figure 4.15 illustrates that this card also has a large prototyping area for custom circuitry, located on the left in front of the screw terminals. Developing custom circuitry that can be interfaced with cards such as this will be covered in detail in Chapter 5.

A GUI has been constructed for the MBB-32 sample application. It is very similar to the GUI developed for the MEM-32 card with the notable exception that it has two sets of checkboxes per byte, one representing the inputs, the other representing the outputs. This GUI has two buttons, one that reads the digital inputs, and one that writes the digital outputs. The GUI is shown in Figure 4.16.

As in the previous applications, first the code obtains the address of the MBB-32 board. Then, the software points to where the software will write the information. Writing the digital outputs is done very similarly to setting the relays on the MEM-32 card. However, reading the digital inputs is an entirely different algorithm.

The following code sets the states of the digital outputs. The only difference is that, instead of activating a relay when a checkbox is checked, the output is set to high (V_{CC} or 5.0 V). Outputs with blank or unchecked checkboxes will be set to low (Gnd or 0V).

```
Public Sub Calc_OutBytes()

Dim ByteNo As Integer

Dim BitNo As Integer

For ByteNo = 0 To 3

   'Set Address Pointer

   Debug.Print (Val(Port_Addr + 1)), (Curr_BAdr + ByteNo)

   DlPortWritePortUchar (Val(Port_Addr + 1)), (Curr_BAdr + ByteNo)
```

```
     Output_DataByte(ByteNo) = 0

        For BitNo = 0 To 7

            If OBit(BitNo + (ByteNo * 8)).Value = 1 Then

                Output_DataByte(ByteNo) = Output_DataByte(ByteNo)
                + 2 ^ BitNo

            End If

        Next BitNo

        Debug.Print "Byte No.;" ByteNo; " = ;"
        Output_DataByte(ByteNo)

        DlPortWritePortUchar Val(Port_Addr),
        Output_DataByte(ByteNo)

  Next ByteNo

  End Sub
```

The real difference lies in reading the digital inputs. The following subroutine illustrates one means of accomplishing this:

```
Public Sub Decode_Inputs()

Dim ByteNo As Integer

Dim BitNo As Integer

Dim Byte_Value As Integer

Dim Check_Byte As Integer

For ByteNo = 0 To 3

   'Set Address Pointer

   Debug.Print (Val(Port_Addr + 1)), (Curr_BAdr + ByteNo)

   DlPortWritePortUchar (Val(Port_Addr + 1)), (Curr_BAdr + ByteNo)

   'Read Byte into Byte_Value

   Byte_Value = DlPortReadPortUchar(Val(Port_Addr))

   Check_Byte = 255 - Byte_Value

   Debug.Print "Check Byte;" ByteNo; " = "; Check_Byte

   For BitNo = 7 To 0 Step -1

       If Check_Byte > = 2 ^ BitNo Then

           Check_Byte = Check_Byte - 2 ^ BitNo

           IBit(BitNo + (ByteNo * 8)).Value = 0

           Else

           IBit(BitNo + (ByteNo * 8)).Value = 1

       End If
```

```
          'Debug.Print Check_Byte

       Next BitNo

     Next ByteNo

   End Sub
```

Similar to the write algorithm, a looping structure is set up to read and analyze all four bytes systematically. The address pointer must be set before attempting to read a byte, as it had to be set before a byte could be written out of the port. Once this has been accomplished, the proper byte is read and stored in the variable Byte_Value using the statement `Byte_Value = DlPortRead-PortUchar(Val(Port_Addr))`. A new variable named Check_Byte is set equal to 255 - Byte_Value. The 255 is the maximum value for an 8 bit byte. A second looping structure is now set up to determine the status of each bit contained within the byte. If a particular bit has contributed a component to the overall value of the Check_Byte, its component is then subtracted from the Check_Byte variable, and that particular bit is shown as low or unchecked. Conversely, if a particular bit has not contributed a component to the overall value, that bit is shown as high or checked.

It is very simple to determine if any particular bit has contributed a component to the overall value of the Check_Byte. If the value of the Check_Byte exceeds 2 raised to the power of the bit number (2^BitNo), then that particular bit has made a contribution that must be subtracted. The Check_Byte is compared against every single bit in the byte from most significant bit (MSB) to least significant bit (LSB). A value is subtracted when possible and the comparison process continues as illustrated in the following example:

BitNo	7	6	5	4	3	2	1	0
Value	128	64	32	16	8	4	2	1

The next chart is an example of 50 as the byte read into the computer:

Byte	Contribution	Result	Comments
50	−128	NO	No contribution
50	−64	NO	No contribution
50	−32	18	Bit 5 is "On"
18	−16	2	Bit 4 is "On"
2	−8	NO	No contribution
2	−4	NO	No contribution
2	−2	0	Bit 1 is "On"
0	−1	NO	No contribution

BitNo	7	6	5	4	3	2	1	0
Value	128	64	32	16	8	4	2	1
Binary	0	0	1	1	0	0	1	0

The binary form of the number 50 is therefore (110010)b. The leading zeros for bits 7 and 6 are omitted from the left side.

One final word of caution, the MBB-32 board shows inputs in one of two possible states, V_{CC} ("1") or Gnd ("0"). However, if an input is left floating, the input could be read as either V_{CC} or Gnd. In practice however, floating inputs are always read as V_{CC}. This is really counterintuitive, since the input is floating and no voltage is applied, the user would expect to read it as ground. In the software supplied to control this card, when an input is left floating it will show up as checked (V_{CC}), when the true voltage at that input may in fact not be high (≈ 5.0 V). Therefore, whenever designing critical circuitry, it is prudent to make sure that an output that will be read by a card

FIGURE 4.17 The MAI-16 analog input board.

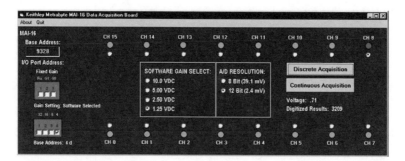

FIGURE 4.18 The MAI-16 application GUI.

such as this will never be allowed to float. In addition, V_{CC} should not represent critical event occurrences, as the absence of any input will trigger the event.

4.5 THE MAI-16 ANALOG INPUT CARD

The order in which the various cards have been presented to the reader is not accidental or by chance. The method of reasoning is as follows. The first card shown, the MEM-08, utilized only a single byte to control its operation. The next card presented, the MEM-32, required four bytes to control its operation. The last card described, the MBB-32, also required four bytes to control its operation, but more significantly; required both reading and writing bytes to the card to perform both input and output operations, whereas the previous cards only utilized writing bytes to perform output type operations.

While the cards presented thus far can add significant value to any application, a major shortcoming of them is that they can only express values in a Boolean condition — on or off. While the digital domain is highly useful in some applications, the real world speaks only one language — analog. Many times in a laboratory it is essential to have the capability of reading an exact voltage. For example, digital cards are not appropriate to monitor the voltage coming off an electrode involved in a titration.

The MAI-16 board is a 16 channel analog input board (Figure 4.17). Each channel can be operated in either 12 bit (higher accuracy, but slower response) or 8 bit (lower accuracy, but quicker response) modes. Each input contains 50 Hz single pole low pass filter to eliminate 60 Hz noise that propagates out of nearly every piece of electronic equipment. For more information see Chapter 5.

Like the MEM-08 board, the MAI-16 board has an annunciator LED to show which of the 16 channels is currently active. This card has two sets of dip switches. One switch selects the gain (or amplification) to be applied to the input signal. The dip switch next to the ribbon connector is the traditional address switch, and each MAI-16 board occupies four consecutive addresses in the Metrabus Address space. As with the previous sample applications, a GUI is set up to reflect the functionality of the board (Figure 4.18).

Unlike the other boards discussed thus far, the MAI-16 requires an additional ± 15.0 V DC power supply. It cannot be powered solely from power drawn off the bus, as the bus is only capable of supplying + 5.0 V DC. An independent power supply can be connected into the bus by means of the MTAP-1 board, or the Metrabus Power Supply (MBUS-PWR) can be purchased from Keithley.

This sample application is several exponents of complexity higher than the previous examples. Like the previous applications, parameters are set up upon initialization of the form. Unlike the previous examples, this application contains several module level variables that can be accessed from any subroutine within the form.

```
Dim CNV12 As Integer 'Start 12 Bit Conversion

Dim CNV8 As Integer 'Start 8 Bit Conversion

Dim SGC As Integer 'Select Gain/Channel

Dim DATAIO As Integer 'Board Address

Dim ADRPTR As Integer 'Address Pointer (Board Address + 1)

' _ _ _ _ _ _ _ _ _ _ _ _

Dim CHSCL As Integer 'Channel & Scale

Dim CHGAIN As Single 'Channel Gain

Dim ADResult As Long

Dim Voltage As Single

Private Sub Form_Load()

DATAIO = Val(BaseAddress.Text)

ADRPTR = DATAIO + 1

Debug.Print "DATAIO$: "; DATAIO; " ADRPTR: "; ADRPTR

Call Gain_Option_Click(1) 'Set Gain to Default

Debug.Print "STARTUP CHGAIN: "; CHGAIN

Call MAI_Address

Call Gain

Call Active_Channel_Click(0) 'Set Default Active Channel

'DlPortWritePortUchar ADRPTR, MAI_Address

End Sub
```

The primary difference between this routine and the previous startup routines is the fact that the gain of the input channel must be set. Although technically correct, gain is really a misleading term as to what is being chosen. By definition, gain indicates that an increase would be applied to the channel. In this particular instance, gain is at best unity (a multiplier of 1) and in all other instances attenuates the applied signal by a factor of $1/2$, $1/4$, and $1/8$. (multipliers of 0.5, 0.25, and 0.125). Input range of the channel is a better term to use because it more effectively conveys what is being chosen. For this particular card, a full scale input range can be chosen from 0 to 10.0 V, 5.0 V, 2.5 V, or 1.25 V.

The input range (gain) of the channel can be selected in one of two ways. The easiest is by means of the dip switch on the MAI-16 board. Table 4.1 illustrates possible settings.

Whenever the Fixed Gain switch (1) is off, the gain for a particular channel is to be selected through software. If the Fixed Gain switch (1) is on, then gain for all channels is fixed at the values dictated by switches G1 (2) and G2 (3).

TABLE 4.1
The Input Range (Gain) Selection Dip Switch

Input Range	Fixed Gain (1)	G1 (2)	G2 (3)
Software Selected	Off	X	X
0 - 10.0 V	On	Off	Off
0 - 5.0 V	On	Off	On
0 - 2.5 V	On	On	Off
0 - 1.25 V	On	On	On

TABLE 4.2
The Software Gain/Channel (SGC) Byte

Bit	7	6	5	4	3	2	1	0
ID	—	—	GS1	GS0	CS3	CS2	CS1	CS0
Value	128	64	32	16	8	4	2	1

Selecting the gain via software is somewhat more complicated but allows for more flexibility in the use of the card in that every channel may have a different gain setting. To select gain by means of software, write the select gain/channel (SGC) byte out to the card. The SGC byte format is illustrated in Table 4.2.

Bits 6 and 7 are not utilized in the SGC byte. The gain is selected through bits 4 and 5 while the channel is chosen by bits 0 thru 3. Table 4.3 and Table 4.4 illustrate the formats for these bits.

For example, to select Channel 10 to have an input range of 2.5 V, the SGC byte would be set to 42. In the following chart, bits 1, 3, and 5 are on yielding contributions of 2 + 8 + 32 = 42.

Bit	7	6	5	4	3	2	1	0
ID	—	—	GS1	GS0	CS3	CS2	CS1	CS0
State	—	—	1	0	1	0	1	0
Value	128	64	32	16	8	4	2	1

At startup, the setting of the dip switch is unknown, which makes it impossible to know if the gain is fixed for each channel, or if the gain is software selectable for each channel. Therefore, a default range (gain) for all channels is chosen with the command: Call Gain_Option_Click(1). The following subroutine sets the range of the current input channel.

```
Private Sub Gain_Option_Click(Index As Integer)

Select Case Index

  Case 0

    CHGAIN = 10

  Case 1

    CHGAIN = 5

  Case 2

    CHGAIN = 2.5

  Case 3

    CHGAIN = 1.25

End Select
```

TABLE 4.3
Ranges Available for Gain Selection of Bit Formats
for Bits 4 and 5 (Input Range Selection) of SGC Byte

GSI	GSO	Full Scale Range
0	0	± 10 V
0	1	± 5 V
1	0	± 2.5 V
1	1	± 1.25 V

Note: These bits define the full-scale range for the selected channel.

TABLE 4.4
Channel Selections for Bit Formats for Bits 0 - 3
(Channel Select) of SGC Byte

CS3	CS2	CS1	CS0	Selected Channel
0	0	0	0	0
0	0	0	1	1
0	0	1	0	2
0	0	1	1	3
0	1	0	0	4
0	1	0	1	5
0	1	1	0	6
0	1	1	1	7
1	0	0	0	8
1	0	0	1	9
1	0	1	0	10
1	0	1	1	11
1	1	0	0	12
1	1	0	1	13
1	1	1	0	14
1	1	1	1	15

Note: These bits select the channel to be programmed.

Source: From *Metrabus User's Guide*, Revision F, Keithley Metrabyte Division, Keithley Instruments, Inc., Cleveland, OH, July, 1994. With permission.

Further along in the startup procedure, the gain dip switch of the GUI is read, and if necessary, an updated gain value is assigned. The range will update if the gain is selected on the card by means of the dip switch. If the gain is to be software selected, the default setting at startup (5.0 V) is used. Of course, the range can be reset at any time by calling the Gain_Option_Click subroutine.

```
Public Function Gain() As Integer

Dim BinStr$

For Switch_No = 1 To 3
```

```
    If Gain_Dip_Switch(Switch_No).Value = 1 Then
        BinStr$ = BinStr$ & "1"
        Else
        BinStr$ = BinStr$ & "0"
    End If
Next Switch_No
Select Case BinStr$
    Case "000", "001", "010", "011"
        Gain_Setting = "Gain Setting: Software Selected"
        Gain_Frame.Visible = True
    Case "100"
        Gain_Setting = "Gain Setting: +10.0 Volts"
        Gain_Frame.Visible = False
        Call Gain_Option_Click(0)
    Case "101"
        Gain_Setting = "Gain Setting: +5.0 Volts"
        Gain_Frame.Visible = False
        Call Gain_Option_Click(1)
    Case "110"
        Gain_Setting = "Gain Setting: +2.50 Volts"
        Gain_Frame.Visible = False
        Call Gain_Option_Click(2)
    Case "111"
        Gain_Setting = "Gain Setting: +1.25 Volts"
        Gain_Frame.Visible = False
        Call Gain_Option_Click(3)
End Select
End Function
```

Now that the card is set up, reading analog data from the card is the next logical step. Data can be read from the MAI-16 card in one of two manners. A discrete time measurement is a single measurement triggered at a particular instant. It could be triggered by hand or by machine when a certain event occurs. A continuous time measurement is the capture of an entire waveform over a specified period. Figure 4.19 illustrates continuous and discrete time signals. Often, a signal will be sampled discretely at evenly spaced intervals in an effort to reproduce the waveform. Such a process is referred to as digitizing a signal, as a digital reproduction is made of the original analog signal. When sampling a signal in this manner, some simple rules must be followed or the signal

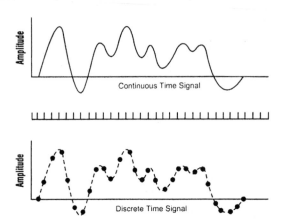

FIGURE 4.19 Examples of continuous and discrete time signals. (From McConnell, E., *Electron. Design Mag.*, June, 1995. With permission.)

reproduced may not be an accurate reproduction of the signal sampled. Section 4.4 focuses on taking single discrete measurements. Continuous time measurements will be covered in Section 4.9.

4.6   DISCRETE TIME MEASUREMENT

Each time an option button is clicked to change the active acquisition channel, a number of events must take place to set up and acquire data from the newly choosen channel. The following subroutine is processed each time a new channel is selected. Each time this subroutine is run, it performs a single discrete time measurement on the new channel selected in order to let the user know the new input value.

```
Private Sub Active_Channel_Click(Index As Integer)

Dim LSB4 As String

Dim MSB2 As String

Dim ENTIRE_BYTE As String

Dim Place As Integer

Dim MSB_12 As Byte, LSB_12 As Byte

Static Last_Channel As Integer

ActiveChannel = Index

Activate(Last_Channel).FillColor = &H8080FF

CHSCL = 0

'Get Gain Value!

Call Gain

'Debug.Print "CHGAIN: "; CHGAIN

Select Case CHGAIN

  Case 10

      MSB2 = "00"
```

```
    Case 5
        MSB2 = "01"
    Case 2.5
        MSB2 = "10"
    Case 1.25
        MSB2 = "11"
End Select
'Debug.Print "MSB2: "; MSB2
Select Case Index
    Case 0
        LSB4 = "0000"
    Case 1
        LSB4 = "0001"
    Case 2
        LSB4 = "0010"
    Case 3
        LSB4 = "0011"
    Case 4
        LSB4 = "0100"
    Case 5
        LSB4 = "0101"
    Case 6
        LSB4 = "0110"
    Case 7
        LSB4 = "0111"
    Case 8
        LSB4 = "1000"
    Case 9
        LSB4 = "1001"
    Case 10
        LSB4 = "1010"
    Case 11
        LSB4 = "1011"
    Case 12
```

```
        LSB4 = "1100"
    Case 13
        LSB4 = "1101"
    Case 14
        LSB4 = "1110"
    Case 15
        LSB4 = "1111"
End Select
ENTIRE_BYTE = MSB2 & LSB4
'Debug.Print "ENTIRE BYTE: ", ENTIRE_BYTE
For Place = 1 To Len(ENTIRE_BYTE)
    If Mid$(ENTIRE_BYTE, Place, 1) = "1" Then
        CHSCL = CHSCL + 2 ^ (6 - Place)
    End If
Next Place
'Debug.Print "CHSCL: ", CHSCL
Activate(Index).FillColor = &HFF&
Last_Channel = Index
'Stop
'Prepare MAI-16 to Acquire Data from Selected Channel
DlPortWritePortUchar ADRPTR, SGC 'Point to SGC
DlPortWritePortUchar DATAIO, CHSCL 'Set Channel and Gain
'Initiate 8 or 12 bit A/D Conversion
Select Case ADRes(0).Value
    Case True '8 Bit Resolution
        DlPortWritePortUchar ADRPTR, CNV8 'Point to 8 Bit
        Conversion
    Case False '12 Bit Resolution
        DlPortWritePortUchar ADRPTR, CNV12 'Point to 12 Bit
        Conversion
End Select
DlPortWritePortUchar DATAIO, 0 'Initiate A/D Conversion
Select Case ADRes(0).Value
    Case True '8 Bit Resolution
```

```
        DlPortWritePortUchar ADRPTR, MAI_Address 'Point to A/D
        Result

        'Wait 1/128th of a Second for Buffer to be filled

        'Bug Fix for X1 Latency

        Call Idle_Time(0.0078125)

        ADResult = DlPortReadPortUchar(DATAIO)

        ADRslt.Caption = "Digitized Results: " & Str(ADResult)

        'Debug.Print "ADResult: "; ADResult

        Voltage = ((ADResult * CHGAIN)/128) - CHGAIN

        Voltage_Label.Caption = "Voltage: " & Str(Format(Voltage,
        "##.00"))

        'Debug.Print "VOLTAGE: "; Voltage

    Case False '12 Bit Resolution

    DlPortWritePortUchar ADRPTR, MAI_Address 'Point to A/D Result

    'Wait 1/128th of a Second for Buffer to be filled

    'Bug Fix for X1 Latency

    Call Idle_Time(0.0078125)

    MSB_12 = DlPortReadPortUchar(DATAIO)

    DlPortWritePortUchar ADRPTR, MAI_Address + 1 'Point to A/D
    Result

    'Wait 1/128th of a Second for Buffer to be filled

    'Bug Fix for X1 Latency

    Call Idle_Time(0.0078125)

    LSB_12 = DlPortReadPortUchar(DATAIO)

    'Combine MSB & LSB bytes

    ADResult = MSB_12 * 16 + LSB_12/16

    ADRslt.Caption = "Digitized Results: " & Str(ADResult)

    'Debug.Print "ADResult: "; ADResult

    Voltage = ((ADResult * CHGAIN)/2048) - CHGAIN

    Voltage_Label.Caption = "Voltage: " & Str(Format(Voltage,
    "##.00"))

    'Debug.Print "VOLTAGE: "; Voltage

End Select

End Sub
```

Recall that before a measurement can take place, both the range and the particular channel of interest must be specified by writing the proper SGC byte. The first action taken by this subroutine is to determine the proper gain with a call to the `Gain` subroutine. Once this has occurred, two sets of select case statements build an SGC byte in the proper format. The SGC byte has been split into two groups of 4 bits, termed LSB4 (the four least significant bits 0 - 3) and MSB2 (the lower two most significant bits 4 and 5). The first select case statement sets MSB2 values according to the range (or gain) desired. The second select case statement sets LSB4 values to reflect the acquisition channel. A "For" loop, then constructs the value of the byte by adding the contributions of the bits that are turned on. The software then points to the SGC and writes the SGC byte value to the MAI-16 board (the SGC byte value is stored in the variable CHSCL).

The subroutine then checks to see which A/D resolution is desired. The 12-bit resolution is accurate within 2.4 mV, while the 8 bit resolution is accurate within 39.1 mV. While 12-bit resolution takes somewhat longer to perform than the 8-bit resolution, for most tasks in the laboratory this is not a consideration. A delay of a few milliseconds is meaningless when performing something such as a titration curve, so go for the accuracy. However, if the data from this A/D converter were being used to guide a missile moving at Mach 2, this delay would no longer be negligible. The software will then point to the proper conversion mode.

The next section of code in the subroutine is the core of the data acquisition process. First, the data acquisition process is initiated by writing a 0 to the board address with the statement: `DlPortWritePortUchar DATAIO, 0`. Once the data acquisition has taken place, one of two methods is chosen to acquire the data based on the resolution chosen (8 bit or 12 bit).

For the 8-bit acquisition process, the subroutine points to the A/D result. After pointing to the result, the program idles for a very short period of time. A propagation delay occurs between when the result is pointed to, and when the result is available to be read. It takes a short period of time to fill the buffer. If this delay is not incorporated, then the result will be missed when the software tries to read the results. This is because the software can execute the next command before the buffer is filled. The results read are in a digitized format that can be converted into voltage by means of the following formula:

```
Voltage = ((Digitized Results * Gain)/128) - Gain
```

The 128 is derived from the MSB, whose contribution is 2^7 which equals 128. The 12-bit acquisition process is a little more difficult, as two bytes must be read and combined to acquire all 12 bits. The two bytes are combined and their digitized results calculated using the following formula:

```
Digitized Results = MSB * 16 + LSB/16
```

The digitized results can then be converted to voltage with the formula:

```
Voltage = ((Digitized Results * Gain)/2048) - Gain
```

The 2048 is derived from the MSB, whose contribution is 2^{11} which equals 2048.

4.7 SAMPLING THEORY AND THE NYQUIST THEOREM

The application discussed in Section 4.4 is great for taking a single reading now and then at intervals on an order of magnitude in seconds. But, when a continuous time signal needs to be captured and an accurate reproduction of what has been acquired must be provided, the input frequency must be sampled at regular intervals. Naturally, as the limit of the interval approaches zero, the reproduced waveform will be an exact replication of the sampled waveform. The maximum length an interval can be spaced between successive samples and still provide an accurate reproduction of the measured waveform is not so intuitive.

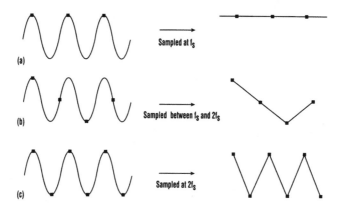

FIGURE 4.20 Periodic signal measurement and the Nyquist frequency. (From McConnell, E., *Electron. Design Mag.*, June, 1995. With permission.)

When reconstructing the original waveform, information is lost between the gaps of each successive sample. This lost information must be estimated in order to reproduce the measured waveform. The simplest way to accomplish this is by linear interpolation. A more robust method would involve some form of smoothing the interpolated values between successive samples, such as FFT smoothing. The puzzler is determining the size of the gap between successive measurements and still maintaining the ability to calculate an accurate reproduction of the original waveform. Figure 4.20 illustrates the solution as it relates to periodic signals. The example uses a sine wave.

Assume the sine wave in Figure 4.20 has a frequency of 1 kHz. If the 1 kHz sine wave is sampled at 1 kHz, the measured waveform will be a straight line (in essence a DC voltage). More importantly, the value of this DC voltage will drift, because the sampling signal and the measured signal are not synchronized. The measured waveform will be reproduced as a DC voltage anywhere between V_{peak} and $-V_{peak}$. Obviously, this reproduction of the input waveform is not accurate, so sampling at an equal frequency to the input waveform is too slow. This concept is illustrated in Figure 4.20a.

If sampling at the input waveform frequency is not fast enough, Figure 4.20c illustrates sampling at double the input frequency. The input frequency (f) is sampled at 2f, and the resulting waveform has the same frequency as the input waveform, but is triangular in shape. This example is true in all cases except when the input waveform happens to be sampled at 2f where the input waveform crosses the zero axis. Should the input waveform be sampled here, the resulting output will be a straight line at ground — no signal. This premise established the fundamental theorem of sampling commonly referred to as the Nyquist theorem.

To reproduce the same frequency as the measured waveform, the sampling rate of the measured waveform must be more than twice the rate of the maximum frequency component in the signal to be measured. The frequency at one half of the sampling frequency is often referred to as the Nyquist frequency.

While adherence to the Nyquist theorem guarantees a correct frequency (the same number of cycles) for the reproduced waveform, it does not guarantee an accurate shape for that waveform. When sampling at slightly above 2f, a sine wave will be reproduced as a triangular wave. While it will be of the right frequency, the shape leaves much to be desired.

For the most accurate reproduction, the input signal should be sampled as fast as possible. Many parameters need to be considered. First, the duration of time in which the signal must be captured is of paramount importance. In some cases, storage capacity may be limited. If samples are taken at high frequencies over longer periods of time there may not be enough memory or disk space to hold the acquired data.

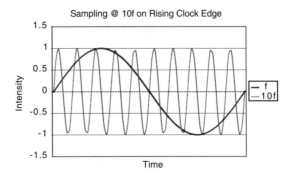

FIGURE 4.21 Sampling waveform of freqency f by 10f.

The speed at which the data can be migrated from the hardware to the storage device is another big consideration. Even if the system has infinite storage capacity, if the acquisition takes places faster than the data can be stored (highly likely), then after some finite amount of time the system buffer will overflow and data will be lost. For many applications, sampling at 10f is a good compromise between accurate waveform reproduction and hardware capacity limitations (Figure 4.21).

4.8 ALIASING

No section on data acquisition could be complete without covering aliasing, which is the inaccurate perception of an event due to the limitations of the observing body. Everyone has looked at a propeller or car wheel rotating in a very rapid manner and observed the car wheel or propeller blades moving very slowly backward. Certainly this is not the true movement of these objects, yet that is how people see them. Actually, these events are happening so quickly that the brain cannot process them all. Since the brain cannot rasterize all the images streaming into it, some of the images are discarded and the perception is given of the object moving slower (and backward) at a fraction of its true speed.

Figure 4.20(b) illustrates this concept rather nicely. When a periodic signal is sampled between f and 2f, the reproduction is aliased down to a lower frequency multiple of the sampled frequency. In fact, the aliased frequency will be the absolute value of the difference between the frequency of the input signal and the closest integer multiple of the sampling rate.

4.9 SAMPLING LIMITATIONS OF VISUAL BASIC

With the fundamental theorems and limitations of data acquisition explained, some basic rules can be established for conducting data acquisition using the tools discussed in previous chapters within a Visual Basic environment. Recall that the Timer control (and function) is only guaranteed to reliably produces intervals as short as $1/18$ of a second. Although in practice this interval can usually be made much shorter, and is to some degree processor dependant.

Well, $1/18$ of a second translates into 18.0 Hz, a very low frequency with a period of only 56 ms. The Nyquist theorem therefore limits the acquisition of periodic signals to a maximum frequency of less than 9.0 Hz ($< 1/2*18.0$ Hz). A reader involved in audio applications or high-speed data acquisition will be laughing because comparatively speaking 9.0 Hz is the equivalent of a turtle on the Autobahn. However, in a laboratory most events do not happen at lightning fast speeds. If performing a titration for example, it is usually adequate to sample the pH of the test solution once every second, or at a frequency of 1.0 Hz.

Also remember that although signals can be sampled as high as 8.9 Hz using the Timer control, an accurate waveform shape will not be reproduced unless the sampling rate is much higher than 2f.

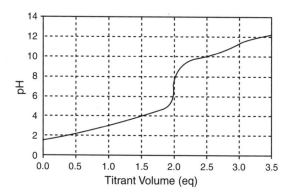

FIGURE 4.22 A typical titration curve.

4.10 SAMPLING OF NONPERIODIC SIGNALS

When discussing the Nyquist theorem in Section 4.5, an example was cited using periodic signals. What happens when the signal to be acquired is not periodic, or its period is unknown? What sampling rate should be used? Often when running a titration (the gradual addition of an acidic solution to a basic solution or vice versa), the signal will remain flat for an extended period of time and then suddenly ramp up, as shown in Figure 4.22.

In Figure 4.22, suppose 0.1 mL of titrant is added to the titration vessel every second, so that at a titrant volume of 0.5 mL, 5.0 seconds have elapsed. If this is a typical titration curve for this kind of experiment, the curve reveals the largest delta y change occurs where the X-axis is equal to 2. Of course, this location will be different for every compound, but the underlying premise for choosing a sampling rate will be the same.

The fundamental choice is to determine the amount of points on the x-axis that should be taken for the region of interest. In this case, the region of interest extends from 1.5 mL to 2.5 mL, a period of time corresponding to 10.0 seconds. If it has been determined that 20 points is sufficient to accurately reproduce the curve in this region, then the sampling rate should be 2.0 Hz, which will sample the pH once every 500 ms or half second. Another way to consider this problem in the region of maximum slope, is to determine the amount of points on the x-axis that are desired for a particular delta on the y-axis. In this example, during an interval of 10 seconds in which the pH rose by about 4 units, 20 points would be necessary to reproduce the titration curve.

4.11 PSEUDO CONTINUOUS TIME MEASUREMENT

True continuous time data using the portio DLL cannot be acquired. However, low frequency waveforms can be sampled enough to produce a reasonable reproduction of the orginal waveform. This section covers the modification of the MAI-16 driver software to perform such a task. Pressing the Continuous Acquisition button will produce the GUI illustrated in Figure 4.23.

This GUI contains some fundamental parameters that will now be explained. The sampling rate is the interval between consecutive samples expressed in milliseconds. The channel is the specific channel on the MAI-16 board (0 to 16) that will be used to acquire the data. Total datapoints is the total number of samples that will be taken. The frequency, which is present only to check for aliasing, should represent the maximum frequency component of the signal to be sampled. The save results to disk checkbox gives the user the option of saving the acquired data to disk in the file c:\temp\mai16data.csv. The parameters stored to disk are shown in Table 4.5 and will be discussed in Section 4.10. If the parameters are not stored to disk they will only be updated on MAI-16 GUI during the interval of acquisition. Calculate the interval of acquisition, by multiplying

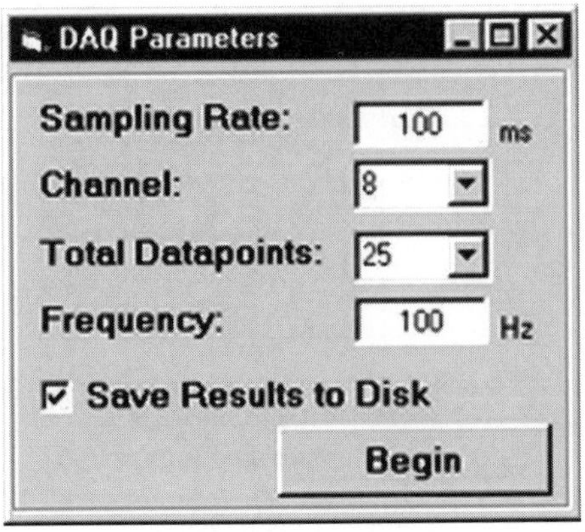

FIGURE 4.23 GUI used to set up pseudo continuous time measurements.

the total datapoints by the sampling rate and dividing by 1000 to yield the interval in seconds. Depress the begin button to start the data acquisition. The code is:

```
Private Sub Begin_Button_Click()

Dim LineHeader$

LineHeader$ = "No.,Sampling Rate,Time,Timer,Delta,Voltage"

'Prepare for Capture to Disk iff necessary

If frmDAQ.SaveCheck.Value = 1 Then

    'If this file should exist from a previous run - erase it !!!

    'This avoids inadvertantly appending data to an existing
    file !!!

    If FileExists ("c:\temp\mai16data.csv") = True Then

    Kill ("c:\temp\mai16data.csv")

    End If

    Open "c:\temp\mai16data.csv" For Append As #1

        Print #1, LineHeader$

    Close 1

End If

'Set Interval

MAI_16.Rate_Timer.Interval = Val(SampRate.Text)

'Enable Timer

MAI_16.Rate_Timer.Enabled = True

End Sub
```

First, this subroutine constructs a header line for the parameters that are to be stored to disk, should that option be chosen. The parameters are:

- No. — the sample number from 1 to n
- Sampling rate in seconds
- Time — the time at which the sample was acquired (hh:mm:ss AM/PM)
- Timer — the number of seconds elapsed since the first sample was acquired
- Delta — the time difference between this sample and the previous sample in seconds
- Voltage — the voltage level of this sample

The timer's interval function is only guaranteed accurate to $1/_{18}$ of a second. The interval can be set lower than $1/_{18}$ of a second however, and on some computers will be accurate to a much higher degree of precision than $1/_{18}$ of a second.

To determine the precise interval, link the timer interval to the system clock, as has been done with this application. Each time the timer event is triggered at interval X, the system time is recorded using the timer function. From the values recorded from the timer functions, the difference in time, between the successive samples, can be calculated and stored in the delta column. If the timer interval is lowered, and the delta intervals do not decrease, the precision of the system has been reached.

Also, when using intervals of less than one second, often the intervals are not quite as accurate as specified. Although the interval may become shorter as its value is lowered, typically as the interval approaches its limit of precision, the interval may in fact not be as short as the period that was specified in the interval property of the control. Shortly after this phenomena occurs, it does not matter how short the interval is set, the duration between successive samples remains the same and the limit of precision has been reached.

The subroutine then checks if the user has in fact selected to store this data to the disk. If data is to be stored to the disk, the subroutine checks to see if the file c:\temp\mai16data.csv exists. If the file does exist, it is deleted because this particular program appends data as it is recorded by the system. If the file already exists, new data will be appended to older data in the same file. The file is then created from scratch and the header line is written.

Finally, the interval of the timer is set, and the timer is enabled starting the interval interruption process. It is not a coincidence that the timer is enabled last in this subroutine. The disk operations that this subroutine performs take a lot of time (deleting and appending files, etc.). If the timer were to be enabled at the beginning of the subroutine, a timer event might not be able to break away during a disk operation such as the deletion of a preexisting file. This in turn could affect the accuracy of the timer interval.

Once the timer is enabled, at the end of each specified interval the Rate_Timer_Timer Subroutine is executed by the following code:

```
Private Sub Rate_Timer_Timer()

Dim delim$, dataline$

Static counter As Integer

Static oldtime As Single

Static zerotime As Single

Dim SampRate As Single

Dim CurrentTime As String, ThisTimer As Single, delta As Single

'Set Parameters Immediately Upon Execution, do not incur delays!

CurrentTime = Time

'Zero Adjust ThisTimer

If counter > 0 Then
```

```
   ThisTimer = Timer - zerotime
   Else
   zerotime = Timer
   ThisTimer = 0
End If
'Calculate delta value
If counter > 0 Then
   delta = ThisTimer - oldtime
End If
SampRate = Val(frmDAQ.SampRate.Text) * 0.001
counter = counter + 1
'Time to stop? How many points Acquired?
If counter > Val(frmDAQ.DPoints.Text) Then
   'Reset Static Variables
   counter = 0: oldtime = 0: zerotime = 0
   Rate_Timer.Enabled = False
   Exit Sub
End If
'Acquire Data
Call Active_Channel_Click(Val(frmDAQ.Chnl.Text))
'Log Analog Input Results to Disk
delim$ = Chr(44) 'Chr(44) = ASCII code for ","
dataline$ = Str(counter) & delim$ & Format(SampRate, "##.000")
& delim$ &_
CurrentTime & delim$ & Format(ThisTimer, "##.000") & delim$ & _
Format(delta, "##.000") & delim$ & Format(Voltage, "##.00")
'Write to File iff required
If frmDAQ.SaveCheck.Value = 1 Then
   Open "c:\temp\mai16data.csv" For Append As #1
        Print #1, dataline$
   Close 1
End If
'Remember old time for delta calculations
```

```
oldtime = ThisTimer

End Sub
```

The timer's value is zeroed out to reflect the time elapsed from the acquisition of the first sample. The timer function ordinarily reflects the amount of time in seconds that has elapsed since midnight. Since the time of interest is the time from start to finish of the acquisitions, the time recorded is zeroed out for readability purposes. Keep in mind that should this subroutine be run and midnight passes during its execution, the timer and delta values will be erroneous. A midnight crossing mechanism has not been built into this subroutine as had been done in Chapter 3 with the Kloehn syringe driver software. Also, the time of the previous acquisition is stored in the variable oldtime and is utilized in calculation of the deltas between successive samples.

4.12 POSSIBLE CAUSES OF INCONSISTENT DATA

To further reinforce and illustrate the concepts behind data acquisition, a 5.0 Hz sine wave is measured at intervals of 1000, 100, 10, and 1 ms. The results obtained from the software and hardware are discussed to clarify the limitations of this system and explain the presence of such limitations. Table 4.5 summarizes the results of the measurements.

In Table 4.5, the data for all four data acquisitions is a periodic wave, in some cases more closely resembling a sine wave than others. Comparing the case with a sample interval of 1000 ms with the case of 1.0 ms yields a gross inconsistency. Both waves are periodic sine waves, yet one has an X-axis scale from 0 to 0.72 and the other has an X-axis scale from 0 to 25.

Figure 4.24 is a perfect example of aliasing. A 5.0 Hz periodic sine wave has a period of 200 ms. The Nyquist theorem dictates that such a wave must be sampled at a minimum of < 100 ms in order to reproduce the correct frequency component of this wave. To sample this waveform at 1000 ms means every 5th cycle of the wave is being sampled. The sampled signal is therefore aliased down to a much lower harmonic of the original frequency, in this case $1/_{40}$th of the original frequency or 0.125 Hz. This is more forcefully illustrated by placing the two waveforms side by side on a graph with the same scale as illustrated in Figure 4.25.

The sine wave looks much cleaner in Figure 4.25 than in previous figures. Figure 4.25 was produced in Origin Lab's Origin and a smoothing function was applied to data. Two sets of data with different x-axis values have been plotted on Figure 4.25. These tasks are much easier to accomplish using Origin as opposed to using Excel. In Origin, a few simple mouse clicks may accomplish the task. In Excel, to accomplish the same task the raw data may have to be recalculated or manipulated (see Chapter 6).

Figure 4.25 illustrates that consistent results are produced only when the sampling rate drops below 100 ms. The values in the timer column illustrate this fact without looking at graphs of the data.

The next point of great interest is the delta column, which represents the time in seconds between successive samples. In each instance this value is different from the true rate specified by the interval property. The interval property has two important limitations: a) not guaranteed accurate for intervals < $1/_{18}$ of a second (≈ 56.0 ms) and b) possibly not in complete sync with the system timer depending upon the state of the background processes.

The true interval for intervals > 56.0 ms (in this case 1000 and 100 ms) is greater than the specified interval by 20.0 ms (0.02 s). On this particular computer, running this specific software, the true time interval will exceed the specified time interval by 20 ms. If the software configuration changes, (background processes eliminated, etc.) this statement may no longer be true. Anti viral software is notorious for slowing down system processes, especially when it is in the process of scanning the system. Of course, if the hardware changes, (different computer used, memory added, etc.) this statement will most certainly no longer be true.

TABLE 4.5
Results of Data Acquisition on 5.0 Hz Sine Wave with Different Sampling Rates

No.	Timer	Delta	Voltage	Timer	Delta	Voltage	Timer	Delta	Voltage	Timer	Delta	Voltage
1	0.000	0.000	-2.69	0.000	0.000	-0.25	0.000	0.000	1.39	0.000	0.000	-3.01
2	1.021	1.021	0.16	0.121	0.121	2.36	0.030	0.030	3.87	0.028	0.028	-4.11
3	2.043	1.022	2.92	0.241	0.120	-4.02	0.060	0.030	2.90	0.058	0.030	-1.95
4	3.064	1.021	3.96	0.361	0.120	3.65	0.090	0.030	-0.56	0.088	0.030	1.73
5	4.086	1.022	2.61	0.481	0.120	-2.22	0.120	0.030	-3.60	0.118	0.030	3.94
6	5.107	1.021	-.024	0.601	0.120	-0.62	0.150	0.030	-3.83	0.148	0.030	2.63
7	6.129	1.022	-2.99	0.722	0.121	2.65	0.180	0.030	-0.95	0.178	0.030	-0.93
8	7.150	1.021	-4.16	0.842	0.120	-4.09	0.210	0.030	2.59	0.208	0.030	-3.78
9	8.172	1.022	-3.08	0.962	0.120	3.54	0.240	0.030	3.94	0.238	0.030	-3.66
10	9.193	1.021	-0.37	1.082	0.120	-2.00	0.270	0.030	1.80	0.268	0.030	-0.63
11	10.215	1.022	2.52	1.202	0.120	-0.92	0.300	0.030	-1.88	0.298	0.030	2.87
12	11.236	1.021	3.97	1.322	0.120	2.83	0.330	0.030	-4.08	0.328	0.030	3.88
13	12.258	1.022	3.00	1.443	0.121	-4.12	0.360	0.030	-3.07	0.358	0.030	1.47
14	13.279	1.021	0.29	1.563	0.120	3.31	0.390	0.030	0.37	0.388	0.030	-2.18
15	14.300	1.021	-2.61	1.683	0.120	-1.69	0.420	0.030	3.49	0.418	0.030	-4.13
16	15.322	1.022	-4.12	1.803	0.120	-1.17	0.450	0.030	3.52	0.448	0.030	-2.82
17	16.343	1.021	-3.38	1.923	0.120	3.04	0.480	0.030	0.49	0.478	0.030	0.72
18	17.365	1.022	-0.79	2.043	0.120	-4.15	0.510	0.030	-2.97	0.508	0.030	3.65
19	18.386	1.021	2.16	2.164	0.121	3.17	0.541	0.031	-4.11	0.538	0.030	3.33
20	19.408	1.022	3.87	2.284	0.120	-1.39	0.571	0.030	-1.98	0.568	0.030	0.12
21	20.429	1.021	3.34	2.404	0.120	-1.51	0.601	0.030	1.70	0.599	0.031	-3.23
22	21.451	1.022	0.86	2.524	0.120	3.27	0.631	0.030	3.93	0.629	0.030	-4.04
23	22.472	1.021	-2.07	2.644	0.120	-4.15	0.661	0.030	2.67	0.659	0.030	-1.64
24	23.494	1.022	-3.96	2.765	0.121	2.94	0.691	0.030	-0.88	0.689	0.030	2.03
25	24.515	1.021	-3.74	2.885	0.120	-1.05	0.721	0.030	-3.75	0.719	0.030	3.98

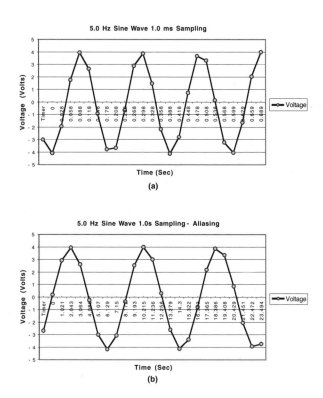

FIGURE 4.24 Identical waveforms captured at different sampling rates yield different results.

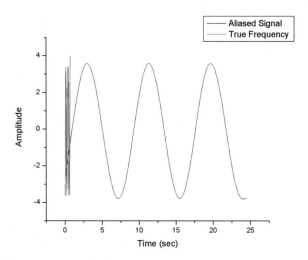

FIGURE 4.25 Original signal (compressed sine wave) aliased down to lower frequency (elongated sine wave).

The interval is reduced below its limit of guaranteed precision of 56 ms (in this case 10 ms and 1 ms) and that the interval is the same value (30 ms). Remember that the interval is only guaranteed to be as fast as 56 ms. The fastest interval on this system is 30 ms. It does not matter how low the interval is set, the shortest interval that can possibly occur on this system is 30 ms. Remember, the true interval is greater than the specified interval by 20 ms. Therefore, a reasonable conjecture is that an interval specified of 10 ms will yield the shortest possible interval for this system.

$$\text{Interval}_{TOTAL} = \text{Interval}_{VALUE} + \text{System Latency}$$

$$30 \text{ ms} = 10 \text{ ms} + 20 \text{ ms}$$

4.13 METHODS FOR FASTER ACQUISITION

Clearly the methods illustrated in Figure 4.25 are somewhat crude in that they do not allow for fast acquisition. However, the methods are simple and very effective when a system needs to be sampled no faster than 3 to 4 times a second, which encompasses many laboratory type experiments. Another great advantage of the techniques demonstrated in Figure 4.25 is that costly software components or add-on controls to complete an application are not required.

To sample at a faster rate, a DriverLINX data acquisition device driver for a 32-bit application can be purchased from Keithley Metrabyte. The device will run under Windows 95/98/NT/2000. This package includes both ActiveX controls that can be placed within a project by placing the component on a form (similar to the timer control) and DLL interfaces that function much like the PortIO DLL. As with the PortIO DLL, the speed at which this software operates is hardware dependant, but is several orders of magnitude higher than the PortIO DLL. Tests on a Pentium-II 233 MHz running Windows NT 4.0 show polled mode operations on the order of 400 µS (2.5 kHz).

However, even faster acquisition can be obtained. A board with hardware that triggers the sampling of the input signal can be purchased. For such systems, the sampling rate is limited only to the speed of the system clock built into the board. Speeds of 20, 50, or 100 MHz are not uncommon. With this type of system, the software program's only task is to tell the hardware when to start acquiring data. The hardware then begins to sample the input and stores the results to its internal memory. The number of samples that can be taken (or conversely the interval) is limited only by the amount of internal memory in the data acquisition board. When the data acquisition process is finished, the software then can request that the data stored in the data acquisition boards memory be transferred into the computer.

Since all the functions in the data acquisition process are carried out at the hardware level (by electronic signals as opposed to software commands), little can be done to slow down the acquisition process. The speed limiting processes such as disk storage and interval limitations within software are gone. Using Visual Basic the interval of data capture will begin within $^1/_{18}$ of a second. Hence, to be safe, issue the command to begin data acquisition $^1/_{18}$ of a second before data would be present at the input of the data acquisition card. To capture data over a particular window of time requires a card with enough memory to store data during the interval of the window, plus $^2/_{18}$ of a second.

4.14 ADDING AUXILIARY POWER TO THE METRABUS SYSTEM

As mentioned earlier, the MAI-16 card requires an external power supply to provide ± 15.0 V power in addition to the + 5.0 V power that all cards require. Other cards may require the ± 15.0 V power supply in addition to the MAI-16, so knowing procedures to supply additional power to the bus is important. If a great number of cards will be connected to the Metrabus system, it would be wise to supplement the PC's internal supply. The 5.0 V power supply provided by the PC to the bus was never intended to drive great loads and it is very easy to overload this fragile supply and cause damage to the host PC. In addition, when using an optically isolated card (highly recommended), an external supply must be used to power cards connected to the bus as no physical connection is present from the bus to the PC.

The easiest way to supply additional power to the Metrabus system is to purchase the Keithley model MBUS-PWR power supply for the Metrabus, connect the ribbon cable, and plug it in. However this supply is costly and for many applications is really overkill. Constructing your own power supply is not difficult.

FIGURE 4.26 The MTAP-1 card provides a convient way to extract or provide power to the bus.

FIGURE 4.27 An assembled Metrabus power supply.

The best way to construct a power supply for the Metrabus system is to use the Keithley MTAP-1 Card shown in Figure 4.26. Although the intended use of MTAP-1 card is to allow power to be drawn from the power supply on the bus, it is equally suited to supply power to the bus. The card also features annunciator LED's that indicate which power supplies are active on the bus. The big advantage to using this card is that it makes it very difficult to connect power to the wrong pin on the bus unless someone is grossly negligent.

Any 3 rail (±15.0 V and +5.0 V) power supply can be connected easily to this card to provide additional power to the Metrabus system. An elegant way to do this (and for safety's sake as well) is to place all the components within a nice enclosure with cutout for the connectors and power cords as shown in Figure 4.27.

The interior connections for assembling such a device are rather simple and can be accomplished easily by anyone with some basic knowledge of electronics who knows how to solder. A typical connection scheme is pictured in Figure 4.28.

FIGURE 4.28 One of many possible power supply type schemes.

4.15 SUMMARY

At least a dozen manufacturers of data acquisition and hardware controller boards are available. To provide an adequate overview for all of them is impossible. However, the methodology behind their inner workings are all very similar. The Keithley system was chosen only because the author has experience using their cards. All the software provided in this chapter has been tested with cards for which it was intended to be used with and found to function properly.

The reader may choose any number of paths based on this new found knowledge. He may choose to enhance the software provided in this text for the Keithley boards or a reader's application may be better suited to adopting the methods shown to control cards from other manufacturers. Whatever the final course of action, it is the mechanics of how hardware interfacing works that is most important. Once a mastery of the underlying principals and concepts has been achieved, the user will have the knowledge to solve the idiosyncratic problems that arise from one card manufacturer to the next.

5   Electronics for Automation

5.1   OVERVIEW

Seldom in life are products exactly the way a customer would want them. Products arrive with features that will be utilized by the majority of consumers. Appealing to the masses is how companies make money and keep costs down. Unfortunately, to achieve a competitive advantage, it is often necessary to have something work exactly the way it is envisioned to work, which is often more robust than its original design.

Instrumentation is no exception. A lot of laboratory instruments were intended only to be operated by hand. They were never intended to be controlled by a computer, or completely automated. They may be capable of providing information or features that were not incorporated into their original design to keep costs down. This chapter is about automating that which was never intended to be automated, and improving that which is already on the market.

Controlling any instrument requires an interface. Some instruments come with a full interface, which allows control of every aspect of the instrument. More often than not, instruments come with limited interfaces. For example, they might allow the data to be collected from a serial port, but the user must set up the instrument parameters by hand. Typically, information is shown to the user by one of three mechanisms: light emitting diodes (LED) that indicate a Boolean type condition (on/off, true/false, etc.), LED binary coded decimal (BCD) displays that show numbers or letters to the user, or liquid crystal displays that can show a variety of information to the operator.

The limited interfaces that so often come with instruments can be problematic if not outright dangerous. If data can be collected by a computer without setting up (or checking) the machine state, how can the user be sure about what is measured? It could be that the instrument is always used in only one state and *should* never change. However, suppose someone wanders into the lab and absentmindedly starts pressing buttons on an instrument for no apparent reason. (Maybe it is "take your son to work" day and little Melvin pushes a button when no one is looking.)

Now the machine starts to measure solubility in European brewery units instead of nephelometric turbidity units. In a best-case scenario the mistake is caught after the first run, and the data for all the compounds run must be done over. In a worst case scenario the mistake is not caught for some time, and erroneous data are sent into the corporate database. The next blockbuster drug is overlooked because poor solubility is erroneously reported, and no one gave the compound a second look.

For the majority of readers, this chapter will prove to be the most difficult to understand in the text. This will especially be true for those who do not have any sort of background in electronics. For those that fall into this category, it would be beneficial to first read a primer on electronics such as *Getting Started in Electronics* by Forrest M. Mims, which can be purchased at any Radio Shack. This material must be covered because it is impossible to have a true discussion on laboratory automation without discussing hardware/software interfacing. Of course, electronic hardware must be discussed.

The first part of this chapter covers some basic but very useful methods commonly utilized by electronics technicians and engineers. It is then shown how to utilize these methods for massaging signal and voltage levels from instruments into states easily interpreted by a common digital interface. The next part of this chapter shows how to find the signals an individual needs from an instrument by "reverse engineering" the instrumentation's hardware. Finally, it is shown by example

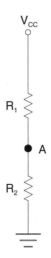

FIGURE 5.1 A simple voltage divider network.

how the methods covered can be used to create an interface for any piece of instrumentation, laboratory equipment, or machinery.

5.2 VOLTAGE DIVISION

Most Boolean values are electrically represented as either 0 V (typically off or false) or 5.0 V (typically on or true). However, when creating an interface it may be necessary to adjust voltages to a desired level, which is where the principal of voltage division comes in. Voltage division takes an input voltage (or reference voltage) and then divides the voltage down to a lower voltage based on the ratio of two or more resistors.

In Figure 5.1 V_{CC} is the input voltage and "A" (V_A) is the voltage that has been divided by means of resistors R_1 and R_2. The voltage at point A (V_A) can be determined by the formula:

$$V_A = V_{CC} \frac{R_2}{R_2 + R_1} \tag{5.1}$$

Example: $V_{CC} = 10$ V, $R_1 = 1.0$ KΩ, $R_2 = 9.0$ KΩ

$$V_A = 9.0 \text{ V}$$

This example is somewhat intuitive; here $^1/_{10}$ of the voltage (1.0 V) is dropped across R_1 (1.0 KΩ) while $^9/_{10}$ of the voltage (9.0 V) is dropped across R_2 (9.0 KΩ). Hence the voltage at point (A) $V_A = 10.0$ V − 1.0 V = 9.0 V

Using this technique, voltages from boards that run off supplies from 0 to 15.0, 20.0, and 25.0 V, or even higher, can be "translated down" to function in a board that uses a transistor to transistor logic (TTL) supply level of 0 to 5.0 V. This technique will be used in the comparator, hysteresis, peak detector, and filtering circuits that follow.

5.3 DIODE WAVEFORM SHAPING

The beauty of the diode (Figure 5.2) is that it only conducts current in one direction. Current flows from positive to negative (usually ground) in the direction of the arrow in this figure. Actually, if

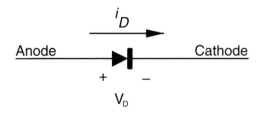

FIGURE 5.2 Basic diode.

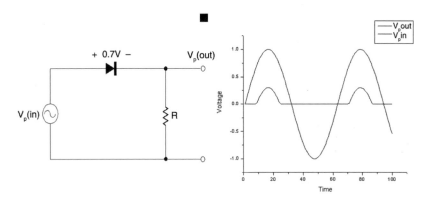

FIGURE 5.3 Half wave rectification.

enough voltage is applied, current will flow in the opposite direction. (If you apply enough voltage to any medium — lightning, for instance — electricity will flow.) This is called avalanche breakdown and this process occurs when a diode is reverse biased (the power is connected to the diode backwards). A practical example of such behavior will be demonstrated later in this chapter.

One very functional aspect of diodes is that they can be used for waveform shaping. In the example in Figure 5.3, the diode is used to convey the positive portion of the signal and suppress the negative portion of the waveform. (Such a circuit is called a half wave rectifier.)

Every diode requires a certain "turn on" voltage or voltage required to overcome the internal mechanism. For a silicon device (most common) this voltage is 0.7 V. For the example in Figure 5.3, when the diode is conducting, the output waveform will be less than the input waveform by its turn-on voltage.

Although rectification is a useful property, diodes can provide even more flexibility to circuits. The ability to clamp waveforms at a certain point is illustrated in Figure 5.4. Here the voltage V_b is applied to the cathode of the diode. The positive portion of the waveform will be "clamped" at $(V_b + V_D)$. This is a useful circuit to have if a signal passed to the input of a device cannot exceed a certain voltage without running the risk of damaging the device.

Earlier it was stated that current would flow through a diode in the reverse direction (cathode to anode) if enough voltage was applied to the cathode of the diode. This process is called avalanche breakdown and when it occurs the diode is operating in the reverse bias region (see Figure 5.5). Base collector junctions of bipolar junction transistors, photodetectors, and Zener diodes are all devices that operate in reverse bias region. The Zener diode is often used in this method to provide a mechanism for over-voltage protection, which protects a circuit from damage should the power supply to the circuit exceed a certain voltage.

In Figure 5.6a, when the input voltage V_{CC} exceeds $V_Z + V_d$, current will flow through the two diodes to ground. The means by which this is accomplished is as follows: when the top Zener diode has a voltage applied to the cathode that exceeds V_Z (the Zener voltage), the diode will "break

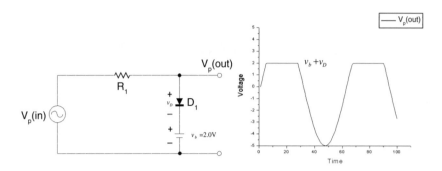

FIGURE 5.4 Voltage clamping using diodes.

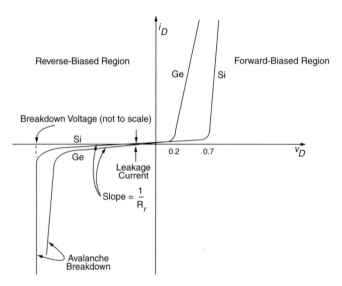

FIGURE 5.5 Diode reverse bias curves. (From Savant, C.J., Roden, M.S., and Carpenter, G.L., *Electronic Design*, 2nd ed., Discovery Press, Los Angeles, CA, 1991. With permission.)

down" and conduct current to the second diode. The second diode functions as a regular diode, and will turn on and conduct current from Vcc straight to ground once its threshold voltage V_d (typically 0.3 to 0.7 V) is reached. If the supply voltage is 5.0 V, a Zener diode would be selected with $V_Z = 5.0$ V. When the voltage exceeds $V_Z + V_d$, the diodes will turn on, protecting the circuit when too much voltage is applied.

Figure 5.6b operates on the same principal, but for a negative input voltage ($-V_{CC}$), the functionality of the diodes is reversed. Here the top diode acts as a regular diode and the bottom diode must have the Zener voltage (V_Z) exceeded at its cathode for diodes to conduct all current to ground.

The two resistors shown in Figure 5.6a and Figure 5.6b are for current limiting. Depending upon the rating of the diodes and the amount of current that could be passing through them in the event of a transient voltage spike, it is possible to "cook" the two Zener diodes if enough current should pass through them.

Typically, the Zener voltage is set to equal the supply voltage for the particular power rail the circuit is employed in. This scheme will prevent damage to circuits from high voltage should the input voltage ever exceed the supply voltage.

(a) Positive Rail   (b) Negative Rail

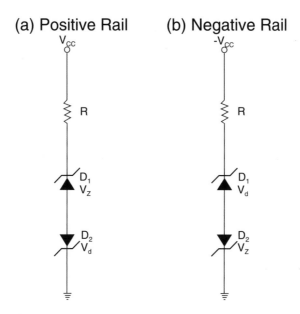

FIGURE 5.6 Zener diode over voltage protection schemes.

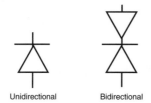

Unidirectional   Bidirectional

FIGURE 5.7 Transient voltage suppressor schematic symbols.

Transient voltage suppression (TVS) diodes (Figure 5.7) are an even better method of over voltage protection. Their reaction time is measured in the picosecond range (nearly instantaneous), and are specifically manufactured for this purpose, as opposed to Zener diodes, which are primarily manufactured for use in power supplies.

5.4 THE TRANSISTOR

A transistor (Figure 5.8) is a semiconductor device with three leads called the emitter, the base, and the collector. The base–emitter junction functions exactly like that of the diode mentioned earlier. Here the base has the same function as the anode on the diode and the emitter serves the same function as the cathode on the diode. Like the diode, the base–emitter junction will only turn on (conduct current) when a high enough potential difference (voltage) exists between the base and the emitter. For most silicon devices (the most common), the turn-on voltage is ≈ 0.7 V.

What is interesting about the transistor is that a very small base–emitter current flow gives rise to a *much larger* collector–emitter current flow. If no current is flowing from the base to the emitter, then no current will be flowing from the collector to the emitter either. In such an instance the transistor is said to be "off." As the base–emitter junction begins to turn on, current begins to flow from the collector to the emitter. As the base–emitter current flow increases, this gives rise to a much larger collector–emitter current flow.

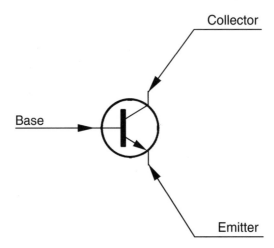

FIGURE 5.8 A bipolar NPN transistor.

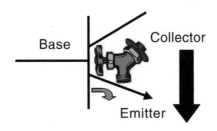

FIGURE 5.9 Transistors function like "electron faucets."

Transistors can be thought of as "electron faucets" (Figure 5.9). Just as a small turn on a faucet knob will allow a great deal of water to flow through the faucet, a small current flowing from the base to the emitter will allow a much larger current to flow from the collector to the emitter.

Transistors are utilized in many applications but are most commonly used as switches and amplifiers, and to sink large amounts of current. Many times the user will wish to use the output of a comparator or operational amplifier (see Section 5.5 to Section 5.7) to drive a relay or some other device that requires a large amount of current. Often the output of a comparator or operational amplifier will not be able to source (pull from the power supply) enough current to drive a high current device such as a relay. However, the output from the comparator or operational amplifier is more than enough to drive the base–emitter junction of a power transistor, which can then source the current through the collector–emitter junction to power the high current device. Figure 5.10 illustrates how to drive a relay using a transistor. The resistor is for current limiting purposes, and V_{in} is where the output of a comparator or opamp would be connected. When V_{in} exceeds 0.7 V the transistor will turn on and the relay will be actuated.

5.5 THE COMPARATOR

This is probably the most useful circuit in terms of converting an analog signal into a digital Boolean value. A basic comparator is illustrated in Figure 5.11. V_{ref} is the reference voltage. Whenever the input voltage (V_{in}) exceeds the reference voltage, the output is equal to the supply voltage +V. Conversely, whenever the V_{in} is less than V_{ref}, the output is equal to ground (0 V). The graph in Figure 5.12 illustrates the output based on the given input signal.

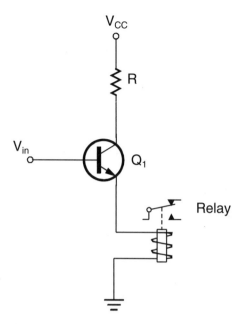

FIGURE 5.10 Using a transistor to drive a relay.

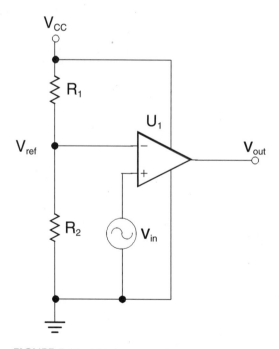

FIGURE 5.11 A basic comparator.

A point worth mentioning is that some operational amplifiers and comparators feature open collector outputs. The advantage of these outputs is that the output voltage is not limited to V_{CC}, but can be set to any level desired. An open collector output means the output of the opamp/comparator is floating as pictured in Figure 5.13. When such a device is used, it will be necessary to connect a "pull up" resistor to the output of the op-amp/comparator, as shown in Figure 5.14.

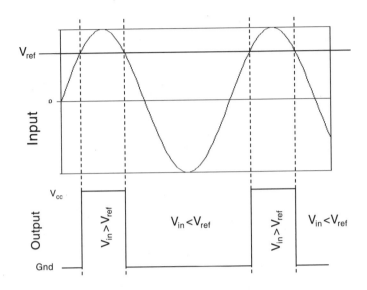

FIGURE 5.12 Input vs. output for the comparator in Figure 5.11.

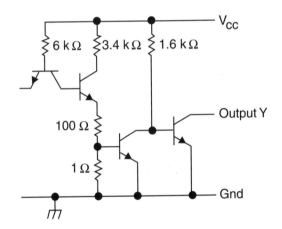

FIGURE 5.13 Partial schematic of open collector type output.

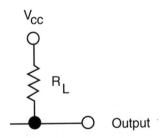

FIGURE 5.14 Open collector outputs require the use of a "pull up" resistor to V_{CC}.

The question now arises about what a comparator circuit is good for. Suppose an instrument can be placed in one of three possible states, S_A, S_B, or S_C. One of three LEDs on the instrument turns on or off for each possible state that the instrument can be in. (Only one of the three LEDs will be lit at a time, because the instrument can only be in one state at a time.)

The instrument has been taken apart and the voltages at the LED drivers on the circuit board have been read when the LEDs are on and off. For the sake of discussion, suppose the voltage level at the LED driver is 2.0 V when the LED is off (V_{off}), and 4.0 V when the LED is on (V_{on}). For a computer to read a digital input and know when the LED is on or off, it must have discrete signal levels of 0 or 5 V for when the LED is on or off (respectively).

This is where the utility of the comparator comes in. First, a voltage divider would be created to set the $V_{ref} = 3$ V. The V_{ref} level would be fed into three operational amplifiers at the (−) inverting input. Next each of the inputs from S_A, S_B, And S_C would be fed into the noninverting input (+) of one of the operational amplifiers. The grounds of the operational amplifiers must be connected to the ground of the instrument *at one point only* (otherwise ground loops will form). The grounds of all the circuits must be the same, or the voltages that are fed into each amplifier will be arbitrary because the reference (or starting point to measure the value of voltage to ground) will be different for every circuit!

The circuit in Figure 5.15 has three different outputs ($V_{SA_{out}}$, $V_{SB_{out}}$, $V_{SC_{out}}$), each corresponding to the state of the instrument in S_A, S_B, or S_C. If the instrument is in state S_n, then $V_{Sn_{out}}$ will be high (V_{CC}). If the instrument is not in state S_n, then $V_{Sn_{out}}$ will be low (Gnd). Table 5.1 shows the possible states.

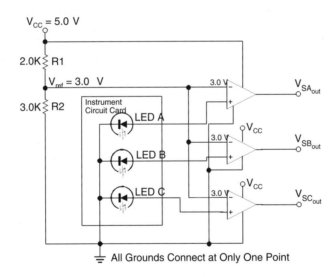

FIGURE 5.15 Reading the state of an LED using comparators.

TABLE 5.1
Truth Table for Circuit in Figure 5.15

	Instrument State			Comparator Output	
S_A	S_B	S_C	$V_{SA_{out}}$	$V_{SB_{out}}$	$V_{SC_{out}}$
0	0	1	Gnd	Gnd	V_{CC}
0	1	0	Gnd	V_{CC}	Gnd
1	0	0	V_{CC}	Gnd	Gnd

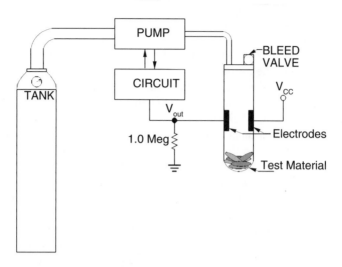

FIGURE 5.16 Sample test apparatus.

The outputs $V_{SA_{out}}$, $V_{SB_{out}}$, and $V_{SC_{out}}$ can be easily read by a digital input board connected to a computer, such as the Keithley Metrabyte MBB-32. Also take note that, by setting $V_{ref} = 3.0$ V, a noise margin of 1.0 V is given to the inputs of the comparators. In other words, noise greater than 1.0 V must be superimposed on the input signal to cause the output to change erroneously. Rarely will a signal have that much noise present; usually a few tenths of a volt would be considered a "large" amount of noise. This example is simple to design for because the differential between the two states is 2.0 V: $V_{LED_{on}}$ (4.0 V) − $V_{LED_{off}}$ (2.0 V) = 2.0 V. Thus, noise factors really do not need to be taken into consideration when designing such a circuit. Unfortunately, noise margins are not always so large, but the next section presents a method of dealing with them.

5.6 COMPENSATING FOR NOISE EFFECTS WITH HYSTERESIS

For a more complicated example with small noise margins, suppose the apparatus illustrated in Figure 5.16 existed. Here a material is placed in a chamber with two electrodes. The chamber has two openings: one to allow a gas from the tank to be pumped in, and one connected to a pressure valve to bleed excess gas off. For the sake of discussion, suppose the material is to undergo a reaction with the gas in the tank. The gas is slightly conductive, and for an optimum reaction to occur the voltage at V_{out} should be 2.77 V. As the reaction occurs the gas changes to become less conductive and more gas must be added to the vessel for the reaction to continue. When more gas is added excess gas is bled off by means of the pressure valve. The gas is only slightly conductive, and if enough "fresh gas" is present, enough current will flow across the electrodes and through the 1 Meg resistor to create an output voltage of 2.77 V. If too much gas is pumped into the vessel, V_{out} will be greater than 2.77 V, and if too little "fresh" gas is present in the vessel, V_{out} will be less than 2.77 V.

Also, the noise margins for this closed system are very low. If V_{out} is greater than 2.97 V, then the reaction will ruined, and if V_{out} is less than 2.57 V, then the reaction will stall. To top it off, the output from the electrode voltage divider is very noisy. The graph in Figure 5.17 shows what would ordinarily be a perfect output of 2.77 V, but noise in the system causes the fluctuations illustrated in the graph.

The graph in Figure 5.18 shows what happens if the noisy output from the electrodes is fed into a comparator. The output oscillates between V_{CC} and Gnd because the noise on the input causes the input signal to rise above and below the threshold. The result of using a circuit like this is that the pump would turn on and off several times a second, which would not only be annoying to

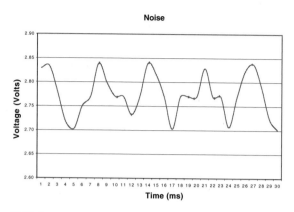

FIGURE 5.17 Input signal to comparator with noise.

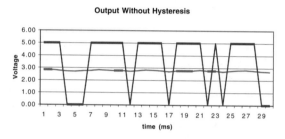

FIGURE 5.18 Output for corresponding input with noise without hysteresis.

anyone sitting next to it, but damaging to the pump as well. (This behavior is similar to the "cheap" window fans sold with a thermostat that turn on and off several times a minute when the thermostat setting is near room temperature.) The question now is how to design a system to compensate for this?

Looking at the noise plot and the process description, Table 5.2 can be produced.

TABLE 5.2
Design Specifications for Sample Process

Desc	Voltage	Comments
Max	2.97	Absolute maximum before ruining reaction
UTP	2.84	Highest V_{out} for ideal + noise
Ideal	2.77	Ideal voltage for reaction (without noise)
LTP	2.70	Lowest V_{out} for ideal – noise
Min	2.57	Absolute minimum before ruining reaction

Notice the abbreviations UTP and LTP, which stand for upper trip (or threshold) point and lower trip (or threshold) point. It is essential to know these points in order to design circuits to compensate for noisy sources. They are taken here from the "noise" graph in Figure 5.17.

Noise is a very common problem in circuit design because it is everywhere and emanates from many sources. Fortunately, a tried and true mechanism called hysteresis can reduce noise in comparator circuits.

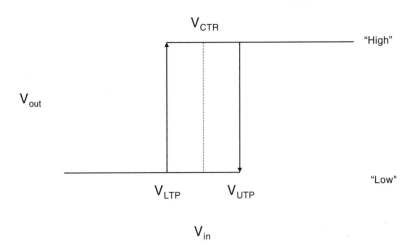

FIGURE 5.19 Output of an inverting comparator with hysteresis.

A comparator with hysteresis (also known as a Schmitt trigger), is a process by which the trip point for the comparator to change states is altered from one discrete voltage point to two discrete voltage points. This is accomplished by means of positive feedback using resistors. The voltage that separates the two discrete voltages is known as the hysteresis voltage (V_H). The highest voltage discrete point is known as the V_{UTP} or voltage upper trip point. The lowest voltage discrete point is known as the V_{LTP} or voltage lower trip point. Together these two points (V_{UTP} and V_{LTP}) create a window, and in order for the state of the comparator to change at the output, the input must rise above or fall below that window. In other words the input must cross the V_{UTP} and the V_{LTP} for the comparator to change states.

When the input voltage exceeds V_{UTP} (the voltage has crossed both V_{UTP} and V_{LTP}), then the comparator will change states at the output. When the input voltage falls below the V_{LTP} (the voltage has crossed both V_{UTP} and V_{LTP}), then the comparator will again change state at the output.

By having two voltage trip points instead of one single voltage trip point, noise immunity is gained because the input voltage change must fall outside the window of the hysteresis voltage (V_H) in order for the output of the comparator to change state. In other words, if the voltage "spikes" from noise at the reference voltage level do not exceed ($V_H/2$), they will not have any effect on the output of the comparator (see Figure 5.19).

Hysteresis has design trade-offs. The larger the hysteresis voltage, the more immunity to noise the circuit has to changing conditions at the expense of sensitivity. By careful application of noise reduction techniques, and judicious use of hysteresis, very robust circuits for different applications can be created.

Figure 5.20 illustrates how to design a circuit to control the pumping of the gas into the test vessel example. Here the reference voltage is set to the "ideal" voltage of 2.77 V. V_{UTP} is set to the highest ideal voltage possible with noise ($V_{UTP} = 2.84$ V). V_{LTP} is set to the lowest ideal voltage possible with noise ($V_{LTP} = 2.70$ V). The hysteresis voltage is: $V_H = V_{UTP} - V_{LTP} = 0.14$ V.

Ideally, the pump should turn off when $V_{in} > V_{UTP}$, and turn on when $V_{in} < V_{LTP}$. The hysteresis desired is accomplished by means of a positive feedback loop consisting of two resistors. To determine the value of the resistors an "n" factor must be computed by means of the following formula:

$$n = \frac{(V_{CC}+) - (V_{CC}-)}{V_H} \tag{5.2}$$

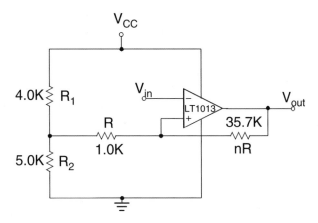

FIGURE 5.20 Schematic for comparator with hysteresis used in sample application.

In this case the supply voltage is 5.0 V to ground (Gnd). ($V_{CC}-$) is therefore 0 V, so the equation is reduced to:

$$\Rightarrow \frac{V_{CC}}{V_H} \tag{5.3}$$

V_{ref} is created by means of a simple voltage divider between V_{CC} and Gnd.

$$V_H = V_{UTP} - V_{LT} = 2.84 - 2.70 = 0.14 \text{ V} \tag{5.4}$$

$$V_{CTR} = \frac{V_{UTP} + V_{LTP}}{2} = \frac{2.84 + 2.70}{2} = 2.77 \text{ V} \tag{5.5}$$

$$V_{CTR} = \text{ideal voltage in chart} \tag{5.6}$$

$$n = \frac{(V_{CC}+) - (V_{CC}-)}{V_H} \tag{5.7}$$

For the case where the supply is between V_{CC} and Gnd this equation reduces to:

$$n = \frac{V_{CC}}{V_H} = \frac{5.0}{0.14} = 35.7 \tag{5.8}$$

$$\text{Choose } R = 1.0 \text{ K} => nR = 35.7 \text{ K} \tag{5.9}$$

To create the reference voltage (2.77 V), use a voltage divider. Choose $R_2 = 5.0$ K.

$$V_A = V_{CC} \left(\frac{R_2}{R_2 + R_1} \right) \tag{5.10}$$

$$2.77 = 5.0\,\text{V}\left(\frac{5.0\,\text{K}}{\left(5.0\,\text{K} + \text{R}_1\right)}\right) \tag{5.11}$$

$$2.77 = \left(\frac{25}{\left(5.0 + \text{R}_1\right)}\right) \tag{5.12}$$

$$2.77\,\text{R}_1 + 13.85 = 25 \tag{5.13}$$

$$\text{R}_1 = \left(\frac{11.5}{2.77}\right) = 4.0\,\text{K} \tag{5.14}$$

One word of caution about setting the reference voltage: a popular method of creating the reference voltage is to create a voltage divider between V_{CC} and Gnd consisting of a variable resistor (pot) and a fixed resistor. The advantage here is that the variable resistor can be adjusted with a screwdriver to yield the exact reference voltage desired; however, this method has some drawbacks.

First, the variable resistor consists of conductive plates that slide along each other to create a varying resistance. While this is fine for many applications, over time electrons arcing across the plates produce a carbon buildup, which will change the resistance of the pot over time.

Second, if the circuit created is to be distributed to a large number of users, the last thing anyone would want on the circuit board is a potentiometer. Why? The second anything goes wrong the first thing some inexperienced person will do is stick a screwdriver into the potentiometer and start turning it hoping to make the circuit work again. This action will uncalibrate the voltage divider and make things worse.

The best way to construct a voltage divider for a comparator, (and the most time consuming as well) is to use precision resistors in parallel or series to obtain exactly the voltage desired for the reference. There is no potentiometer to go out of calibration, and there are no components for unknowledgeable persons to adjust with a screwdriver.

5.7 WINDOW COMPARATOR

Some measurements do not focus on the value of one discrete point; an entire range of values may be acceptable, and anything outside that range is deemed unacceptable. Such a range is referred to as a window, and a window comparator will determine if the input voltage is within a range specified by two reference voltages. Figure 5.21 shows a window comparator with a window from 1.0 to 4.0 V ($V_{ref_{HIGH}}$ = 4.0 V, $V_{ref_{LOW}}$ = 1.0 V).

Window comparators are often used to make "idiot lights." For the preceding example, if V_{in} is within the window of 1.0 to 4.0 V, then V_{out} is low (Gnd). If V_{in} is outside the window, then V_{out} is high (V_{CC}). An LED could be connected to V_{out} that would light anytime the input value was outside the window. This is how the idiot lights on an automobile work.

Here is an example related to the pharmaceutical industry. Suppose a machine is to dispense powdered samples into vials. Also assume that each powdered sample has the potential to be different (different powder size, electrostatic potential, material, etc.). Some powders will be easy to shake into a vial. Others will want to clump together. The different powders are blown into the vials at the same rate. Therefore, the powders with larger flake sizes will fill up the vials faster.

With all these different parameters for all the different types of powders, the fill rate for each particular powder will be different. Therefore, a range that is an acceptable fill level for each vial could be developed. Suppose that each vial should have between 95.0 and 105.0 mg of powder in it. A pressure sensor could be used to weigh each vial during the filling process. Once the vial is

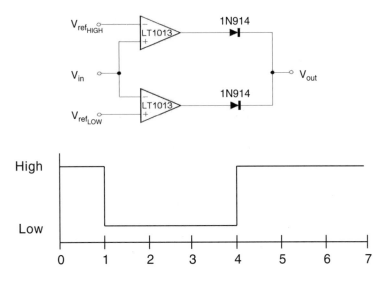

FIGURE 5.21 Schematic for window comparator.

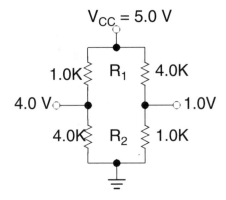

FIGURE 5.22 Two separate voltage dividers.

filled to within the acceptable window the filling process should stop. If the pressure sensor's output is >1.0 V for weights above 95.0 mg and <4.0 V for weights below 105.0 mg, then the preceding circuit will produce a low output when the vial is filled within the acceptable tolerances. Such a circuit could be used first to stop the filling process once the amount of material falls within the window, and second to check the final result to see if the vial is still within the required tolerance once the filling process has stopped.

The two reference voltages can be set by means of a voltage divider network as described earlier. One method would be to use two voltage dividers such as pictured in Figure 5.22. However, a much better method would be to use three resistors as the shown in Figure 5.23.

This method is superior because it uses one less component (resistor). However, this method is superior for even deeper reasons. When two voltage dividers are used and placed on a circuit board, they will inevitably be placed in different areas of the circuit board, drawing their power from different places on the circuit board. This is a problem because one area of the board may harbor more induced noise than another area of the board, and the inputs to the operational amplifiers will not fluctuate due to noise in a consistent manner. By using only one voltage divider network,

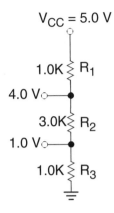

FIGURE 5.23 Using three resistors to provide two voltage references — superior method.

the noise present on the reference supply lines should be nearly the same. Additionally, the size of the window will remain the same, although it may be offset somewhat due to noise fluctuations on the inputs.

5.8 PEAK DETECTION

Oftentimes voltage at a point being monitored will rise and fall over time depending upon what is being measured. In such circumstances it is sometimes convenient to have a circuit that will "remember" the highest voltage at that point for a limited amount of time. This is especially true if the voltage point can only be polled at long intervals during which the value at the voltage point may change substantially. Another such use would be if the highest voltage at a given point over a specified interval is desired, but the user does not want to poll the point and make constant comparisons over that interval. Polling and comparing voltages can easily be done by computer, but it is taxing on system resources, which could probably be put to better use. For such applications a peak detection circuit is handy (see Figure 5.24).

The voltage from any given point on a circuit board can be applied to V_{in}. The output at V_{out} will be the highest voltage applied at V_{in}. This is accomplished by means of the capacitor that stores the voltage applied at the input. Be advised that the highest voltage will not be stored indefinitely. The output V_{out} will drift lower at a rate of about 10 mV/sec. As long as V_{out} is sampled every few seconds, an accurate reading of the highest level applied to V_{in} is assured. To reset the acquisition of the highest voltage level stored, the reset switch is closed, thus shorting the capacitor to ground and draining any voltage it had stored in it.

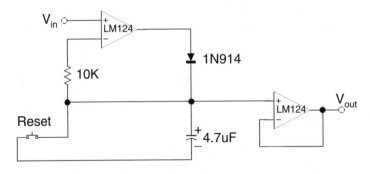

FIGURE 5.24 Basic peak detector (requires leaky capacitor — not an electrolytic).

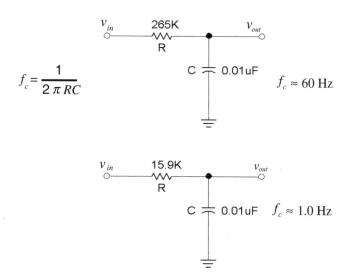

$$f_c = \frac{1}{2\pi RC}$$

$f_c \approx 60\ Hz$

$f_c \approx 1.0\ Hz$

FIGURE 5.25 Examples of passive single pole low pass filters — rolloff 20 dB/decade.

A typical application would be to have a computer connected to an analog input board acquire the input voltage from V_{out}. After the voltage had been acquired, a computer-controlled relay could connect the reset switch restarting the acquisition process. Such circuits are very useful in instances when a pulse on a line is indicative of a state change or process condition. The peak detector is capable of storing the highest voltage value present on a line over an interval of time. In this way an interface can monitor whether or not a pulse actually occurred by checking the voltage output of the peak detector. If the output of the peak detector is at ground, no pulse has occurred since the peak detector was last reset.

5.9 LOW PASS FILTERING

Sometimes when looking at a particular point on a circuit card from an instrument, both DC and AC voltages will be present. More often than not, the DC voltage is all that is needed to be monitored, and having the superimposed AC signal will make monitoring the steady state DC signal all but impossible.

AC signals on circuit cards come from a variety of sources. The most obvious would be noise. AC signals also emanate from signal lines that run close together, causing crosstalk. Many boards have AC signal lines running throughout them controlling counters, multiplexors, latches, etc.

To remove the AC component from a point being monitored on a circuit card, a device called a low pass filter is used. A very simple low pass filter is illustrated in Figure 5.25. This circuit is referred to as a passive low pass filter, so called because no active components (such as operational amplifiers) are used in its construction. As a signal propagates from v_{in} to v_{out}, the AC component of the v_{in} signal is routed to ground through the capacitor.

However, the AC component of the signal is only routed to ground above the cutoff frequency (f_c). At frequencies less than the cutoff frequency, most of the AC component of the signal will be passed through to v_{out}. At the cutoff frequency, the AC component is reduced by half its power level (or 3 dB). As the frequency continues to rise above the cutoff frequency, the AC component of the signal continues to diminish until it is attenuated beyond measurement.

The cutoff frequency for the single pole passive low pass filter is:

$$f_c = \frac{1}{2\pi RC} \tag{5.15}$$

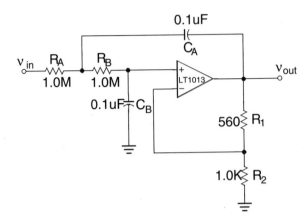

FIGURE 5.26 Double pole low pass filter constructed to have Butterworth response.

TABLE 5.3
Cutoff Frequencies for Sample R and C Values Using the Butterworth Response

Cutoff Frequency f_c	R	C
67.76 Hz	500.0 KΩ	0.0047 μF
1.59 Hz	1.0 MegΩ	0.1 μF

A more robust filter (called an active filter) can be constructed using operational amplifiers. Next is an example of a double pole, active, low pass filter with a Butterworth response. The rolloff, or rate of attenuation a filter has after the cutoff point, is 20 dB/decade for each pole that a filter has. The Butterworth response is one of three major filter characteristic responses. The main advantage to this response is that it has a maximally flat response in the pass band. For the filter in Figure 5.26 to have the Butterworth response, set the ratio of $R_1/R_2 = 0.586$. The cutoff frequency is again given by:

$$f_c = \frac{1}{2\pi RC}$$

However, notice that the values used in this equation do not set the cutoff frequency at 60 Hz. To use the filter to reduce AC noise, the cutoff frequency must be set well before 60 Hz because the amplitude of the input frequency is still passed through the filter at the cutoff frequency. To eliminate any signals present at 60 Hz, the cutoff frequency must be set well below 60 Hz, in this case 1.59 Hz (see Table 5.3).

The preceding circuit is excellent for reducing typical AC noise sources at and above 60 Hz. Noise has a tendency to propagate down long lines that may connect instruments to computer interfaces. If a long line is being used to supply an instrument signal to an interface, it is good design practice to filter the line prior to passing the signal to an opamp. Ribbon cables are especially prone to *crosstalk*, a condition where neighboring signals on different wires are absorbed at a lower intensity and then propagated down the adjacent lines.

5.10 DECOUPLING CAPACITORS

This section deals with how to prevent noise when designing interfacing circuits by decoupling circuits with capacitors. Proper decoupling ensures that noise generated inadvertently from an interface will not propagate back to the instrument it controls and cause problems to the instrument's native control circuitry, or to the interface itself. As the frequency of operation increases, decoupling becomes ever more important. As frequency increases noise increases, and undesirable characteristics such as crosstalk begin to emerge as signals "bleed" from one line to another. Chances are that any interface designed to control an instrument will not have any high frequency (>1.0 MHz) signals on it; however, it is best to follow good design practices on any project.

The gold standard of decoupling is: at every integrated circuit there should be a 0.1 µF ceramic capacitor (to remove high frequency components) and a 1.0 µF tantalum capacitor (to remove low frequency components) connected from each power lead to ground. At every point on the circuit board where power is routed onto the circuit board, this combination should be repeated. Also, randomly distribute this combination of capacitors throughout the circuit board (between power and ground) as space permits. If this rule is followed noise should be kept at bay.

One cardinal rule should be followed when decoupling is implemented to achieve maximum benefit. When a capacitor is mounted on a board, lead lengths and traces from the board to the capacitor should be kept as short as possible because they are a major source of inductance; inductance must be minimized to obtain good decoupling performance under high speed transient conditions. Using capacitors to reduce line noise that emanates from internal switching within an IC chip requires low inherent inductance within the decoupling capacitor. As a general rule, a multilayer ceramic capacitor has about four times less inductance than a tantalum capacitor. See Decoupling Basics in Appendix E for more details.)

The 0.1 µF ceramic capacitor is used to route high frequency noise to ground. The 1.0 µF tantalum capacitor is used to route low frequency noise (such as 60 Hz, which emanates from every lighting fixture near the circuit board) to ground. Decoupling capacitors essentially couple AC noise to ground before it has a chance to propagate into components on the circuit board and wreak havoc.

5.11 CREATING CUSTOM INSTRUMENT INTERFACES

Here is where knowing all the preceding information pays off. If an instrument does not have the functionality required to automate a process desired, the user can attempt to build in such functionality with a custom interface. For example, if an instrument was meant to be operated by hand and has only touch-key buttons and LEDs, an interface could be constructed in the following way.

First a bank of computer controlled relays such as the MEM-32 from Keithley Metrabyte would be connected to each and every touch key button. When a button press is required, the software should trigger a contact closure on the relay connected to that button for about $^1/_2$ second. This would mimic a person momentarily pushing the button.

Determining the state an instrument is in is much more problematic because a unique state (usually a voltage level somewhere on the instrument circuit board) must be found to correspond with each possible machine state. That means the voltage level at the chosen point must be checked and known to be unique for every different machine state. Sometimes that is not possible, and voltage levels must be taken from several points to determine a machine state. The only saving grace for taking voltages from several points is that the logic to check the instrument state can be done in software rather than hardware when reading the value of the points.

LEDs are probably the easiest objects for which to construct a logical interface. Merely probe for the DC voltage at the anode of the diode when it is lit and unlit and then construct a comparator to return a low or high state using the voltage at the anode of the LED for the comparator input. Sometimes there may be a group of LEDs indicating a variety of conditions. In the case where only one condition in the group can be true, it is a good idea to check that the interface only indicates one LED in the group is lit. This is a method of having the software check for a potential problem with the hardware.

Decoding seven-segment LED displays (Figure 5.27) is also a relatively simple task. There are two ways to go about doing this. If all that is required is the return of the numeral displayed 0 to 9, a seven-segment to BCD decoder IC can be used. Using such a chip, the outputs to the seven-segment LED (usually labeled a to g) are connected to the corresponding inputs (a to g) on the seven segment to BCD decoder. The seven segment to BCD decoder will then generate four digital outputs that return the number (in binary form) currently displayed on the LED. As an alternative to this, if the site where the BCD signals originate can be located, an individual can pick off the signals at this point and decode them from there.

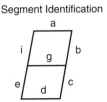

Segment Identification

FIGURE 5.27 Typical seven-segment LED display.

Sometimes, the seven-segment LEDs will display "special characters" whose BCD value may not be known or defined. In such a case a series of seven comparators can be constructed to determine which segments of the LED display are on and which are off. A computer program can then read the output of each comparator (via an input/output board such as the Keithley MBB-32) and make the determination as to which segments are lit and what exactly they mean.

Liquid crystal displays are by far the hardest to read. First, the connections to them tend to be very small so it is difficult to read their state with a meter or scope. Second, there is nothing intuitive about which connection controls which area of the liquid crystal display, so a lot of poking around has to be done to determine which inputs control which areas of the display. Finally, it is very difficult to solder connections to these types of displays because the connections to them are so small, and the liquid crystal display is a very delicate device, which can easily be damaged by the heat of a soldering iron. In addition, LCDs that use serial input or are multiplexed will be impossible to read in this manner. Other more sophisticated methods must be utilized to read such displays.

As a side note, when soldering small wires to small delicate connections, it is good practice to:

- Use a very hot iron to minimize the time the tip has to stay in contact with the connection.
- Clean the connections after soldering with an acid brush and a mixture of 50% MeOH and 50% acetone.
- Wash off the dry white powder that is formed, using soap and water. (This is what the rosin and flux reduce to after being exposed to the MeOH/acetone mix.)
- Let the circuit board dry. Then place a bead of clear nonconductive silicon or any commercially available epoxy or urethane designed for encapsulation and protection of electrical components across the wires where they are soldered to the connection. (Note: RTV silicone will produce acetic acid while curing, a source of ionic contaminants.)

Cleaning the connections, which are very close together, eliminates any shorts from conductive materials that may have bridged the gap between the connections. Also, by removing the rosin and flux no parasitic capacitances can form between the fine connections. Finally, the adhesive provides material strength to the connection between wire and the circuit board. If a wire undergoes stress for whatever reason, the stress is placed between the wire and the board (via the adhesive), not between the wire and the soldered connection.

When scouting for places to look for voltage changes that can determine a change of state in a machine, it is highly likely that an AC as well as a DC voltage may be encountered. Using the low pass filtering techniques, it is very easy to strip away the AC component of the signal at any point on the board and provide a pure DC output voltage, which is easily read by a comparator.

Also be aware that most circuit boards contain multiple layers. A connection can be soldered to anywhere on the top or bottom layers of the board. If for some reason it is not desirable to make a connection to a solder pad, an individual can follow the trace to anywhere on the board and make a connection. Some fine sandpaper can be used to remove the green insulating material from any trace on the board and then a connection can be soldered at that point with relative ease.

Often, the ground and power traces (or "rails") will be located on much thicker traces, and power for a custom-designed interface can be pulled directly from the native circuit board in this manner. If for some reason a connection from point A to point B cannot be traced out on an instruments circuit card, be aware that signals jump from layer to layer by interlayer connections called "vias" that look like small connections, with holes surrounded by copper too small in diameter to accept a component lead.

As a safety consideration, when probing points on a live circuit card it is prudent to employ the "one hand rule." In other words, the scope lead should be in one hand and the other hand should be in your lap. It is often easy to lose track while probing a live circuit, and should an errant hand touch 115 VAC, it will lead to a severe shock.

Once a set of discrete outputs has been obtained using the techniques illustrated previously, a digital input/output board can be used to read the output states of the comparators used in the interface.

5.12 LIQUID LEVEL SENSING CIRCUITS

Mechanical and electrical are two common means to sense the level of a liquid in a vessel. A float and level system is an example of a mechanical means. If the liquid being used is conductive (and it only has to be slightly conductive), a liquids level can be sensed by electronic means.

Mechanical devices have a lot of disadvantages to them. First, they are bulky in size and will only fit in the confines of a limited number of containers. Second, being mechanical, they will wear out over time and are prone to all the nasty things that can happen to any machine that relies upon moving parts. Third, they are easily disabled by environmental factors such as ice, debris in the fluid, and corrosive liquids. Fourth, mechanical devices can be falsely triggered in environments where external stresses are applied, such as centrifuges or mixers. Last, they are impractical if the differential in liquid height is very large or very small within the container.

Electrical devices can overcome almost all of the shortcomings of mechanical devices. One type of electrical liquid level-sensing circuit works as follows. Two probes are immersed in the liquid at a specific height. When the liquid reaches the level of the probes, current begins to flow across the probes. The liquid that carries the current from probe to probe acts as a resistor. This resistor can be incorporated into a voltage divider. The output of this voltage divider can be used as the input voltage to a comparator. The reference voltage of the comparator can be set such that the output of the comparator will change as soon as current begins to flow across the electrodes.

The design is quite simple; what is not so simple is selecting the proper material for the probes. If the liquid to be measured within the container will always be the same, this is usually not a problem. A suitable conducting probe that will not interact with the liquid can usually be found. The problem arises when a variety of liquids will be filling a container, some acidic, some basic, etc. Then it becomes quite difficult to find a probe that will not interact with all the different kinds of liquids used in the system (in fact sometimes it is impossible). If the liquids do not stay in the container for a very long period of time, it may be possible to flush the container out and wash the electrodes after each use before harm can come to them.

Another factor to watch for is the effect the electrodes have on the liquid in the container. This is especially true when doing spectroscopic calculations over a range of wavelengths. A good test is to take an electrode and immerse it in a flask of the liquid to be measured for about a week, and then run a scanning spectrum on the liquid over a range of wavelengths. Compare this to a spectrum of the liquid taken in its pure form right out of the bottle. Any differences are probably from some

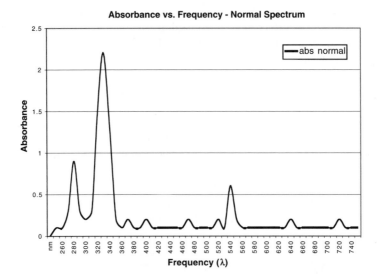

FIGURE 5.28 Sample optical spectrum of liquid to be used in testing vessel.

kind of material leeching out of the probe and into the liquid. If the differences are significant, then that type of material should be avoided when measuring absorbances within those wavelengths.

For example, suppose the liquid to be used in the vessel has an optical spectrum that looks like the one in Figure 5.28. Now suppose the "region of interest" is between 390 to 490 nm. Any absorbance greater than 0.5 (absorbance is usually measured in optical density [OD] units) is indicative of a reaction. Looking at the spectrum of the liquid in Figure 5.28, it is nearly flat in this region and the liquid will not interfere with the substance to be measured.

The electrode to be used is then soaked for 48 hours in a clean container filled with the liquid having the preceding spectrum. A spectrum is then taken of the liquid that was in contact with the electrode for 48 hours, which produces the plot shown in Figure 5.29. In this spectrum two additional peaks are observed, one at 450 nm and one at 560 nm. The 560 nm peak is outside the region of interest for this experiment, so its existence is not an issue. However, the 450 nm peak is in the region of interest and will trigger a false positive. Therefore, this electrode material is unsuitable for the liquid used in this experiment because it is leaching some impurity, which will adversely affect the experiment being run. This is more clearly illustrated on the plot in Figure 5.30, which overlays the spectra on top of each other.

It is also important to remember that any adhesive or glue used to affix the electrode to the vessel must also be immersed in the test liquid and have its optical spectrum taken. Adhesives are very prone to having substances leach out of them. However, sometimes only a finite amount of impurities will leach out in a short period. So it may be prudent to soak the adhesive for 48 hours, take a spectrum, and then repeat the process. The impurities excreted the first time may be nearly imperceptible the second time around.

A peak within the region of interest does not necessarily negate using that particular material for an electrode, provided the peak generated by the material is consistent in location and intensity. If, for example, it is known that the material immersed in the liquid will generate a peak no greater than 1.5 OD units at 490 to 492 nm, it is very easy to compensate for this in the analysis software. Simply remove from consideration all peaks that fit this profile in the analysis. If only 0.5% of the experiments will yield a peak that matches this profile, then a very small number of the compounds assayed will produce an inconclusive result.

With the process of selecting a suitable electrode finished, a liquid level circuit can be constructed such as the one pictured in Figure 5.31 using an LT-1013 opamp. Notice that this circuit contains hysteresis, implemented by using the positive feedback resistor R_H (55 K), which prevents

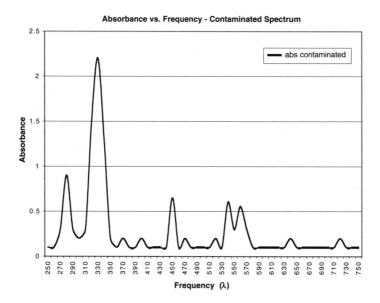

FIGURE 5.29 Optical spectrum of liquid after exposure to contaminants that have leeched from electrodes.

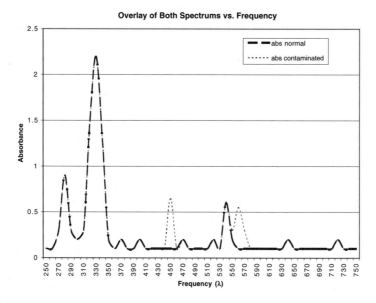

FIGURE 5.30 Comparison of previous two spectra, with and without electrode contaminants.

the pump from oscillating on and off from voltage differentials as the current flows from one electrode to the other. Hysteresis is a very important design criteria in liquid level-sensing circuits because often the liquid to be measured is being agitated in some manner. (In a laboratory environment this is often done by a magnetic stir bar.) Therefore the amount of liquid in contact with the electrodes will fluctuate by some degree and the resistance between the electrodes will fluctuate to some extent. As the resistance varies so too will the input voltage (V_{in}) to the comparator vary, which would cause the pump triggered by V_{out} to oscillate on and off.

This circuit works very well in many applications; however, one problem will arise that is neither obvious nor intuitive in nature. Unfortunately, when a DC current is passed through a liquid

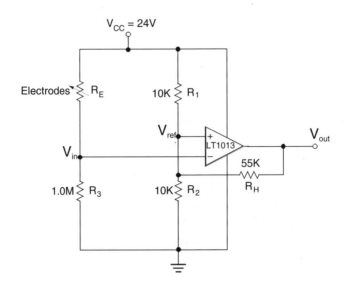

FIGURE 5.31 Using the LT1013 high precision opamp as a comparator to detect liquid levels.

between two electrodes, it can sometimes cause the electrodes to corrode depending upon the material used as electrodes and the liquid they are immersed in. The first time this circuit was utilized it was for a methanol rinse in which a septum piercing needle was rinsed off after each use. Because the needle was immersed in many compound samples, all of which had varying structures, the methanol rinse was contaminated with a variety of substances. This led to the electrodes corroding over time. The electrodes were prone to corrosion when they came in contact with certain compounds and had a current flowing through them.

The solution to this problem is to use an AC current in the liquid because alternating current at a relatively high frequency will not cause corrosion of the electrodes. When a constant current (DC) is passed through an electrode in a liquid, the molecules will be polarized by the direction of the current and tend to migrate toward the voltage sources. This is how electroplating, a beneficial use of this phenomenon, is done. Unfortunately, this property also causes most electrodes in chemical solutions to corrode with impurities because the impurities will migrate toward the voltage sources.

An AC current does not provide a constant voltage source; because the voltage level is in constant fluctuation, the molecules may start to line up but once the voltage level drops the molecules will resume their normal chaotic behavior. An AC current can easily be generated by a 555 timer integrated circuit, as shown in Figure 5.32.

The values of components R_A, R_B, and C determine the frequency and duty cycle of oscillation. The output will be a square wave whose duty cycle can be calculated by using the following formulas:

$$t_1 = 0.693\,R_A C \tag{5.16}$$

$$t_2 = \left[\frac{\left(R_A R_B\right)}{\left(R_A + R_B\right)}\right] C \ln\left[\frac{\left(R_B - 2\,R_A\right)}{\left(2\,R_B - R_A\right)}\right] \tag{5.17}$$

$$f = \frac{1}{t_1 + t_2} \tag{5.18}$$

To maintain the duty cycle of the square wave at 50%, it is best to keep the values of the resistors set close to the values given in the example and change the capacitance to vary the

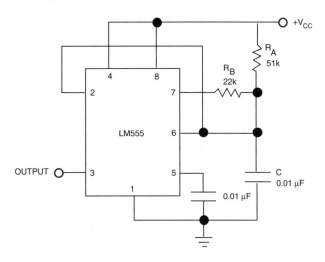

FIGURE 5.32 Using the 555 timer to create an AC signal.

TABLE 5.4
Calculating the 555 Timer Duty Cycle Parameters

RA	RB	C	t_1	t_2	f
51 K	22 K	.01 µF	0.353 ms	0.374 ms	1.374 kHz
56 K	22 K	.0047 µF	0.182 ms	0.750 ms	3.012 kHz

frequency of operation. Note that oscillations will not occur if $R_B > 0.5$ (R_A), as pin 2 will not be brought down to a low enough voltage level to trigger the second comparator in the 555 timer.

A 3.0 kHz square wave will do nicely for this application, and can be generated using the values: $R_A = 56$ K, $R_B = 22$ K, and C = 0.0047 µF. Table 5.4 illustrates the calculated values to determine the duty cycle and frequency of operation.

The example given in the 555 timer datasheet runs at 1.374 kHz, and has nearly a 50% duty cycle. A schematic to set up the 3.012 kHz square wave is pictured in Figure 5.33. With this circuit a 3 kHz square wave can be passed through a liquid such as methanol from one electrode to another without corrosion occurring. The prior design used a comparator to measure the DC voltage coming from the electrode to determine the level of the liquid. Here, a square wave will emanate from the electrode whose maximum voltage will be dependent upon the level of liquid in the vessel with the electrodes. A comparator can still be used, but the waveform must be massaged into a DC voltage. The circuit shown in Figure 5.34 accomplishes this.

The square wave from the 555 timer is input at v_{in} into the first 1N914 diode. Both diodes conduct when the square wave voltage is "high." While the square wave is high, the 10 µF capacitor charges up to peak value of the square wave. When the square wave goes low on the second half of the duty cycle, the 10 µF capacitor discharges through the second diode into the noninverting input of the LT1013 opamp, thus maintaining a "high" voltage level even during the "low" portion of the duty cycle of the square wave.

The 100 K resistor at the noninverting input pulls that input to ground once the capacitor has discharged. If the 100 K resistor is omitted, the output will always be high because the noninverting input will be "floating" when the duty cycle of the square wave goes low. Once the square wave input stops or its maximum value drops below 2.5 V, the output of the LT1013 comparator will go low.

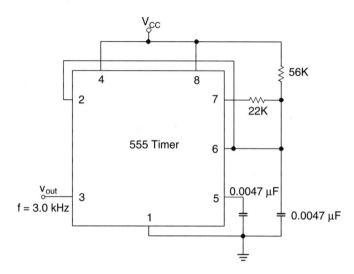

FIGURE 5.33 An approximate 3 kHz AC signal generator with 50% duty cycle using the 555 timer.

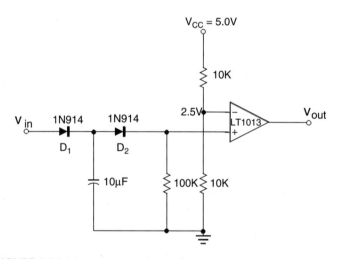

FIGURE 5.34 Massaging an AC waveform to a steady state DC signal.

The diodes between the capacitor serve one purpose: to isolate the capacitor from high imped-ance or potential grounding points. If D2 is omitted, the capacitor will see the high input impedance of the amplifier, and the time constant of the capacitor will be so large that it will take too long for it to discharge and trigger a low output on the comparator. If D1 is omitted, the capacitor could be inadvertently discharged to ground through the electrode in contact with the liquid to ground. This would allow the noninverting input to drop to low when the duty cycle of the square wave went low, and the output of the comparator would be identical to the output of the 555 timer.

Using the capacitor isolated by the diodes, the circuit triggers a low output only when the square wave is no longer present, or has a peak value less than 2.5 V. Whenever a square wave with a peak value greater than 2.5 V is present on the input of this circuit, its output will be high. This output can easily be read by a computer using a digital I/O board. The input/output scheme of the overall circuit is illustrated in the diagram in Figure 5.35.

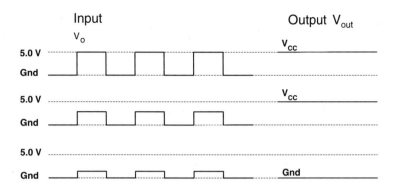

FIGURE 5.35 AC input vs. DC output using previous circuit.

5.13 SOLENOID VALVE DRIVER CIRCUITS

For an assay of any complexity, valves must be utilized to switch lines between various liquids, solvents, and buffers. Values typically come in two operating voltages: 12 and 24 VDC. Most valves can be operated in one of two methods. The first method is to drive the valve with its operating voltage for the duration in which the valve is to be actuated. This method works quite well when the valve only has to be driven for a few seconds at a time and is then given a long interval to rest before being actuated again (i.e., a short duty cycle). However, this method is not desirable when the valve has to be driven for extended periods of time (>60 sec).

When a valve is driven at its operating voltage for extended periods of time it tends to become very hot — hot enough that it will burn skin when it is touched. The heat causes many problems. First, it is hard on the internal mechanism of the valve. Second, it can change the physicochemical properties of the liquid passing through the valve.

The second method of driving a valve is more sophisticated, but its use will increase the overall reliability and accuracy of the system in which it is deployed. This method works as follows. First the driver sets a "pull-in voltage," which actuates the valve. The pull-in voltage is the operating voltage of the valve, typically 12 or 24 VDC. After the valve is actuated, the voltage output of the driver drops to the "hold-in voltage," which is typically 3, 5, or 8 VDC. The hold-in voltage allows less current to flow through the valve in its energized state, which in turn reduces the heat generated by the valve and does not cause unnecessary internal stresses to the valve.

Most laboratory valves, such as those from General Valve Corporation, allow the use of a reduced hold-in voltage. However, not all valves support this feature, and some may not function properly if it is implemented. It is best to check the manufacturer's specification before using a valve driver of any sort. Also note that using a lower hold-in voltage than specified may lead to intermittent valve failure that is not mechanical in nature.

Many valve drivers on the market will accommodate a variety of different solenoid valves. General Valve Corporation's model #90–30–100 is an excellent valve driver, which will accommodate 12 and 24 VDC valves, and has user selectable hold in voltages via circuit board jumpers.

If a valve does not function properly with a driver, the first thing to check is the pull-in voltage, which is often less than the operating voltage of the valve. To test the pull in voltage hook the valve up to a variable power supply and increase the voltage until the valve can be heard to "click" on. If the voltage at which the valve clicks on exceeds the pull-in voltage, the valve is defective and should be returned to the manufacturer for repair. (Often manufacturers are sloppy about quality control regarding the pull-in voltages because many users power the valves at their operating voltage, not caring if doing so causes other adverse effects.)

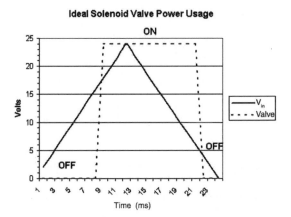

FIGURE 5.36 Valve state vs. input voltage.

The hold-in voltage can be checked by reducing the voltage after the valve clicks on. At some point the valve will click off when the voltage has been reduced sufficiently. The minimum hold-in voltage should be the lowest voltage setting that exceeds the voltage at which the valve clicks off when the input voltage is reduced from the pull-in voltage. Figure 5.36 illustrates this concept with a 24 V valve that has a pull in voltage of 18.0 V, and a hold in voltage of 5.0 V.

Another great cause of valve failure, not electrical in nature but suitable to mention while on the subject of troubleshooting valves, is passing material through the valve that will make its seals swell. Most base models of valves come with a synthetic seal that will swell when exposed to chemicals such as methanol or other solvents. When the seal swells it mechanically prevents the actuator from moving when the valve is energized. Most quality valve manufacturers offer upgraded seals on valves (which usually triples the price). The material most resistant to swelling and degradation from common laboratory liquids is Kalrez.

Another common cause of valve failure comes from using a valve on a line to which vacuum is applied. If a valve is not rated to take the vacuum applied to a line (20 and 100 psi are common ratings), then the solenoid within the valve may not be strong enough to close or open the given valve. This problem may occur intermittently, which makes it difficult to diagnose the failure as either electrical or mechanical in nature.

5.14 SUMMARY

It would be impossible for any text covering as many topics as this one to bring the reader from ground zero to a full-scale understanding of electronic design. Therefore, some compromises had to be made. This chapter presents methods found most useful when designing circuits to control various forms of instrumentation. It presents how to use those methods, as well as sample applications on how to incorporate those methods to create a working interface. Those in need of more background information should seek additional reference material, such as those texts suggested in Appendix I, Further Reading.

6 Robust Data Analysis Using Origin Lab's Origin

Without question, Microsoft Excel has attributes that make it a tremendously useful program. It is easy to use for simple data analysis and incorporates the features most likely to be used by individuals in a variety of disciplines such as business, science, and engineering. It has a macro language that is easy to learn and comprehensive in terms of its ability to automate the functionality of the mother program (Excel). The features that make the program broadly applicable and easy to use are also self-limiting when a user wishes to perform a very specific task.

A very wise engineer named Bob Blank at Lockheed Martin once said there are only two kinds of problems: the kind money can solve, and the kind money cannot solve. While it is true that a variety of aftermarket add-ins can be purchased for Excel that will expand its utility, nothing can be done about the poor graphics the program uses to display plots and graphs. True, the data could be displayed in a third-party utility like Jandel Table Curve 3-D, but this adds yet another layer of complexity when the focus should be on streamlining the overall process.

The capabilities of Origin Lab's Origin program far exceed those of Excel. It offers a variety of very sophisticated mechanisms for curve fitting, graphics, filtering, matrices, statistics, and transform functions. Like Excel, Origin also has a scripting language called "Labtalk," which is based on "C." In contrast, Excel's macro language is modeled after Visual Basic. Origin also has a steep learning curve to accompany its sophistication. However, a sound philosophy on this is that once someone learns Origin, he can forever harness its computational power for all future projects.

There are many programs written that can accomplish the same tasks for which Origin is used in this chapter. They include Matlab, Mathematica, MathCad, Maple, and others. Nothing is wrong with or inferior about these programs. However, most data acquired in a laboratory are acquired in a spreadsheet type of format; Origin is the best suited of these programs to handle data in a spreadsheet type of format.

Origin's scripting language is a bastardized version of "C" that has "C" -like syntax (looping structures, etc.), but also contains non"C" like syntax used to reference Origin worksheets and plot windows. "C", by its very nature, is a more difficult language to program in than Basic, and hence Labtalk is more difficult to learn to program in than Excel Visual Basic for Applications (VBA).

6.1 LABTALK FUNDAMENTALS

Like all programming languages, Labtalk has its limitations. A good place to begin this discussion would be with Labtalk's system variables because Labtalk has some fundamental limitations on variable definitions and use. Many variables are utilized by the system and can be overwritten by Origin's internal processing if a user should unwittingly choose to incorporate a system variable within a Labtalk script.

Scripts are limited to using only 26 *string variables*. Each string variable is preceded by a "%" followed by a letter of the alphabet. However, some of these string variables are utilized (or served) by the system, which makes it easy for a programmer to get into trouble. Table 6.1 shows only the reserved string variables.

TABLE 6.1
Reserved String Variables in Labtalk

String Variable	Description
%A	Used extensively by Origin and commonly used in development scripts.
%B	Used extensively by Origin and commonly used in development scripts.
%C	The name of the current active dataset.
%D	The name of the last dataset set as active with the set command.
%E	The name of the window containing the latest worksheet selection.
%F	The name of the dataset currently in the fitting session.
%G	The current project name.
%H	The current active window title.
%I	The current baseline dataset.
%X	The DRIVEPATH of the current project. The '\' character can be displayed from the %X or %Y system string variable. For example, if the following text is typed in the text control dialog box (with 'Link to Variables' enabled in the label control dialog box): The file path for this document is \v(%X). Origin displays: The file path for this project is CAMyDiAMyRle.
%Y	The DRIVEPATH for the ORIGIN60.INI file. To display the '\' character, see %X.
%oZ	A long string for temporary storage.

Source: From *Labtalk Manual*, Version 6, Origin Lab Corporation, Northampton, MA, 1999. With permission.

Keep in mind that string variables can only contain *one* letter in their descriptor. For example, the following script:

```
%K = "Brian";

%KK = "Bissett";

type -a%K;

type -a%KK;
```

will print the following to the script window:

```
Brian

BrianK
```

Notice what happened here: %KK is not identified as a string variable. Instead, %K is recognized and printed to the screen and then the letter K is printed following the value assigned to %K (%K + "K" = "Brian" + "K"). In general, avoid using %C, %E, and %H because they are often in use by the parent program (Origin).

It is also possible to parse out sections of string variables. This is not done using traditional types of functions common to "C." The following script illustrates how to accomplish this:

```
%M = "How to parse those _Strings!";

type -a Original String:%M;

//Return all text to the left of "_"

type -a%[%M,'_'];

# Return all text to the right of "_"

type -a%[%M,>'_'];
```

```
//Return all text to the left of 6th character
type -a %[%M,6];
//Return all text to the right of 6th character
type -a %[%M,>6];
//Return "parse"
type -a %[%M,8:12];
//Return Token 2
type -a %[%M,#2];
//How Long is this String?
type -a %[%M];
```

When run, this code will produce the following output to the script window. (The command "type –a" instructs the script to dump the result to the script window.)

```
Original String: How to parse those _Strings!
How to parse those
Strings!
How t
o parse those _Strings!
parse
to
28
```

The end of each command is terminated with the ";" character just as in "C." If a ";" is not used to terminate the end of a command, an error or an unintended result will occur.

Comments precede each command in the previous script. The characters "//" and "#" denote that comments are to follow and any further information on that line should be ignored. Previous versions of Labtalk also recognized the old "C" method of commenting "/*", but the more recent versions no longer support this method of adding comments.

One last comment is in order on parsing strings in relation to tokenizing strings. It is possible to use a delimiter other than a space when separating sections of string variables. In the case of tab delimited text the syntax would be as follows:

```
type -a%[%M, #2, \t];
//here the \t denotes a tab for a delimiter
```

The \t is used to represent the tab character, which is a *nonprintable* character. Table 3.4 in Chapter 3 displays ASCII codes that provide the hexadecimal codes for all nonprintable characters, some of which are used in developing device drivers in Visual Basic. Origin has shortcuts for nonprintable characters, such as:

```
\n New line.
\r Return.
\d Delete/backspace.
\t Tab.
```

However, any nonprintable character can be specified by using the notation \xhh, where hh are two hex digits representing the character code desired.

When tokenizing a string, anything contained within any kind of brackets — (), [], etc. — will be considered a single token, even if the delimiter is contained within the brackets. For example:

```
%M = "a (unique kind) or [special token]";

//Return Token 2

type -a%[%M,#2];

//Return Token 4

type -a%[%M,#4];
```

will display the following output to the script window:

```
unique kind

special token
```

Numeric variables are more flexible than string variables because no limit is placed on the number of them (within reason). The only real limitation on numeric variables is that they cannot start with a number and must not contain any noninteger or alphanumeric characters. As with string variables, some numeric variables are reserved for use by Origin.

System variables have a specific meaning in Origin; for example, X1, X2 define the endpoints of the x-axis. The following system variables should be avoided when constructing Labtalk scripts:

```
i,j,k

x,y,z,t

X1,X2,X3,X4

Y1,Y2,Y3,Y4

Z1,X2,Z3,Z4

V0,V1,V2,V3,etc.
```

Operation variables are variables used in scripts called by Origin during various operations. User-defined scripts may also use these variables, but with no guarantee that the variable will remain unchanged if an Origin operation is performed (such as curve fitting). The following operation variables should be avoided in Labtalk scripts:

```
xx,yy,zz,tt

ii,jj,kk

temp,e,tiny

beg,begin,start,startRow,

end,endRow,endRows,numRows
```

A numeric variable can be converted to a string variable by using the form $(). This can be especially handy if a portion of a string (such as a title of a plot) needs to have a numeric component in it taken from a calculation.

```
//Number to String Conversion Examples

//Using C Conventions
```

```
//Assign Numeric Value
flow_rate = 43.6792;
%M = Sewage Flow Rate: $(flow_rate);
type -s%M;
//Return only Integer Portion
%M = Sewage Flow Rate: $(flow_rate,%d);
type -s%M;
//Set Number of Decimal Places to 2
%M = Sewage Flow Rate: $(flow_rate,.2);
type -s%M;
//Set Engineering Format
//4.3f = 4 total Digits 3 Decimal Places!
%M = Sewage Flow Rate: $(flow_rate,E%4.3f);
type -s%M;
```

In the preceding script it is conceivable that %M would be used in the title of a plot and requires the flow_rate be converted to a string and possibly to a certain format in terms of decimal places. The output of this script is as follows:

```
Sewage Flow Rate: 43.6792
Sewage Flow Rate: 43
Sewage Flow Rate: 43.68
Sewage Flow Rate: 43.679
```

Because of some limitations on variable use in Origin, someone might question how data (especially string data) should be stored on a long-term basis in Origin. The answer to this question is within a dataset. A dataset is a one-dimensional array that can contain numeric and/or string data. Whenever a worksheet is created within Origin, a dataset is created for each column in the worksheet (to hold its data). The dataset names follow the convention of "worksheet name"_"column name". Datasets can also be created independently of worksheets using the create command. An independent dataset can also be viewed and edited as a worksheet using the edit command.

```
//Datasets - Creation and use
//Create a STRING Data Set with 25 Elements
create StringSamp_A -s 25;
//Populate the first few elements
StringSamp_A[1]$ = "one";
StringSamp_A[2]$ = "two";
StringSamp_A[3]$ = "last";
//Look at this dataset using a WorkSheet
edit StringSamp_A;
```

The preceding script creates a string dataset with 25 elements. It then populates the first three elements of the dataset with the text "one," "two," and "last." This dataset can then be viewed (and altered) by means of a worksheet using the edit command. Notice that after this command is executed a worksheet named "StringSamp" with a column named "A" appears with 25 rows, as each row represents an element in the dataset StringSamp. If the column "A" had not been explicitly declared when the create command was issued (by using "StringSamp_A"), a worksheet still would have been created but Origin would have chosen and named a column to place the data in.

Conversely, data can be read out from the worksheet into a string variable using the methods shown in the following script:

```
//Read Information From a Dataset
%M = col(A)[1]$;
type -a %M;
%M = StringSamp_A[2]$;
type -a %M;
%M = col(A)[3]$;
type -a %M;
```

which produces the following output to the script window:

```
one
two
last
```

The real advantage to using datasets to store data is that any function that Origin can perform on a worksheet — sorting, Fourier transform, etc. — can be utilized on the members of a dataset. In the case of string data, the user has no real choice for long-term string storage because the number of string variables within Labtalk is somewhat limited.

Numeric datasets operate in a similar fashion to string datasets but have a little more flexibility. Take, for example, the script:

```
//Demonstration on Creating Numeric DataSets
//Create with Discrete Values
NumbSamp_A = {2,4,6,8,10,12};
//Create Range {start,stop,increment}
NumbSamp_B = data(1,3.5,0.5);
edit NumbSamp_A;
```

In this example two datasets are created. The first specifies discrete values in order by element (row) number. The second uses the data function to create a range of data linearly spaced at a specified interval (in this case 0.5). Numerical data sets can also be created by creating and populating columns of a worksheet, or by using the create command with -c or -d options.

In Excel VBA specialized syntax was available for accessing items in worksheets and OriginLab offers this as well. Having the ability to access worksheet data contained within a particular cell is of paramount importance in any application. This is done by means of the following command:

```
%(worksheetname,ColumnNumber,RowNumber);
```

Although it is true that any data contained within the worksheet can be accessed by utilizing the dataset name, this command makes it possible to access data using numeric variables to represent row and column positions, which is often done using loops. The full dataset name of any column within a worksheet can be returned using the syntax:

```
%(WorksheetName,ColumnNumber);
```

and a column label can be returned with:

```
%(ColumnNumber,@L);
```

A great deal of additional information can be returned about a particular worksheet column using the command:

```
%(WorksheetName,@ Option, ColumnNumber);
```

The @ Option choices are listed in Table 6.2.

TABLE 6.2
Column Option @ Return Codes

Option	Return Value
@#	Returns the total number of worksheet columns. *ColumnNumber* may be omitted.
@C	Returns the column name.
@D	Returns the dataset name.
@E#	If *columnNumber* = 1, returns the number of Y error columns in the worksheet. If *columnNumber* = 2, returns the number of Y error columns in the current selection range. If *columnNumber* is omitted, Labtalk defaults to *columnNumber* = 1.
@H#	If *columnNumber* = 1, returns the number of X error columns in the worksheet. If *columnNumber* = 2, returns the number of X error columns in the current selection range. If *columnNumber* is omitted, Labtalk defaults to *columnNumber* = 1.
@O	Returns the offset from the left-most selected column to the *columnNumber* column in the current selection.
@OY	Returns the offset from the left-most selected Y column to the *columnNumber* column in the current selection.
@OYX	Returns the offset from the left-most selected Y column to the *columnNumber* Y column counting on Y columns in the current selection.
@OYY	Returns the offset from the left-most selected Y column to the *columnNumber* X column counting on X columns in the current selection.
@T	Returns the column type. 1 = Y, 2 = disregarded, 3 = Y error, 4 = X, 5 = label, 6 = Z, and 7 = X error.
@X	Returns the number of the worksheet's X column. Columns are enumerated from left to right, starting at 1. Use the syntax: %(worksheetName,@X);
@Xn	Returns the name of the worksheet's X column. Use the syntax: %(worksheetName,@Xn);
@Y-	Returns the column number of the first Y column to the left. Returns *columnNumber* if the column is a Y column, or returns 0 when the Y column does not exist. Use the syntax: % (worksheetName, @Y-,ColumnNumber);
@Y#	If *columnNumber* = 1, returns the number of Y columns in the worksheet. If *columnNumber* = 2, returns the number of Y columns in the current selection range. If *columnNumber* is omitted, Labtalk defaults to *columnNumber* = 1.
@Y+	Returns the column number of the first Y column to the right. Returns *columnNumber* if the column is a Y column, or returns 0 when the Y column does not exist. Use the syntax: %(worksheetName, @Y+,ColumnNumber);
@YS	Returns the number of the first selected Y column to the right of (and including) the *columnNumber* column.
@Z#	If *columnNumber* = 1, returns the number of Z columns in the worksheet. If *columnNumber* = 2, returns the number of Z columns in the current selection range. If *columnNumber* is omitted, Labtalk defaults to *columnNumber* = 1.

Source: From *Labtalk Manual*, Version 6, Origin Lab Corporation, Northampton, MA, 1999. With permission.

TABLE 6.3
Arithmetic Operators in Origin

Operator	Description
+	Addition
-	Subtraction
*	Multiplication
/	Division
^	Exponentiate (X^Y raises X to the Y'h power)
&	Bitwise And operator
|	Bitwise Or operator

Source: From *Labtalk Manual*, Version 6, Origin Lab Corporation, Northampton, MA, 1999. With permission.

TABLE 6.4
Origin Assignment Operators

Operator	Description
	Assign the argument to the variable.
+=	Add the argument to the variable contents and assign to the variable.
-=	Subtract the argument from the variable contents and assign to the variable.
*=	Multiply the argument by the variable contents and assign to the variable.
/=	Divide the variable by the argument and assign the result to the variable.
^=	Raise the variable contents to the argument and assign to the variable.
++	Increment variable contents by 1.
--	Decrement variable contents by 1.
--O	Perform interpolation outside the domain of the second dataset for all X values of the first dataset. For example: datal b --O data2 b;
#=	Copy the contents of a vector to another vector. The destination is overwritten. For example: Destination Vector #= SourceVector
	To get the size of a vector, use the following notation: NumericVar=Vector+
#+=	Add the contents of a vector to another vector. The destination vector is not overwritten. For example: Destination Vector #+= SourceVector
	To get the size of a vector, use the following notation: NumericVar=Vector+

Source: From *Labtalk Manual*, Version 6, Origin Lab Corporation, Northampton, MA, 1999. With permission.

As in any programming language, Labtalk must have operators to perform arithmetic and logic operations on objects. Origin utilizes the arithmetic operators shown in Table 6.3, which are fairly universal in terms of use.

In terms of assigning values to variables, Origin uses the syntax shown in Table 6.4 for assignment operation, some of which is identical to "C" and some of which is unique to Origin.

For logic and relational operators, Origin utilizes the standard "C" operators listed in Table 6.5.

The = = operator in Table 6.5 is in boldface font. This is because this operator is one of the greatest pitfalls in programming in Labtalk or "C", especially if the programmer is used to programming in another language such as Basic.

TABLE 6.5
Logic and Relational Operators in Origin

Operator	Description
>	Greater than
>=	Greater than or equal to
<	Less than
<=	Less than or equal to
==	**Equal to**
!-	Not equal to
&&	And
||	Or

Source: From *Labtalk Manual*, Version 6, Origin Lab Corporation, Northampton, MA, 1999. With permission.

Use the = = operator only when trying to determine if one operator is equal to another operator, as is done with the *if* statement. When setting one operator equal to a value or another operator, use the = operator. For example:

```
%M = ConditionX;

potential = 32.78;

if (potential == 32.78){

type -s "equal";

};
```

This is such a killer bug because, if a statement was written as:

```
if (potential = 32.78){//NO NO NO THIS IS WRONG!

type -s "equal";

};
```

nothing stands out as seemingly incorrect.

Intuitively, when someone glances at this statement, nothing appears logically or syntactically wrong. Only a trained programmer will look at this statement and instantly see a warning flag. Even a hard core programmer who works with several languages might miss this error because in some languages such syntax is OK.

Another nasty pitfall in Labtalk occurs when comparing string variables. If the string variables are surrounded by quotation marks (for example, "%M"), then Origin compares the string characters. If the string variables are not surrounded by quotation marks (for example, %M), then Origin compares the values of the string variables. This is not the norm for programming languages; usually, a string must be converted to a number using a function before comparing a string to a number. For example:

```
aaa=4;

bbb=4;

%M=aaa;

%N=bbb;
```

```
if  (%M==%N)

   type  "YES";

else

   type  "NO"
```

The result will be YES, because in this case the values of the strings (rather than the characters) are compared, and aaa==bbb==4. However, in the following example:

```
aaa=4;

bbb=4;

%M=aaa;

%N=bbb;

if  ("%M"=="%N")

   type  "YES";

else

   type  "NO";
```

the result will be NO, because the strings are compared and "aaa" ! = "bbb"!

Sometimes, despite a person's best efforts and use of correct syntax, programs will not function the way they were intended to. For such cases, knowing how to utilize the following commands makes debugging Labtalk scripts much easier.

The Echo command allows the user to see what is happening "behind the scenes" while commands are issued and scripts are executing in real time. The chart in Table 6.6 shows the "bit number" for each option available for use in the echo command. Commands can be combined by adding them. The combination Echo = 7 is particularly useful because it allows the user to see what is happening as commands are selected and executed from the menu bar. The echo command can be turned off by setting Echo = 0.

If a user wishes to build a particular functionality into a script, but is unsure how to do it, Echo can be set to 7 and the script necessary to implement the functionality will be printed to the script window as the user manually executes the command. This is much like recording a macro in Excel using VBA and then looking at the resultant code.

Another very useful tool in debugging scripts is the #! statement. Statements following this command are ignored *unless* the system variable system.debug (or @B) is set to 1. The beauty of this statement is that debugging statements that print out vital parameters at critical points in a

TABLE 6.6
Echo Command Bit Options

Echo #	Result
1	Display commands that generate an error.
2	Display scripts that have been sent to the queue for delayed execution.
4	Display scripts that involve commands.
8	Display scripts that involve assignments.
16	Display macros.

Source: From *Labtalk Manual*, Version 6, Origin Lab Corporation, Northampton, MA, 1999. With permission.

script's execution can be left in a script and not hinder its performance when it is run in a normal fashion. However, when it comes time to add to the script and debug it further, these parameters can be set to execute at will.

Some questions concerning additional common mistakes in Labtalk programming:

- Does each statement end with a semicolon? Remember that commands such as *if*, *define*, and *for* should not have a semicolon between the command and the script.
- Is there an equal number of open brackets and closing brackets for each statement?
- Are $() and %() utilized appropriately when accessing numbers and worksheet information?

6.2 BUTTON SCRIPTS

The simplest way to execute small scripts is to associate them with a button on a worksheet. When the button is pressed, the script is executed. What could be simpler? Although this method does not bode well for large scripts, it is great for small scripts to be utilized within worksheet templates.

A logical place to put a button is at the top of the worksheet. Here it is instantly identifiable, and if the script is to perform actions on a single column, the button can be placed directly above the column. To do this, a gap should be put between the beginning of the worksheet and the top of the window frame. This can be accomplished by selecting from the menus Format->Worksheet->GapFromTop, which will produce the dialogue box illustrated in Figure 6.1.

A Gap From Top value of 20 is usually sufficient to place a number of buttons. Once a gap has been established within the worksheet, create a button to run a script. A button can easily be created using the text button (T). (Note: if the tools toolbar is not visible, the text button (T) will not be shown. The tools toolbar can be activated from the view menu.) Click on the text button and then click on the worksheet where the intended button is to be placed. Type some text and hit enter when finished. There will now be a text item on the worksheet. Select the text item on the worksheet, right click on it, and choose label control. Figure 6.2 shows the dialog sheet that will appear.

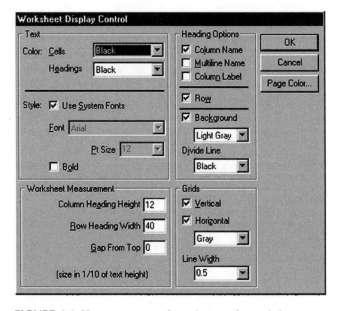

FIGURE 6.1 How to set a gap from the top of a worksheet.

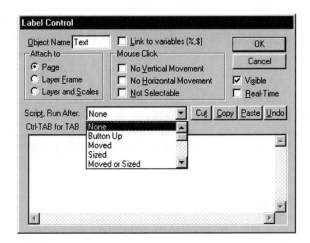

FIGURE 6.2 Creating a button-activated script.

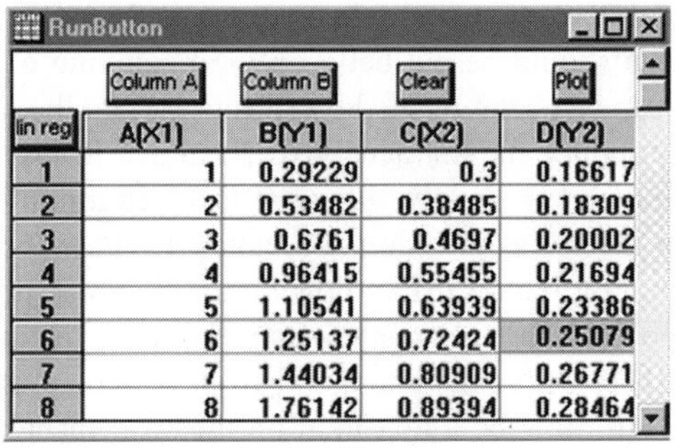

FIGURE 6.3 The RunButton worksheet with five button scripts.

In the "script run after" drop down box, select "button up." In the large textbox at the bottom of the dialogue sheet a script may be typed that will be executed when this button is pressed. After exiting out of the dialog box, toggle in and out of the button edit mode by CTRL+B or Edit->Button edit mode. (*Note:* CTRL+B no longer toggles in and out of button edit mode if SP2 for Origin 7.0 is installed.) When out of the button edit mode the text will appear as a button that will run the script when pressed. When in the button edit mode, edit the script held by the label control (text) by right clicking on it, and choosing label control.

The script can be selected to run after a variety of things, such as moving or resizing, are done to the text. It is difficult to imagine an instance when such behavior would be useful for running a script, but the functionality is there, should it be required. It might be interesting to add some code to low level format the boot sector of someone's hard drive if he chooses to move or resize a label. *Too* much functionality built into macro languages is how viruses get started, however. The button option for most users is the most flexible.

Now that it is known how to make a button from which script can be executed, it is time to discuss a sample application. Opening the file "ButtonExample.OGW" will bring up a worksheet in Origin that has five button scripts and serves to illustrate not only utilizing button scripts but also linear regression in Origin (Figure 6.3).

The button labeled "Column A" populates Column A(X1) with the integers 1 through 8. This is accomplished by means of the following script:

```
type -s%H;

for (ii=1; ii<=8;ii+=1) {
```

```
type "ii=$(ii)";

%(%H,1,ii)=ii;

};
```

Here a simple "C" type looping structure populates the first eight rows of column A(X1) using the %(worksheetname,ColumnNumber,RowNumber) syntax to populate the individual cells. Remember that %H is always the name of the current worksheet.

The button labeled "Column B" populates Column B(Y1) with a series of nearly linearly spaced values, with a spacing of approximately 0.2 units. Notice that a random number between 0 and 1 is multiplied by 0.2 and added to result on each **pass**, which gives the dataset some variability when each dataset is generated for the demonstration:

```
for (ii=1; ii<=8;ii++) {

//type "ii=$(ii)";

%(%H,2,ii)=ii*(0.2)+ rnd()*0.2;

};
```

Now it is time to tackle the area that gives so many people difficulty — actually performing a linear regression using script. This difficulty stems from the fact that when the Origin manual on Labtalk script for linear regression and nonlinear curve fitting is read, there are an overwhelming number of functions and commands. It becomes very confusing to know which commands must be used, as well as in what order they should be used. Complicating matters more is the fact that many commands serve the same purpose within Origin.

The button labeled "lin reg" populates columns C(X2) and D(Y2) with a 100 point linear regression of the data contained within columns A(X1) and B(Y1). This button contains the following script (to be explained further):

```
stat.data$ = %H_B;

stat.fitxData$ = %H_C;

stat.fityData$ = %H_D;

stat.makeX.fitnpts = 100;

stat.makeX.fitX1 = %(%H,1,1);

stat.makeX.fitX2 = %(%H,1,8);

stat.makeX();

stat.lr();

stat.lr.a=;

//Display intercept in Script window

stat.lr.b=;

//Display slope in Script window

stat.cod=;

//Display R^2 Value in Script Window
```

First the data to be fit are specified with .data$, in this case column B of the RunButton worksheet. The columns with the X and Y data of the fitted function are then specified with

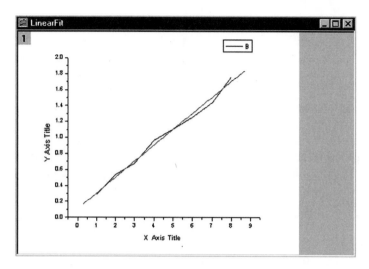

FIGURE 6.4 Original and linear fitted data (bold) in Origin plot window.

.fitxData$ and .fityData$. The number of points in the fitted curve is then set to 100. Initial values for the start and end points of the fitted function are then specified via .fitX1 and .fitX2. The .makeX() command then makes the required dataset and the .lr command performs the actual linear regression.

From a common sense perspective it would seem that the linear regression (.lr) should first be performed before an actual dataset (.makeX()) could be produced. This is not the case and if these commands are reversed the D(Y2) column will not be populated.

Finally, the last portion of the script displays the intercept, slope, and R^2 (COD) value to the script window.

Although it is great to have all this new found information, it does not mean very much unless it can be presented in a manner that conveys the true value of its content. The button labeled "plot" plots the original data in black and the corresponding linear fit in red. The code to accomplish this is:

```
Worksheet -s 1 1 4 100;

Worksheet -p 200;

Window -r %H LinearFit;
```

The first line selects the rows and columns to be plotted. The second line instructs Origin to plot the selected data. (200 is the type of plot, in this instance a line plot.) The last line renames the window of the plotted data to something appropriate, in this case "LinearFit". This code will generate a sample plot such as the one shown in Figure 6.4. The button labeled clear is done for convenience. It allows the user to erase the contents of the worksheet with the press of a button, as opposed to having to select the data with a mouse and delete it. The code to accomplish this is:

```
ClearWorksheet RunButton;
```

6.3 SCRIPT FILES

The button scripts illustrated earlier are useful for relatively short tasks that require the use of a number of keystrokes. They are especially helpful within templates, where data will be brought into a worksheet in the same form every time. In such cases button scripts can save the user from having to initialize initial conditions for curve fitting and similar tasks that would ordinarily need to be repeated for each dataset.

When it comes to lengthy complicated tasks, however, button scripts are not appropriate. First of all, it is impossible to construct a neat button script of any length because the tab function does not work within the button script editor. In addition, the button script editor offers no real debugging tools, and long scripts are difficult to access and read within the textbox where they must be written. However, using an Origin script file (*.ogs) is a better way to write scripts of a more complicated nature.

The concept behind the Origin script file is straightforward. Scripts are stored within separate files from Origin projects where they may be accessed and debugged using the Labtalk editor, which is capable of coloring text based on its context, much like the editor used in Visual Basic. The editor may be launched by selecting File->New->LabTalk Script.

A very simple Labtalk Origin script file (SimpSamp.ogs) follows:

```
//A very simple Origin Graphic Script to Illustrate the

//Script File Concept

[First Section]

//Print Message to Status Bar

type -q First Section has executed!;

[2nd Section]

//Print Message to Script Window

type -a Second Section has Run!;

[Last Section]

//Print Message with OK dialog Box

type -b You are a bumbling fool!;

[Clear Status]

//Clear the Status Bar

type -q;
```

This script file is subdivided into sections. The demarcation of each section is preceded by the name of the section in square brackets []. This example script has four sections, each of which can be thought of as a separate script or subroutine within the script file. When a section is called, only the lines of code within its section are executed.

Running script within an Origin script file is accomplished by means of the run object. For example, to run the [last section] of the previous script file one would use the script:

```
run.section(SimpSamp,Last Section);
```

If the file SimpSamp.ogs did not contain any sections it would be run using the script:

```
run.section(SimpSamp);
```

To execute scripts type these commands from the script window or create a button to execute them on a template; however, these methods are rather tedious. A better mechanism for the execution of application-specific scripts is by having menu options that can run any particular script.

Setting up Origin to invoke scripts from the menu is a multistep process. This process will be demonstrated to run the sections in the SimpSamp script. First, a script must be written to set up the menu items. This script must be saved as a configuration file with the extension (*.cnf). This file must be saved to the root Origin directory (for version 6.1 this is: \Origin61) or the full path must be specified in the run.section statement. Listed next is the code for a sample configuration file (saved as "TestCNF.cnf" on the website http://www.pharmalabauto.com):

```
//Refer to Menus for Worksheets & Graphs
menu -wg;
//Insert the Menu TestScripts in Position 7
//(7th from Left - Between Analysis & Tools)
menu 7 TestScripts;
//Add 1st Menu Item to Active Menu (TestScripts)
menu (Section 1) {
run.section(SimpSamp,First Section);
};
//Add 2nd Menu Item to Active Menu (TestScripts)
menu (Section 2) {
run.section(SimpSamp,2nd Section);
};
//Insert Divider ("_____")
menu;
//Add 3rd Menu Item to Active Menu (TestScripts)
menu (Last Section) {
run.section(SimpSamp,Last Section);
};
//Make Edit Menu Active
menu ?Ed;
//Insert Divider ("_____")
menu;
menu (Clear Status Bar) {
run.section(SimpSamp,Clear Status);
};
```

The mechanics of this code will be covered later in this section. The ideal situation would be to have this code execute at the startup of Origin so that the menus specified within it would be available to the user immediately. This can be accomplished rather easily by including this file in the Origin .ini file. In the [Config] section of the Origin .ini file, modify the following statement from:

```
File1=Macros FullMen
```

to:

```
File1=Macros FullMenu TestCNF
```

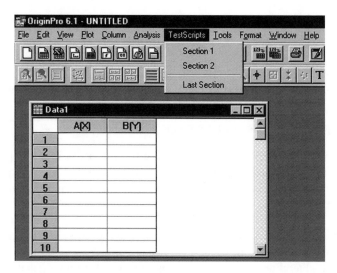

FIGURE 6.5 A new menu added via a configuration file (*.cnf).

This revision tells Origin to run the TestCNF.cnf script upon startup. This script will create a new menu called "TestScripts" between the analysis and tools menus and add an additional menu item to the end of the edit menu called "Clear Status Bar" (Figure 6.5).

Now it is necessary to explain the mechanics of the (*.cnf) file that makes these additional menu items. It is the menu command that makes customized menus possible. When working with the menu command, the first thing to do is to specify the active menu bar. This essentially means the objects to which the active menu commands will apply. The choices are: graphs (-g), worksheets (-w), layout page (-p), or matrix window (-m). These options can be combined to specify more than one object. For example, most user-defined menu items will apply to worksheets and graphs, which can be denoted by:

```
menu -wg;
```

With this accomplished, the menu to be worked on (or the active menu) must be specified. For built-in menus it is best to use the ?Reference notation, where Reference = the first two digits of the menu name. For example, to reference the Analysis menu the command would be:

```
menu ?An;
```

Only built-in menus can be referenced this way. User-defined menu items must be referenced by *position*, where the position is defined as the number from the left that the chosen menu item resides. Be advised, however, that the menu item position can change when user-defined menus are added. Therefore, it is always a good idea to reference built-in menus using their reference notation — it will never change.

On many menus a divider separates and groups menu commands that are alike. This can be done in Origin by using the command:

```
menu;
```

A divider will be placed after the currently referenced position. Also note that submenu positions can be referenced by the (.n) notation. For example, the script to reference the fourth menu item in the second menu is:

```
menu 2.4;
```

When a menu item is created, script can be assigned to that menu item using the form:

```
menu (label in menu bar) {

//Script to Execute here

};
```

It should now be clear how the "TestCNF.cnf" script constructs these menu items.

6.4 PASSING PARAMETERS IN LABTALK

This chapter would not be complete without covering parameter passing. It is possible to pass up to five parameters in Labtalk script. This can be done in either script file sections or macros. The passed arguments must always be %1, %2, %3, %4, and %5. When passing parameters, keep the following in mind:

- Passed parameters behave as string variables in that they are always substituted prior to execution of the command in which they reside
- When passing literal text or string variables, surround them by quotation marks
- Passing numbers or numeric variables does not require quotation marks *except for negative numbers*
- To pass a numeric value *by reference*, list the variable
- To pass a numeric variable *by value*, enclose the argument with the $() notation

This list shows that it is possible to pass numeric values by value and by reference. However, the manner in which this is done is really a quasi by reference and by value method. Programmers used to working in "C" or Visual Basic may be confused by this approach. The following script serves to illustrate the differences:

```
//Passing Parameters Example Script

//To run type at script window: run.section(passparms,main)

[Main]

numbpassed = 10;

%M = TestString;

//Pass By Reference

type -a By Ref Before: $(numbpassed);

run.section(passparms,ByRef,numbpassed);

type -a By Ref After: $(numbpassed);

//Pass By Value

type -a By Val Before: $(numbpassed);

run.section(passparms,ByVal,$(numbpassed));

type -a By Val After: $(numbpassed);

//Passing String Data (Note "" around var%M)

//%M cannot be changed by called subroutine

type -a String Before:%M;

run.section(passparms,String,"%M");
```

```
type -a String After:%M;
[ByRef]
//type -a $(%1);
%1 = %1 * 5;
//type -a $(%1);
[ByVal]
//The statement below is illegal when a parameter is passed
//By Value!
//%1 = %1 * 5;
[String]
//type -a%1;
%1 = "New Value";
//type -a%1;
```

When run, this script produces the following output to the script window (to be explained next):

```
run.section(passparms,main);
By Ref Before: 10
By Ref After: 50
By Val Before: 50
By Val After: 50
String Before: TestString
String After: TestString
list v;
 1                    X   3   0
 2                    Y   3   0
 3                    T   3   4.26
 4                    Z   3   0
 5                    I   3   0
 6                    J   3   0
 7                    E   3   0
 8                 ECHO   3   0
 9                COUNT   3   0
10                   PI   5   3.14159
11               FITNPTS  5   60
12                FITX1   5   1
```

13	FITX2	5	10
14	TIMERCYCLE	5	5
15	ACCEPTFIT	5	0.05
16	FITWTMODE	5	0
17	INITNFITPTS	5	60
18	AVESTEP	5	3
19	MV_BASELINE_PTS	5	12
20	NUMBPASSED	1	50
21	TESTSTRING	1	—

The ByRef section actually changes the value of the variable numbpassed from 10 to 50 by means of the statement "%1 = %1 * 5;". The action of this statement in the ByRef section carries back to the main section because %1 is a direct reference to the passed parameter, in this case the variable numbpassed.

The ByVal section is where things get confusing for those accustomed to programming in a "traditional" language. When the variable numbpassed is passed using the $() notation the parameter %1 actually becomes the value assigned to the variable numbpassed (in this case 50). In a language such as Visual Basic or "C," the value of numbpassed would be transferred into the variable %1, and the value of %1 could then be altered any way the programmer sees fit, but the value of %1 would not propagate back and change the value of the variable numbpassed because it was passed by value. Such is not the case in Labtalk. Because %1 *is* 50, %1 cannot be set to equal any other number because it would be analogous to saying 50 = 3. When a variable is passed by value its value cannot be changed by the recipient subroutine. Therefore, statements such as %1 = ... are illegal when a value is passed to %1 by value using the $() notation.

The string section is even more unusual. Here "%M" passes the literal string "TestString", and %1 becomes the string "TestString". As counterintuitive as it may seem, when the statement %1 = "New Value"; is executed in the [string] section, a new variable named "TestString" is created and is assigned to be equal to "New Value". However, since the designation "TestString" is representative of a numeric variable in Labtalk, the new variable is created but no value is assigned to it because the string "New Value" cannot be assigned to a numeric variable. This can be checked by typing "list -v"; in the script window as shown earlier.

The preceding two cases are difficult to rationalize for those accustomed to dealing with traditional by-reference and by-value parameter passing. It is critical, however, that someone understand the underlying mechanism as to how Labtalk handles passed parameters because parameters must be passed in a program with any degree of utility or complexity.

6.5 CURVE FITTING USING THE "NLSF" COMMANDS

The nonlinear least squares fitter ("nlsf") is one of the most versatile tools within Origin, and also one of the least well explained within the documentation. The fitter has a number of nonlinear functions built in for immediate use. However, it is also possible to create custom functions for any application. Fitting may be done using the Levenberg–Marquardt (LM) algorithm, or the Simplex minimization method. The Levenberg-Marquardt method is much faster, but the Simplex minimization method will never crash the fitter (i.e., it will never have a 0 matrix pivot point). Most importantly, it is possible to automate the fitting of experimental data by using Labtalk script. To illustrate how to make use of this robust tool, a sample application will be built to fit absorbance vs. pH data to the mono and the bis pKa equations.

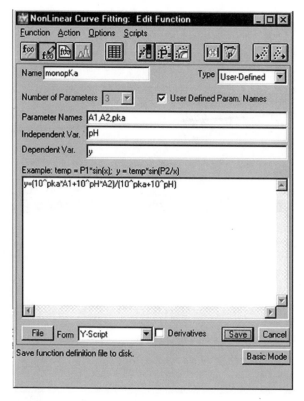

FIGURE 6.6 Setting up the mono pKa curve fitting function.

The first step is to create the fitting functions for the mono and bis pKa equations. First, the mono pKa equation:

$$y = (10^{\wedge}pka*A1+10^{\wedge}pH*A2)/(10^{\wedge}pka+10^{\wedge}pH)$$

where:

   y = measured absorbance (dependent variable)
   pH = measured pH buffer (independent variable)
   pka = the fitted mono pKa
   A1 = limiting acidic absorbance value
   A2 = limiting basic absorbance value

To set this equation up, bring up the nonlinear curve fit GUI from Analysis->NonLinear Curve Fit by typing in Alt+A+L (or Ctrl+Y). The best category to place these functions in would be pharmacology, so click on this item in the list box on the left-hand side of the GUI. Then select Function->New from the menu or hit the third button from the left to define a new function. Fill in the GUI as illustrated in Figure 6.6 and push the save button.

One last item to be set correctly is of paramount importance. In the nonlinear curve fit GUI brings up the "after fit" GUI screen by hitting the last button on the right or selecting from the menu: Scripts->After Fit. The menu shown in Figure 6.7 will appear. This menu allows the user to set up events that are to occur after the curve fitting is complete (including running custom scripts). Notice the box near the top labeled "fit curve." If the checkbox "generate fit curve" is not checked, no curve will be generated for the fitting session. If the box is checked, make sure the

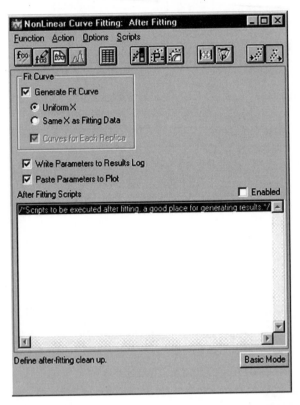

FIGURE 6.7 The after fitting dialog box — choose generate fit curve — uniform X.

option "uniform X" is selected. This is the most useful because uniform X creates a fitted curve where the X values are evenly spaced across the interval of the dependant variable. The number of X values used is determined by the nlsf.xpoints method. (If this method is not set, a default number of evenly spaced points will be created; this default value is set in the nlsf.ini file and can be changed by the user.)

If "same X as fitting data" is checked, then as many points as there are X values in the dataset will be generated. This will lead to a very roughly shaped curve because the interval between successive fitted points will be the same as the sample data. This option should be used to solve only for the points contained in the X dataset. This is not the correct choice to obtain a dataset from which a nice smooth curve can be plotted. In addition, if this item is chosen, when the fitting session is complete a new column of data will be appended to the end of the worksheet containing the fitted dataset. This column will contain the Y-axis values corresponding to the discrete X-axis points in the fitted dataset. This is a good way to clutter up a perfectly good worksheet.

One particularly nasty bug is present in versions 6.0 and 7.0 of Origin. When a user-defined curve fitting function is saved, the "generate fit curve" in the after fitting dialog box always defaults to "same X as fitting data" when the function is saved. For instance, if the function is edited and "uniform X" is chosen in the "after fitting" dialog box and then the user switches to the "edit function" dialog box and saves the fitting function, the function will be saved with the option of "same X as fitting data" instead of "uniform X". Somehow the act of selecting the edit function dialog box changes the "generate fit curve" to a default state of "same X as fitting data". This bug only occurs when the Form = Y-Script is chosen from the drop down box in the edit function dialog box.

The only way around this seems to be to edit the fitting function definition file by hand in a text editor, but even this will not help if the function is inadvertently saved again. This bug can

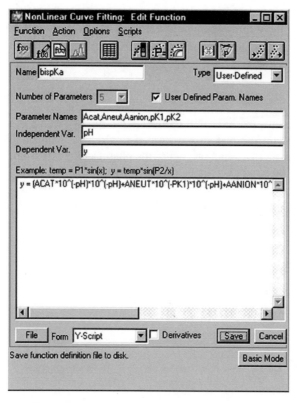

FIGURE 6.8 Setting up the bis pKa curve fitting function.

now be fixed by applying Service Release 1 to Origin 7.0. The same procedure should be followed for the bis pKa function (Figure 6.8) defined as:

$$y = (A_{CAT}*10^{(-PH)}*10^{(-PH)}+A_{NEUT}*10^{(-PK1)}*10^{(-PH)}+A_{ANION}*10^{(-PK1)}*10^{(-PK2)})/(10^{(-PH)}*10^{(-PH)}+10^{(-PH)}*10^{(-PK1)}+10^{(-PK1)}*10^{(-PK2)})$$

where:

y = measured absorbance (dependent variable)
pH = measured pH buffer (independent variable)
pk1 = the first fitted bis pKa
pk2 = the second fitted bis pKa
A_{CAT} = absorbance of cation
A_{NEUT} = absorbance of neutral species
A_{ANION} = absorbance of anion

One downside to saving custom curve fitting functions is that the files are by default named user1.fdf, user2.fdf, …, user*n*.fdf, where user1.fdf would be the first custom function saved, etc. This can be a problem if the user has a lot of custom fitting functions and wishes to distribute or edit one from the \fitfunc directory. It is nearly impossible to know which file references the fitting function under search.

However, there is a way to remedy this. First change the name of the *.fdf file to a specific name (for example, from user1.fdf to monopKa.fdf). Then the user must manually edit the *.fdf file and change the brief description to equal the filename without the extension. (All the fitting functions are stored in the \fitfunc subdirectory under Origin's main directory.) For example:

```
[GENERAL INFORMATION]

Function Name = monopKa

Brief Description = monopKa
```

If the user does not do this, the fitter will not be able to find the fitting function.

The last step is to edit the nlsf.ini file located in Origin's main directory. In this case, it is desired to have the custom nlsf functions under the pharmacology grouping, so the pharmacology section of the nlsf.ini file would be changed to look like:

```
[Pharmacology]

Default Function=

BiPhasic=BIPHASIC

DoseResp=DRESP

OneSiteBind=BIND1

OneSiteComp=COMP1

TwoSiteBind=BIND2

TwoSiteComp=COMP2

monopKa=monopKa

bispKa=bispKa
```

Before the entries would have looked like:

```
monopKa=user1.fdf

bispKa=user2.fdf
```

Although it is a little bit more work to set things up like this, it is well worth it in the long run, especially if curve fitting functions will be shared with other users. If everyone uses the default user*n*.fdf format, chances are that, when someone distributes a user*n*.fdf file, the recipient of the curve fitting file will already have a user*n*.fdf file — making things more complicated for everyone involved. It is best to have unique names for curve fitting function files.

It is possible to take this process one step further. If, by chance, no built-in category is suitable for the set of functions to be defined, it is possible to create a custom category. In this example a custom category called pKa will be created to store a variety of pKa curve fitting functions (mono, di, and tri cases).

To create a custom category within the nonlinear curve fit advanced fitting tool, the nlsf.ini file in Origin's root directory must be edited. To add a custom category termed "pKa", the first step is to edit the [category] section of the nlsf.ini file (located near the top of the file).

For this instance, the category section of the nlsf.ini file would need to be altered to look like:

```
[Category]

Origin Basic Functions=fitfunc

Chromatography=FitFunc

Exponential=FitFunc
```

```
Growth/Sigmoidal=FitFunc

Hyperbola=FitFunc

Logarithm=FitFunc

Peak Functions=FitFunc

Pharmacology=FitFunc

Polynomial=fitfunc

Power=FitFunc

Rational=FitFunc

Spectroscopy=FitFunc

Waveform=FitFunc

pKa=FitFunc
```

At the end of the [category] section add the name of the new category with " = FitFunc" appended to the end of the category. In this instance "pKa = FitFunc" is appended to the end of the category section.

As was done previously when adding custom fitting items to an existing category, the function name and the function filename (minus the extension) must be added under the category heading in the "nlsf.ini" file. In this example the following pKa functions were added to the [pKa] section of the "nlsf.ini" file:

```
[pKa]

monopKa=monopKa

bispKa=bispKa

monobase=monobase

dibase=dibase

monoacid=monoacid

diacid=diacid

tribase=tribase

monoacidmonobasicampholyte=monoacidmonobasicampholyte

monoaciddibaseampholyte=monoaciddibaseampholyte

diacidmonobaseampholyte=diacidmonobaseampholyte

triacid=triacid

Default Function=monoacidmonobasicampholyte
```

The result of these modifications is shown in Figure 6.9.

A word of caution is in order when controlling Origin's nlsf curve fitter from script. A lot of the nlsf commands are very similar in nature, and in fact some really do the same thing. However, in the course of using the fitter, it has been determined that some commands will work in some instances and others will not. In other words, the nlsf curve fitting functions are *very* fussy to code. If they are not exact, the curve fit will not work. In fact, reversing some statements, which should have no apparent effect, will cause a useless nlsf script to begin working. The key is not to get

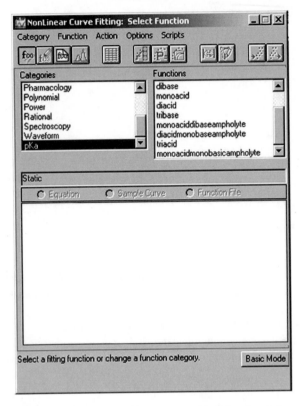

FIGURE 6.9 A custom pKa category with fitting functions.

frustrated when coding nlsf scripts, and to have a solid knowledge of the nlsf commands and the patience to substitute one command for another or move commands around when things are not working.

In particular, the following blocks of commands are redundant:

nlsf.init	nlsf.uninit
nlsf.begin	nlsf.end
nlsf.cleanupfitdata	
nlsf.x$	nlsf.y$
nlsf.indep	nlsf.setdepend
nlsf.fitdata$	
nlsf.fit	nlsf.iterate

The redundancy issue stems from several years of upgrading the fitter to include new features. Old nlsf commands were often scrapped in favor of new ones that could handle the new implementations better. The old commands still work and are retained for backward compatibility, but really should not be listed in the newer version's documentation.

When setting up an nlsf fitting session, the nlsf.cleanupfitdata should always be used at the onset. Actually, it could also be used at the end of each fitting session to give the subsequent fitting session a clean start, but if this is not added at the end of the code of the last fitting session, the results from the prior fitting session will not be cleaned.

The following commands are optional:

- Add the following in the beginning to make sure the fitting session entered is completely clean:

```
nlsf.init();//load nlsf dll and allocate memory for fitting
```

- Add the following in-between nlsf.init() and nlsf.iterate(10) to produce a more refined fit:

```
nlsf.tolerance = 1e-5;
```

- Add the following at the end of the script after nlsf.end(); to rescale the graph and show the whole plot:

```
rescale;

//Or alternatively this command can also be used:

layer -at;
```

- Consider performing the iterations in a loop so that small sets are performed at a time rather than one large set all at once. This more often results in a better fit. In other words, do something like:

```
loop(passno,1,5)

{

nlsf.iterate(10);

}
```

The redundancy issue is especially true with the assignment of independent and dependent variables. The preferred properties at this point are nlsf.y$ and nlsf.x$ because these are the easiest to use when scripting a multiple dataset fit (a feature that was not a part of the original fitter and therefore not possible with nlsf.indep, nlsf.setdepend, and nlsf.fitdata$).

The explanation of the nlsf.fit command is misleading in the Origin Labtalk manual. It implies that a FitFunc dataset will automatically be generated when the nlsf.fit command is executed. (The FitFunc dataset is supposed to contain the x and y curve fitted values to the selected dataset having the number of points in the curve fit determined by nlsf.xpoints.) This is not true. The FitFunc dataset will only be generated if the FitFunc dataset already exists and its associated nlsf.xmode properties have been set. The nlsf.iterate(n) differs from the nlsf.fit command in that nlsf.iterate(n) performs iterations without producing any visible results. The nlsf.end produces the results and ends the fitting session.

The best way to demonstrate coding and use of the nlsf curve fitter is to provide a working example. Pictured in Figure 6.10 is a workbook with two buttons and a status object that will perform mono and bis pKa curve fits on absorbance vs. pH data from a spectrophotometer. The X column contains the values of the pH buffer contained row wise in a 96 well microplate. The mono (Y) column contains the absorbance data for a mono pKa curve. The bis (Y) column contains the absorbance data for a bis (or dual) pKa curve. Pressing either button produces a dialog box that prompts the user to select the dataset to be fit. Notice the status object in the upper left corner of the worksheet that stores the fitted parameters. Origin has many GUI tools like the status object that can be placed on worksheets while in the button edit mode. The Labtalk code to perform the mono pKa curve fit contained in the button monofit is:

```
//Mono pKa Fitting example

//Save the name of the current worksheet
```

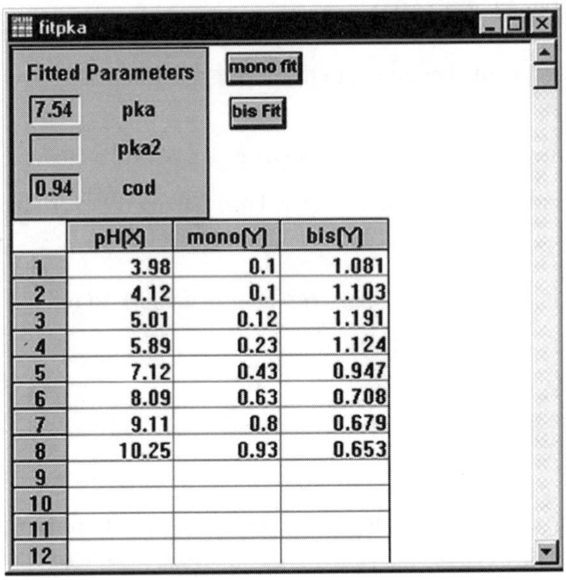

	pH(X)	mono(Y)	bis(Y)
1	3.98	0.1	1.081
2	4.12	0.1	1.103
3	5.01	0.12	1.191
4	5.89	0.23	1.124
5	7.12	0.43	0.947
6	8.09	0.63	0.708
7	9.11	0.8	0.679
8	10.25	0.93	0.653
9			
10			
11			
12			

FIGURE 6.10 The fitpKa worksheet

```
%W=%H;
%Z="";doc -e DY {%Z=%Z%C};dependataset=2;
getnumber (mono pKa Curve Fit) dependataset:Z (Select Dataset
to be fitted);
dependataset+=1;
//Create a plot window named "monopKa"
GetEnumWin monopKa;
//Select the Custom Fitting Function
nlsf.func$=monopKa;
//Clean Up all internal Objects Used in nlsf
nlsf.cleanupfitdata();
//initialize parameter values
A1=%(%W,dependataset,1);//.p1
A2=%(%W,dependataset,8);//.p2
pka=5;//.p3
//Print Parameters to Screen
#! type -a $(A1) $(A2) $(pKa);
//Dependent Data (Y-Axis) to be Fit
nlsf.y$ = %(%W,dependataset);
//Independent Data (X-Axis) to be Fit
nlsf.x$ = %(%W,1);
```

```
//Total Number of Points for Fitted Curve

nlsf.xpoints = 60;

//Improve accuracy of fit

nlsf.tolerance = 1e-5;

//Perform Curve Fitting

nlsf.fit(20);

//Add the original Data to the plot

layer -i%(%W,2);

//Rescale X & Y Axis to show All data

layer -at;

//Update "FitPams" Status Object

%W!FitPams.v1 = $(nlsf.p3,.2);

%W!FitPams.v2$ = "";

%W!FitPams.v3 = $(nlsf.cod,.2);

//Rename Axis Labels

label -xb pH Buffer Values;

label -yl Absorbance;
```

Running this script will produce a plot like that shown in Figure 6.11. The bis pka curve fitting script is similar, but has some very important differences in nlsf coding commands. First its code is listed, followed by an explanation of its differentiation.

```
//Bis pKa Fitting example

//Save the name of the current worksheet
```

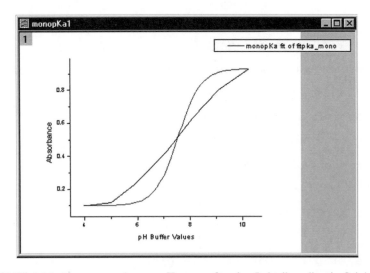

FIGURE 6.11 An automated mono pKa curve fit using Labtalk coding in Origin.

```
%W = %H;

%Z="";doc -e DY {%Z=%Z%C};dependataset=2;

getnumber (bis pKa Curve Fit) dependataset:Z (Select Dataset
to be fitted);

dependataset+=1;

//Create a plot window named "bispKa"

GetEnumWin bispKa;

//Select the Custom Fitting Function

nlsf.func$=bispKa;

//Clean Up all internal Objects Used in nlsf

nlsf.cleanupfitdata();

//initialize parameter values

Acat=%(%W,dependataset,1);//.p1

Aneut=%(%W,dependataset,4);//.p2

Aanion=%(%W,dependataset,8);//.p3

pk1 = 5;//.p4

pk2 = 8;//.p5

//Print Parameters to Screen

#! type -a $(Acat) $(Aneut) $(pk1);

//Do Not Allow Acat or Aanion to Vary!

nlsf.v1=0;

nlsf.v3=0;

//Dependent Data (Y-Axis) to be Fit

nlsf.y$ = %(%W,dependataset);

//Independent Data (X-Axis) to be Fit

nlsf.x$ = %(%W,1);

//Total Number of Points for Fitted Curve

nlsf.xpoints = 60;

//Improve Accuracy of Fit

nlsf.tolerance = 1e-5;

//Perform Curve Fitting

nlsf.fit(20);

//Add the original Data to the plot

layer -i%(%W,dependataset);
```

```
//Rescale X & Y Axes to show All data

rescale;

//Update "FitPams" Status Object

%W!FitPams.v1 = $(nlsf.p4,.2);

%W!FitPams.v2$ = $(nlsf.p5,.2);

%W!FitPams.v3 = $(nlsf.cod,.2);

//Rename Axis Labels

label -xb pH Buffer Values;

label -yl Absorbance;
```

The first major difference to notice in this code is the two additional parameters contained in the fitting function, which must also be set with initial values prior to starting a curve fitting session. Also notice the statements nlsf.v1 = 0 and nlsf.v3 = 0. These tell the fitter not to let parameters 1 (A_{Cat}) and 3 (A_{Anion}) vary during the fit. Why are these commands necessary? Looking at the following plot, the pKa's are located at 4.45 and 7.21. Intuitively this seems correct because they are located at the centers of the two sloping curves.

Notice, however, the shape of this curve. Unlike the monotonic pKa curve that slopes from high to low, this curve is more complex because it first swings up and then down, giving it two sloping regions. (In fact some would argue that a third sloping region exists on the end tail.) A bis pKa can even be U-shaped, giving it a slope up and a slope down.

Because of the more complex geometry of the bis pKa curve, it is necessary to tie or fix the two endpoints of the curve when fitting it. If the endpoints are allowed to wander, they may wander all over the place and a satisfactory fit will never be achieved. If the statements nlsf.v1 = 0 and nlsf.v3 = 0 are commented out of the bis pKa script, the curve fitter will crash because it cannot settle on a solution. Try it (see Figure 6.12).

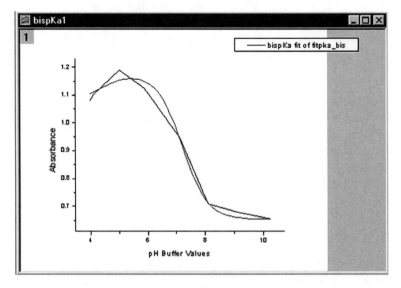

FIGURE 6.12 An automated bis pKa curve fit using Labtalk coding in Origin.

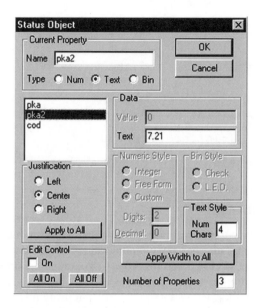

FIGURE 6.13 Setting the status objects' properties.

One last point to make concerns the population of the status object, which shows the solutions to the first pKa, the second pKa (if it exists), and the R^2 (or COD) value, indicating the "goodness" of the fitted curve. (COD = 1.0 is a perfect correlation.) Notice the statements:

```
%W!FitPams.v1 = $(nlsf.p4,.2);

%W!FitPams.v2$ = $(nlsf.p5,.2);

%W!FitPams.v3 = $(nlsf.cod,.2);
```

FitPams.v2$ is treated as a string because, for a mono pKa fit, the second pKa does not exist. If FitPams.v2$ were treated as a numeric variable, setting it to equal a null would fill the status object with a zero, which is not the value of the second pKa. It is desirable to leave this field blank when a second pKa does not exist (the monotonic case.) This can then be accomplished by the statement:

```
%W!FitPams.v2$ = "";
```

which is used only in the monotonic pKa script; however, such syntax will only produce the desired results if the item in the status object is treated as string data.

Also look at the convention = $(nlsf.p4,.2). This tells the compiler to assign the contents of nlsf.p4 to some variable with only two decimal places (the standard "C" convention). However, this statement will not fix the number of decimal places or digits shown in the status object. This must be done by setting the properties of the status object (Properties->Special) using the GUI shown in Figure 6.13.

6.6 HARNESSING ORIGIN REMOTELY

As has been previously shown, Origin is capable of many complex and robust mathematical algorithms from within the Origin environment. However, what happens when it becomes necessary to harness this power from another application that does not possess the functionality to perform such computations on its own?

Two principal methods of interapplication data communication under windows are dynamic data exchange (DDE) and object linking and embedding (OLE [pronounced "olay"]) automation. DDE is the older method and is rapidly being replaced by OLE in many applications. Many applications, including Origin, still support only DDE.

Both methods are difficult to use for a variety of reasons that will be discussed briefly. The principal difficulty in using either method is documentation. Commands, properties, and methods accessible from DDE or OLE vary from application to application. To compound the problem, syntax may also differ among applications as well. For example, many applications (including Microsoft Excel) require that each command received through a DDE channel be enclosed in brackets ([]); others do not have this requirement. To make this distinction, try it one way and if it does not work, try it the other way. This pragmatic solution is often the only way to make such a determination, especially when documentation is lacking.

Another intricacy involved with DDE and OLE is that, in order for it to work, *both* applications must be function correctly and accept each other's requests. Typically, whenever a problem occurs when using DDE or OLE and technical support is called, the answer is that the communication problem is the other application's fault. It is amazing that among Microsoft's own applications (such as the Office Suite), very few problems occur. This is probably because Microsoft has fixed the problems as they arose.

Despite the difficulties, DDE and OLE provide the means to automate applications in ways that would not be possible without their existence. While troublesome in many ways, the benefit often outweighs the cost. Usually, a workaround can be found for the various incompatibilities that do exist.

6.7 DYNAMIC DATA EXCHANGE (DDE) WITH VB

The best option for communicating with Origin is through standalone Visual Basic (VB). The method described in Figure 6.14 works without any glitches.

An application has been set up to utilize the bis pKa curve fitting function outlined in Section 6.5. Here a worksheet named "DDEfit.ogw", which has the mono and bis pKa curve fitting functions but contains no data, is utilized. This application will pass the pH and absorbance data to the Origin worksheet, push the bis pKa curve fit button, and retrieve the bis pKa curve fitting results to VB. When the bis pKa curve fit button on the GUI in Figure 6.14 is pushed, the code in the VB_DDE_Sample sub is executed:

```
Sub VB_DDE_Sample()

'YOU MAY NEED TO CHANGE THE PATH TO THE DDEFIT.OGW WORKSHEET !!!

Call Start_Origin("C:\Program
Files\OriginLab\Origin61\Origin61.exe", _
"G:\Author\PharmaLabAuto\Section6\Origin\DDEFit.ogw",
vbMinimizedNoFocus)
```

FIGURE 6.14 GUI for VB DDE bis curve fitting example using Origin.

```
Call  Initiate_OrgConversation

Call  Send_pH_Data

Call  Send_Bis_Data

Call  Click_Bis_Button_LX

Call  Extract_Parameters

Call  Terminate_Conversations

'End

End Sub
```

The first thing to do is to start Origin:

```
Sub Start_Origin(apppath$, projpath$, winstyle)

'Starts Origin Remotely, Stores its Task ID to Origin_TaskID

'apppath$ = Full path to Origin61.exe

'projpath$ = Full path of Project to Open when Origin Starts
(Optional)

'winstyle% - Window Focus and Style of Opened Object (Origin)

Dim pathname$

'A space must be placed between apppath$ and projpath$ and

'projectpath$ must be in "" (Chr$(34) = ")

pathname$ = apppath$ + " " + Chr(34) + projpath$ + Chr(34)

'WD98: WindowStyle Argument of Shell Function Ignored (Q178328)

'If the pathname argument is set to a document instead of a
program,

'the windowstyle argument will have no effect.

Origin_TaskID = Shell(pathname$, winstyle)

'Since Shell windowstyle is ignored

'Use AppActivate to set focus back to Excel

AppActivate "DDETEST"

End Sub
```

Here the shell command is used to start Origin and the task ID of the Origin application started is saved to Origin_TaskID. The AppActivate statement returns focus to Excel. This really should not be necessary but the Winstyle argument of the shell command does not function correctly in this instance (see Microsoft article Q178328, www.microsoft.com).

To provide DDE communication under standalone VB, the.LinkXXXXX methods will be used. These methods must be associated (or bound) to a label, textbox, or picturebox control. For this sample, textboxes were used.

Whenever using DDE, a communication topic supported by the server application must be referenced. (In this example VB is the client and Origin is the server. Origin will *serve* the client by providing curve fitting of data passed to it by VB, and allowing the results of the fit to be extracted to VB.) Origin only supports two topics: *variable* and *org*. The variable topic is used for Labtalk scripts and/or variables while org is used to reference worksheets. Some degree of overlap exists between the two topics.

Two textboxes have been added to the form on the application. The LinkLabelVar textbox is used solely to transmit topics under the variable domain. The LinkLabelOrg textbox is used only to transmit topics under the org domain.

A conversation is then initiated by this subroutine:

```
Sub Initiate_OrgConversation()

'This will initiate a DDE "Conversation" Channel from VB to
Origin

ORIGIN_DDE!LinkLabelOrg.LinkTopic = "Origin|ORG"

ORIGIN_DDE!LinkLabelOrg.LinkItem = "DDEfit"

ORIGIN_DDE!LinkLabelOrg.LinkMode = COLD

End Sub
```

Here a link is established using the org topic. The item linked to is the Origin worksheet DDEfit. COLD is a global variable equal to 2.

With the link established, it is now time to send over the pH data. This is accomplished with:

```
Sub Send_pH_Data()

'Send pH data to Origin - Excel Column 1

Dim ii As Integer, datastring$

Dim pHData As Variant

datastring$ = "3.98,4.12,5.01,5.89,7.12,8.09,9.11,10.25"

pHData = Split(datastring$, ",")

For ii = 1 To 8

   Call Send_Data_Org("DDEfit", "pH", (ii), (pHData(ii - 1)))

Next ii

End Sub

Sub Send_Data_Org(Worksheet$, column$, row&, data!)

'Send Data From Excel to a Specific Origin Worksheet/Cell

Dim SendString$

SendString$ = Worksheet$ & "_" & column$ & "[" &
Trim$(Str(row&)) & "]=" & Trim$(Str(data!)) & ";"

Call Send_Command_Org(SendString$)

End Sub

Sub Send_Command_Org(cmd$)

'Execute the command passed in Origin

ORIGIN_DDE!LinkLabelOrg.LinkExecute cmd$

End Sub
```

Any Labtalk command can be sent to Origin via DDE using the .LinkExecute method in VB. For example, the command:

```
ORIGIN_DDE!LinkLabelOrg.LinkExecute "DDEfit_pH[3]=5.01;"
```

sets the third row of the pH column in the DDEfit worksheet equal to 5.01. The subroutine Send_Data_Org needs only to have the worksheet name, column name, row, and data passed to it. It will then format the information into a convention understood by Origin.

The command does not have to be present inside the textbox for it to be sent to the server application. The textbox merely serves as a link to the server application.

The pH values are stored in a comma-delimited format in the variable datastring$, and are parsed out by means of the split function. The pH data could just as easily have been read from a file or spreadsheet object.

With the pH data in the DDEfit worksheet, the absorbance data can be sent over using the same algorithm.

```
Sub Send_Bis_Data()

'Send bis pKa data to Origin

Dim ii As Integer, datastring$

Dim bisdata As Variant

datastring$ =
"1.081,1.103,1.191,1.124,0.947,0.708,0.679,0.653"

bisdata = Split(datastring$, ",")

For ii = 1 To 8

   Call Send_Data_Org("DDEfit", "bis", (ii), (bisdata(ii - 1)))

Next ii

End Sub
```

At this point the user would then push the "bis Fit" button in order to have Origin run the bis pKa curve fitting script. The big question is how to press a button from script. The next subroutine shows how to accomplish this:

```
Sub Click_Bis_Button_LX()

'"PUSH" the bis pka fit button on the worksheet using Labtalk
Code

Send_Command_Org ("DDEfit!Text1.run();")

End Sub
```

(The "bis Fit" button's name is "Text1" in Origin. Enter button edit mode to see this.)

With the curve fitting completed, it is time to extract the fitted values from Origin and bring them into the VBA. For this process the variable topic must be utilized. In the previous example the client application was sending commands to the server application. Now the client application is requesting information from the server application.

```
Sub Extract_Parameters()

'Extract Fitting Parameters Using the Variable Topic

Call Initiate_VarConversation
```

```
Call SetLinkVar("nlsf.p4")

ORIGIN_DDE!pKa1Label.Caption = FetchVar

'Debug.Print "pKa1="; FetchVar

Call SetLinkVar("nlsf.p5")

ORIGIN_DDE!pKa2Label.Caption = FetchVar

'Debug.Print "pKa2="; FetchVar

Call SetLinkVar("nlsf.cod")

ORIGIN_DDE!CODLabel.Caption = FetchVar

'Debug.Print "R^2="; FetchVar

End Sub

Sub Initiate_VarConversation()

'This will initiate a DDE "Conversation" Channel from VB to
Origin

ORIGIN_DDE!LinkLabelVar.LinkTopic = "Origin|Variable"

ORIGIN_DDE!LinkLabelVar.LinkMode = COLD

End Sub

Sub SetLinkVar(OriginVar$)

'Estabish What Variable from Origin is to be Retrieved

ORIGIN_DDE!LinkLabelVar.LinkItem = OriginVar$

End Sub

Function FetchVar() As String

'Extract the specified Variable from Origin

ORIGIN_DDE!LinkLabelVar.LinkRequest

FetchVar = ORIGIN_DDE!LinkLabelVar.Text

End Function
```

As before, the first thing to do is to establish a DDE link. This time, however, the LinkLabelVar textbox is used, and the topic is variable as opposed to org. Next the link must know what variable in Origin is to be retrieved. This is accomplished with the SetLinkVar subroutine, which sets the .LinkItem property to the name of the variable of interest. Finally, the variable is retrieved with the FetchVar function, which extracts the variable's value from Origin using the .LinkRequest method. As shown previously, the Extract_Parameters subroutine resets the .LinkItem property to retrieve a different parameter in Origin. In this example, pKa1, pKa2, and COD (R^2) are the variables retrieved.

With the analysis complete and the results returned, the final thing to do is shut down the DDE connections. To do this, set the LinkMode property to 0 by entering the following code:

```
Sub Terminate_Conversations()

'Terminate all DDE Conversations
```

```
Call Terminate_VarConversation

Call Terminate_OrgConversation

End Sub

Sub Terminate_VarConversation()

'Terminate the Given DDE Variable Link

ORIGIN_DDE!LinkLabelVar.LinkMode = 0

End Sub

Sub Terminate_OrgConversation()

'Terminate the Given DDE Org Link

ORIGIN_DDE!LinkLabelOrg.LinkMode = 0

End Sub
```

6.8 DYNAMIC DATA EXCHANGE (DDE) WITH VBA

It is also possible, although much more problematic, to communicate with Origin in Excel using VBA. To provide DDE communication under VBA, functions such as DDEInitiate, DDERequest, and DDEPoke must be used. The .LinkXXXXX methods *are not* supported in VBA. The following DDE functions exist within VBA. Some of these functions require more than one parameter, and may encompass more than one .LinkXXXX property/method. However, the chart in Table 6.7 shows approximate equivalence.

An example macro has been constructed to harness the fitpKa worksheet developed in Origin earlier in Section 6.5 as the "engine" to provide robust mono and pKa curve fitting ability within Excel's environment. The macro will start Origin and establish a DDE connection between Excel and Origin, pass the appropriate data to Origin, "push" the appropriate button to execute the mono or bis pKa curve fitting script, return the results calculated by Origin to Excel, terminate the DDE connection, and close Origin. First, some basic parameters must be set up:

```
'Excel -> Origin Dynamic Data Exchange DDE Example

'Will Send Data to Origin for Calculations

'Return results to Excel

Option Explicit
```

TABLE 6.7
VBA DDE Functions and Their Nearest
VB Equivalent

VBA Function		VB Equivalent
DDEExecute	⇔	.LinkExecute
DDERequest	⇔	.LinkRequest
DDEPoke	⇔	.LinkPoke
DDEInitiate	⇔	.LinkTopic
DDETerminate	⇔	.LinkMode

```
Dim Origin_TaskID As Variant

Dim Channel As Variant

'These paths are Machine Specific and Will Need to be Changed!

Const origin_path$ = "C:\Program
Files\OriginLab\OriginPro70\Origin70.exe"

Const origin_sheet_path$ = _

"C:\Author\Books\PharmaLabAuto\PharmaLabAutoText\CDROM\Origin\
workssheets\DDEFit.ogw"
```

Here two variables that are global across the module in which they reside: Origin_TaskID and Channel. Origin_TaskID is the specific ID number assigned to the Origin program when it is started by the shell command. Channel is the channel number returned upon initiation of a DDE connection. Notice also that two paths are set up that point to the Origin program and the DDEFit (.ogw) worksheet locations. These paths will, in all likelihood, need to be changed in order for the example to run on a different machine.

The following subroutine is called when the user selects from the Menu Automation->Section 6->Origin DDE:

```
Sub DDE_Sample()

    Windows("Origin7DDE.xls").Visible = True

    DDEForm.Show

    'These statements which would "push" buttons on a form in VB

    'Do Not Work in Excel VBA!

    'DDEForm!VisibleOptButton.Value = True

    'DDEForm!DDERunButton.Value = True

End Sub
```

The execution of this subroutine launches a GUI that the end user will use to control the DDE communication example (Figure 6.15). Notice that the GUI allows the user to show and hide the Origin7DDE.xls workbook at will using the option buttons in the "workbook status" frame. Another point worth mentioning is that in Excel VBA it is not possible to push a button on an Excel form (GUI) by means of the .Value property as can be done in stand-alone Visual Basic. Hence commands such as

```
DDEForm!DDERunButton.Value = True
```

would not function in an Excel VBA macro.

When the user presses the "Run DDE Demo" button in the lower right-hand corner of the GUI, the following subroutine, which controls the DDE communication process between Excel and Origin, is executed:

```
Sub Run_Sample()

Call Start_Origin(origin_path$, origin_sheet_path$,
vbMinimizedNoFocus)

Call Initiate_Conversation

Call Send_pH_Data
```

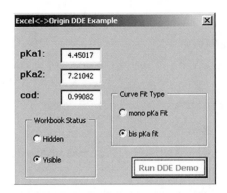

FIGURE 6.15 GUI for the Excel<->Origin DDE sample application.

```
Call Send_Fitting_Data

Call Click_Button

Call Get_Parameters

Call Terminate_Conversation

'ActiveWindow.Visible = False

End Sub
```

The first task to be accomplished is to start Origin in such a way that Excel will have control as the "calling" application. The following subroutine will accomplish this:

```
Sub Start_Origin(apppath$, projpath$, winstyle)

'Starts Origin Remotely, Stores its Task ID to Origin_TaskID

'apppath$ = Full path to Origin61.exe

'projpath$ = Full path of Project to Open when Origin Starts
(Optional)

'winstyle% - Window Focus and Style of Opened Object (Origin)

Dim pathname$

'A space must be placed between apppath$ and projpath$ and

'projectpath$ must be in "" (Chr$(34) = ")

pathname$ = apppath$ + " " + Chr(34) + projpath$ + Chr(34)

'WD98: WindowStyle Argument of Shell Function Ignored (Q178328)

'If the pathname argument is set to a document instead of a
program,

'the windowstyle argument will have no effect.

Origin_TaskID = Shell(pathname$, winstyle)

'Since Shell windowstyle is ignored

'Use AppActivate to set focus back to Excel

AppActivate "Microsoft Excel"

End Sub
```

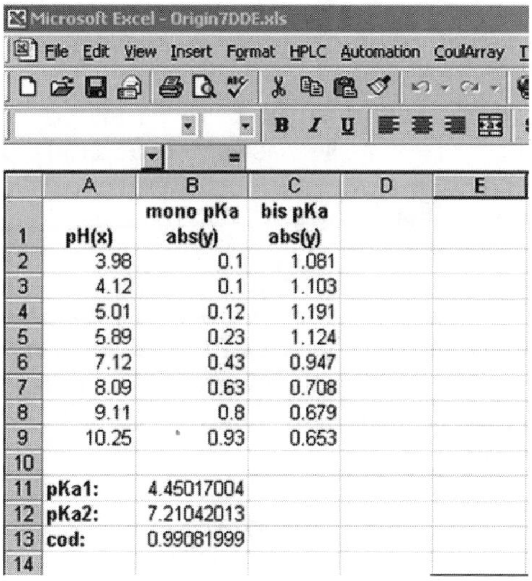

	A	B	C	D	E
1	pH(x)	mono pKa abs(y)	bis pKa abs(y)		
2	3.98	0.1	1.081		
3	4.12	0.1	1.103		
4	5.01	0.12	1.191		
5	5.89	0.23	1.124		
6	7.12	0.43	0.947		
7	8.09	0.63	0.708		
8	9.11	0.8	0.679		
9	10.25	0.93	0.653		
10					
11	pKa1:	4.45017004			
12	pKa2:	7.21042013			
13	cod:	0.99081999			
14					

FIGURE 6.16 The Excel–Origin DDE example worksheet.

A point worth mentioning in the preceding code is that under some circumstances (see previous coding comments), the Window-style argument of the shell command will have no effect. Because this occurs in the preceding coding, the AppActivate statement is utilized to return focus to the calling application, which in this case is Excel.

As in the previous example, a DDE link must be established:

```
Sub Initiate_Conversation()

'This will initiate a DDE "Conversation" Channel from Excel
to Origin

   Channel = DDEInitiate("Origin", "Variable")

End Sub
```

Unlike the previous example, the DDEInitiate function returns a specific channel number, which must be referenced by all DDE functions that wish to use the channel opened by the DDEInitiate command.

Looking at Figure 6.16, the pH data in this instance is stored in the first column of the OriginDDE worksheet. It is sent to Origin by means of the following subroutines:

```
Sub Send_pH_Data()

'Send pH data to Origin - Excel Column 1

Dim ii As Integer

For ii = 1 To 8

   Call Send_Data("DDEfit", "pH", (ii), ActiveSheet.Cells(ii,
   1).Value)

Next ii

End Sub

Sub Send_Data(Worksheet$, column$, row&, data!)
```

```
'Send Data From Excel to a Specific Origin Worksheet/Cell

Dim SendString$

SendString$ = Worksheet$ & "_" & column$ & "[" &
Trim$(Str(row&)) & "]=" & Trim$(Str(data!)) & ";"

DDEExecute Channel, SendString$

End Sub
```

The Send_Data subroutine is the real workhorse of this macro. It is within this subroutine that data are transferred from Excel to Origin using the DDEExecute command.

The absorbance data stored in the Excel worksheet in Figure 6.16 can be transmitted to the Origin DDEFit worksheet in the same manner:

```
Sub Send_Fitting_Data()

'Send appropriate pKa data to Origin

Dim ii As Integer

For ii = 1 To 8

    Select Case DDEForm!monoOptionButton.Value

        Case True

            Call Send_Data("DDEfit", "mono", (ii),
            ActiveSheet.Cells(ii + 1, 2).Value)

        Case False

            Call Send_Data("DDEfit", "bis", (ii),
            ActiveSheet.Cells(ii + 1, 3).Value)

    End Select

Next ii

End Sub
```

Once all the data have been transmitted to Origin, the user must be able to execute the code associated with the correct fitting function — in this instance the code executed when the "mono fit" or "bis Fit" buttons are pressed. Simulate a button press via DDE; the object name that the button code resides in must first be determined. The object name can readily be determined by going into button edit mode (Ctrl-B), left clicking, and selecting label control. Doing so shows the correspondence in Table 6.8.

TABLE 6.8
Button Name and Corresponding Object Containing Code

Button Name	Object Name:
mono fit	text
bis fit	text1

This subroutine will execute the code associated with the button corresponding to the proper pKa curve fitting function:

```
Sub Click_Button()

Select Case DDEForm!monoOptionButton.Value

   Case True

      '"PUSH" the mono pka fit button on the worksheet using
      Labtalk Code

      DDEExecute Channel, "DDEfit!Text.run();"

   Case False

      '"PUSH" the bis pka fit button on the worksheet using
      Labtalk Code

      DDEExecute Channel, "DDEfit!Text1.run();"

End Select

End Sub
```

Once this subroutine has triggered Origin to perform the indicated curve fitting function, a means must be devised for Excel to read the parameters calculated by Origin. This can be accomplished by using the function:

```
Function RequestData(Channel, param$) As Single

Dim returnList As Variant, i As Integer

'The stronger typecasting requires the returnList in DDE to be

'(1) cast as Variant (2) treated as array EVEN IF ONLY
REQUESTING A SINGLE ELEMENT

returnList = Application.DDERequest(Channel, param$)

For i = LBound(returnList) To UBound(returnList)

   RequestData = Val(returnList(i))

Next i

End Function
```

In earlier versions of Excel, it was possible to use DDERequest with a variant to extract a single parameter. (For example, MyData = DDERequest(VariableId, "cod") would function in earlier versions of Excel. This code will no longer work in Excel.) However, the more recent stronger typing of VBA code requires that all variables used with DDERequest be treated as an array, even when only a single element is to be requested. This is the reason the variant returnList is treated as an array even though only one parameter is returned.

With this function it is now possible to construct a subroutine to get the most recent curve fitting parameters. Any old or previously calculated parameters are erased prior to writing in the new parameters. If a bis pKa fit followed by a mono pKa fit is performed, the second pKa parameter calculated in the bis pKa fit will not remain in the worksheet when the results for a mono pKa fit are produced:

```
Sub Get_Parameters()

'Erase any pre-existing parameters in Excel Workbook and GUI

Workbooks("Origin7DDE.xls").Worksheets(1).Cells(11, 2).Value = _
""

Workbooks("Origin7DDE.xls").Worksheets(1).Cells(12, 2).Value = _
""

Workbooks("Origin7DDE.xls").Worksheets(1).Cells(13, 2).Value = _
""

DDEForm!pKa1TextBox.Text = ""

DDEForm!pKa2TextBox.Text = ""

DDEForm!codTextBox.Text = ""

'Fetch Most Current Curve Fit Parameters - Write to Worksheet
& Form (GUI)

Select Case DDEForm!monoOptionButton.Value

    Case True

        Workbooks("Origin7DDE.xls").Worksheets(1).Cells(11,
        2).Value = _

        RequestData(Channel, "nlsf.p3")

        Workbooks("Origin7DDE.xls").Worksheets(1).Cells(13,
        2).Value = _

        RequestData(Channel, "nlsf.cod")

        DDEForm!pKa1TextBox.Text = RequestData(Channel,
        "nlsf.p3")

        DDEForm!codTextBox.Text = RequestData(Channel,
        "nlsf.cod")

    Case False

        Workbooks("Origin7DDE.xls").Worksheets(1).Cells(11,
        2).Value = _

        RequestData(Channel, "nlsf.p4")

        Workbooks("Origin7DDE.xls").Worksheets(1).Cells(12,
        2).Value = _

        RequestData(Channel, "nlsf.p5")

        Workbooks("Origin7DDE.xls").Worksheets(1).Cells(13,
        2).Value = _

        RequestData(Channel, "nlsf.cod")
```

```
      DDEForm!pKa1TextBox.Text  =  RequestData(Channel,
      "nlsf.p4")

      DDEForm!pKa2TextBox.Text  =  RequestData(Channel,
      "nlsf.p5")

      DDEForm!codTextBox.Text  =  RequestData(Channel,
      "nlsf.cod")
```

```
End Select
```

```
End Sub
```

When the curve fit is complete and the parameters of interest have been extracted, it is time to close down Origin and terminate the DDE connection. This can be done using the code:

```
Sub Terminate_Conversation()

'Terminate the Excel-Origin DDE "Conversation" Link

'Exit Without Changing

DDEExecute Channel, "doc -s;dde -q 1;exit;"

DDETerminate Channel

End Sub
```

The tricky thing about closing down an application using DDE is that when designing an automated task, closing down an application without the need for user intervention can be difficult. In most instances, when an application is closed down using DDE, the application will (1) prompt the user to save changes and (2) warn the user that if the application closes, an active DDE connection will be terminated.

In the preceding example these problems are eliminated by sending a series of Labtalk commands. The command "doc –s;" will disable the save flag so that no prompt will appear upon exiting Origin. The command "dde –q 1;" will disable all DDE warning message boxes in Origin. The command "exit;" will terminate the current session of Origin and close the program down. These commands can be sent in a series within a single DDE statement, as is done in the preceding subroutine in the line

```
DDEExecute Channel, "doc -s;dde -q 1;exit;"
```

The completed application can be run by selecting Automation->Section 6->Origin DDE from Excel's menu bar. Doing so will bring up the Origin DDE example worksheet along with the GUI depicted in the earlier illustrations.

The most important point of this example is to understand the mechanism by which an Excel <-> Origin DDE connection can be established and controlled. Although the prededing sample application is trivial in that each communication must be triggered manually (by pressing a button on the "Excel <-> Origin DDE example" form) and that only one parameter is calculated and returned, it is important to keep in mind that this application could be modified with a minimal amount of effort to calculate numerous parameters automatically and return several parameters with each DDE request.

7 Agilent (HP) Chemstation Macro Writing

7.1 BASICS OF CHEMSTATION MACRO WRITING

Chemstation macros can be written in any text editor such as notepad or wordpad. Instead of using the default extension (*.txt), Chemstation macros use the extension (*.mac). Macros written in a standard text editor require that the user manually modify the extension prior to saving with the default (*.txt) to the required (*.mac). All macros should be saved to the path…\hpchem\core\ (which by default is c:\hpchem\core). Macro filenames must also not exceed the old Windows 3.1 filename convention of 8.3 (eight-character name, three-character extension " --------.---"). A macro with a filename of nine characters or more cannot be loaded into the Chemstation environment. The Chemstation command line is identified by the bold text command line in Figure 7.1. All commands to load, reload, and run macros are typed in the command line.

Before any macro can be run, it must first be loaded into the system. To load the macro into the system, the file (*.mac) must reside in the \core directory. Suppose a macro is created with the name "usermacro.mac". Once the file has been saved into the \core directory it can be loaded into the Chemstation software by typing the following at the command line and hitting return:

```
macro "usermacro.mac"
```

If the macro is edited at any time and saved during a Chemstation session, the file must be reloaded prior to execution for the changes to take effect. It is not enough to make the changes in a text editor and then rerun the macro. The macro must be saved to the \core directory and reloaded for the changes to take effect. A macro that has been loaded into memory can be run by typing the following at the command line:

```
macro usermacro.mac, go
```

or, more simply,

```
usermacro
```

A very important distinction must be made here. Each *.mac file can contain several macros. If a macro is changed, the file that it resides in must be resaved and reloaded. The macro can then be run by typing its name at the command line. The confusion occurs because sometimes a macro name and a file name will be the same (typically when only one macro is in a file). In that instance, to load the macro into memory type:

```
macro "macroname.mac"
```

However, if the preceding is typed, and "macroname" is the name of a macro residing within a file named "othername.mac", the command would produce an error because the Chemstation software would be looking for a file named "macroname.mac" that would not exist. Instead, reload othername.mac, which contains the macro macroname. Bottom line — when reloading a macro into memory after making changes, reload the file in which the macro resides; do not try to reload the macro itself.

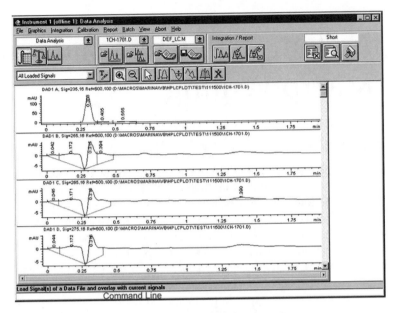

FIGURE 7.1 The Chemstation software GUI interface.

Admittedly, loading macros by hand from the command line gets old very quickly. However, after a macro has been perfected and debugged, there is a way to load it automatically once the Chemstation software starts. Each time the Chemstation starts, it looks for a file in \core called "user.mac". If it finds this file, it loads the macros listed in it. In this way, user-created macros can be automatically loaded and available for use in every Chemstation session. The syntax within the "user.mac" file should look like the following:

```
macro firstmacro.mac

macro lastmacro.mac
```

Here two macros, "firstmacro" and "lastmacro", would automatically be loaded upon the startup of the Chemstation software.

A number of functions and commands built into the Chemstation software are not documented, but are nonetheless available for use by programmers. By typing the show command at the command line, a dialog box will be brought up that shows the user all the available functions and commands, and the parameters required for their use. To access this dialog box, type the following at the command line and the dialog box shown in Figure 7.2 will appear:

```
show
```

7.2 CHEMSTATION MACRO VARIABLES

All variables used within Chemstation macros are global unless they are declared local. All string variables end with a "$". For example:

```
Compound$ = "acetaminophen"

Laboratory$ = "Drug Metabolism"

Chemist$ = "J. Smith"
```

All numeric variables do not end with a "$". For example:

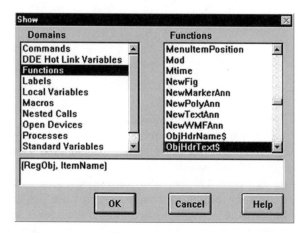

FIGURE 7.2 Using the show command to display all available functions.

```
NoOfPeaks = 5

piByTwo = 6.28

MaxArea = 205
```

If a variable is to be utilized in only one macro, it should be declared a local variable with the local command. Local variables do not retain their values outside the particular macro within which they are declared. For example:

```
Local NoOfPeaks = 3

Local Chemist$ = "J. Doe"
```

The examples contain upper- and lower-case letters. This is for readability only. The macros do not distinguish between upper- and lower-case letters in variable names. Therefore, "MaxArea" will be equivalent to "maxarea" anywhere in the macro. Speaking of readability, macros are commented using the exclamation point (!). For example:

```
! Programmer: Al Gore

! Also inventor of the internet

MaxArea = 205 ! Maximum Area defined here
```

It is a good idea to make as many variables as possible local variables. Too many global variables will inevitably lead to conflicts because macros overwrite the values of global variables, which are assigned values in other macros. For small applications this is not critical, but for large applications it exponentially becomes more difficult to keep track of all the global variables.

Although local variables have scope only within the function in which they reside, their value can be passed to other functions via the parameter command. For example:

```
name passer

! Macro demonstrates how to pass local variables

local compound$, numbpeaks

compound$ = "aspirin"

numbpeaks = 3

ShowLocals compound$, numbpeaks
```

```
endmacro

name ShowLocals

Parameter dummy$, numb

Print dummy$, numb

endmacro
```

Typing "passer" at the command line will execute the macro passer, which defines the local variables compound$ and numbpeaks. The macro passer then calls the macro showlocals, passing the local variables into the defined parameter variables (compound$->dummy$: numbpeaks->numb). These values are then printed to the status bar using the print command.

One point is worth mentioning for those accustomed to programming in the different forms of Basic. The print command in the Chemstation macro language does not support the operator ";" (which is usually used to separate items to be printed without putting any spaces between them). Only commas can be used to separate items to be printed, and when a comma is used as the separator it will cause the next item to be printed adjacent to the previous item, with one space separating them.

A final thought on variables in the Chemstation macro programming language is with regard to strings. In most versions of Basic, functions such as Mid$, Left$, and Right$ will return portions of strings. Unfortunately, the Chemstation macro language does not have these functions built in. However, methods to parse apart sections of strings, albeit crude, do exist.

Any portion of a string variable can be extracted and assigned to a new string variable by the use of indexing. To extract a portion of a string, enclose in brackets next to the string to be parsed the start and stop positions of the data to be extracted separated by colons — for example:

```
name$ = "Cindy Maclelland"

firstname$ = name$[1:5]
```

will set

```
firstname$ = "Cindy".
```

All strings are concatenated or appended using the "+" function — for example:

```
temp$ = "Pency " + "Prep " + "School".

temp$ = "Pency Prep School".
```

The Chemstation macro language supports the following string commands:

Chr$(number) will return the character represented by the number from an ASCII table.
Instr(str1$, str2$) returns the starting position of str2$ in str1$.
Len(string$) returns the number of characters that make up string$.
Val("10") returns the numerical version of the given string, in this case 10.
Val$(55) returns a string form of the given number.

Here is a macro that demonstrates the use of the various string functions:

```
name stringdemo

! macro will demonstrate use of string functions

! available in the Chemstation Macro Programming

s1$ = "u"
```

```
name$ = "Rudolf Schmidt"

s2$ = "5"

numb = 12

Print Chr$(65)

Print INSTR(name$,s1$)

Print LEN(name$)

Print Val(s2$)*5

temp$ = "ADAM-" + Val$(numb)

Print temp$

endmacro
```

This macro will produce the following output in the list messages window:

```
Loading Macros from D:\HPCHEM\CORE\LAB.MAC...

A

2

14

25

ADAM-12
```

Output explanation:

65 is the ASCII character code for "A".
"u" is the second character from the left in "Rudolf Schmidt".
"Rudolf Schmidt" is 14 characters long.
5 * 5 = 25. Here the string "5" is converted to a number.
"ADAM-12" The number 12 is made into the string "12" and appended to "ADAM-".

The following macro demonstrates how to extract substrings from the string named "wholestring$":

```
name showindex

! Subroutine to demonstrate string parsing

wholestring$ = "one two three"

endloc = Len(wholestring$)

one$ = wholestring$[1:3]

two$ = wholestring$[5:7]

three$ = wholestring$[9:Len(wholestring$)]

print one$

print two$

print three$
```

```
print endloc

endmacro
```

One very useful function is the SubString$ function. It will take any string that has numerous items separated by a delimiter (usually a comma or a space) and return whatever token (item number) is desired. The SubString$ function has the following parameters:

```
SubString$(String$,index,UseQuotationMarks,Delimiter)
```

For example:

```
Print Substring$("one,two,three," 2,,  ",")
```

This statement will print "two" to the command line. If the index is changed to 1, it will print "one". If the index is changed to 3, it will print "three".

Understanding how to search for and extract information from string variables is a crucial part of Chemstation macro programming because, as will be shown later, all Chemstation parameters are stored in objects and tables. Objects and tables access objects via registers, which lump parameters together delimited by some character, usually a comma. To gain the value of an individual parameter, it is often necessary to parse through a string register to get it.

7.3   USING MACROS AS FUNCTIONS AND SUBROUTINES

The macros shown previously are small and limited in usefulness. However, macros can call other macros and in this manner macros can be used as subroutines for other macros. Unfortunately, macros used as subroutines or functions are not distinguished from ordinary macros by means of any special syntax, with the exception of functions that have parentheses around passed parameters. Therefore, it is good practice to place comments at the beginning of every macro called as a subroutine to denote its purpose and place in the overall scheme of things. Every macro called as a subroutine must have a return statement for it to return upon completion to the point from which it was called.

Macros can also be called as functions that will return a single value (string or numerical) to the calling macro. Unlike macros called like subroutines, which have passed parameters following the macro name separated by commas, all passed parameters must be enclosed by parentheses and separated by commas.

The following is an example of how to call macros in both a function and subroutine type form:

```
name ShowLocals

Parameter dummy$, numb

Print dummy$,numb

return

endmacro

name callsubfuncs

! Macro demonstrates how to call other macros

! as Subroutines and Functions

! Call a subroutine

subrot1

! Call the macro "function" cube
```

```
! Note the use of parentheses around passed
! parameters when dealing with functions
result = cube(3)
Print "Three Cubed is = ",result
! Call a subroutine and pass parameters
! Note No parentheses for passed parameters
! when macros called like subroutines
ShowLocals "Three", 3
endmacro
name cube
! This macro is a function because it returns a value
Print "Macro cube called as a function"
Parameter numb
local result
result = numb * numb * numb
return result
endmacro
name subrot1
Print "Macro subrot1 called like a subroutine"
! Returns Execution to the point in the calling
! macro immediately after which it was called
return
endmacro
```

7.4 TOOLS TO AID MACRO WRITERS

The only written reference to aid Chemstation macro writers is published by Agilent (formerly HP) and titled *HP Chemstation Macro Programming Guide*. In the words of Don Grothen, the Agilent programmer who specializes in writing macros, this manual is about as clear as mud. However, some useful sample macros and tools can aid those who endeavor to write their own custom macros.

The first tool anyone should install is the macropad. This is basically a spiffed up version of notepad customized for writing, loading, and debugging macros. The file is included on the website http://www.pharmalabauto.com and is titled "macropad.exe". Unzip the file to a preferred directory (…\Hpchem\macropad\ is a good choice). The file is run by double clicking or running "macro-pad.exe". Use this environment when constructing macros.

The second tool that is an absolute must is the debugger. It too is included on the website and is named "macdebug.exe". The contents of this file can be unzipped to any folder chosen. The readme file contains important information, which someone took the time to write for a reason. To get started quickly using the debugger, place all the files with the extension (*.mac) into the folder…\hpchem\core. The debug window can be brought up by typing the following command:

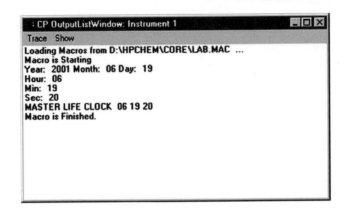

FIGURE 7.3 Using the macro debug tool to display various parameters.

"listmessages on,,400" in the command line. In fact, it is a good idea to place the code "listmessages on,,400", in the user.mac file in the core directory, which will automatically load the debugger each time the Chemstation software is opened.

The debugger is a critical tool because it allows the programmer to see what is happening through the entire run cycle of a macro. It logs all errors, actions, and messages that occur as the macro is running. Even more important, it allows the programmer to print out the values of variables at different points within the program. To illustrate this, copy the file "lab.mac" from the website http://www.pharmalabauto.com to...\hpchem\core and type the following commands from the command line:

```
macro lab.mac
```

```
datentime
```

The information depicted in Figure 7.3 should be displayed on the screen.

Several macros are contained by the lab.mac file, which is first loaded into the Chemstation's memory. The macro "datentime" is one macro in the file lab.mac. It is then executed once it is loaded. The macro "datentime" shows how to extract the date and time from the system using the system functions Date$ and Time$. The extracted information is then printed to the screen using the print command. The syntax of the "datentime" macro is:

```
name datentime

! Macro Shows how to Extract Date and Time Information

! From Date$() and Time$() Variables

local yyyy$, mo$, dd$

local hh$, mm$, ss$

local SDate$, STime$

Print "Macro is Starting"

!On Error GoTo ErrorCatch

On Error ErrorCatch

! EXTRACT DATE INFORMATION HERE INTO STRING VARIABLES

SDate$ = Date$(1)

STime$ = Time$(1)
```

```
! EXTRACT DATE PARAMETERS
yyyy$ = SDate$[7:10]
mo$ = SDate$[1:2]
dd$ = SDate$[4:5]
! EXTRACT TIME PARAMETERS
hh$ = SDate$[1:2]
mm$ = SDate$[4:5]
ss$ = SDate$[7:8]
! PRINT EXTRACTED VALUES
Print "Year: ",yyyy$, "Month: ",mo$, "Day: ",dd$
Print "Hour: ", hh$
Print "Min: ", mm$
Print "Sec: ", ss$
Print "MASTER LIFE CLOCK ",hh$,mm$,ss$
Print "Macro is Finished."
endmacro
```

7.5 REGISTERS, OBJECTS, AND TABLES

Data acquired by the Chemstation software are stored in a hierarchical fashion that descends from registers. A register comprises a contents list, a header, and one or more sections called objects. Each object has summary information followed by detailed information. Every register has a unique name, which must be used to access the information contained within the register.

Objects are one "level" below registers and are contained within registers; each object is identified by a unique number. Objects consist of headers, a data block, tables, and annotations. The Chemstation has four different types of objects: 1) chromatogram, 2) spectrum, 3) matrix, and 4) user defined.

Each chromatographic signal is loaded into a separate object. If a data file contains five different signals (each at a different wavelength from a different detector), the ChromReg register will contain five objects. ChromReg[3] would provide access to the third object. If no index is specified, the default index is 1. The ChromReg register is the working register for all chromatograms. A utility will be shown later that allows the user to access all the register, object, and table names.

Tables are contained within objects, which can contain more than one table. Tables hold data relating to the objects in which they are contained. Data can be in both string and numerical formats. Tables also have headers that contain information valid for all cells in the table, such as table dimensions, method names, etc.

In summary, data are accessed in this hierarchy: register -> object -> table. Figure 7.4 is a good illustration of how to rationalize the way that Chemstation software stores and accesses data. Each register is equivalent to a filing cabinet, and each filing cabinet has many drawers, which are equivalent to objects. Each drawer in the filing cabinet has many files, which are equivalent to tables stored within objects. Note that it is possible to have multiple registers or file cabinets.

The user now determines the names of the registers, which is required knowledge in order to extract any data. The easiest way to accomplish this is through a user-contributed macro named

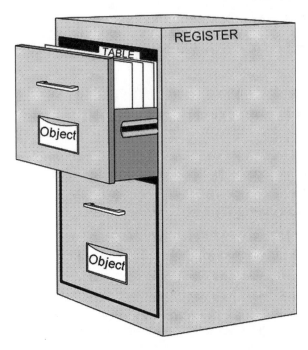

FIGURE 7.4 Understanding registers, objects, and tables.

#	Name	Objects
1	_CONFIG	6
2	_DAMETHOD	1
3	_MENU	1
4	_PREPBATCHREF	2
5	_SEQUENCE	2
6	CALVIEW	3
7	CHROMREG	5
8	CHROMRES	1
9	DADIAG	1
10	EEV_REG	1
11	FRPT_CONFIG	1
12	GUI	1
13	HOOK	1
14	NLSBATCH	1
15	NLSCORE	1
16	NLSLALS	3
17	NLSLCLU	0
18	NLSLCMNU	4
19	NLSLDAD	2
20	NLSLFLD	1
21	NLSLMWD	2

FIGURE 7.5 Using the "mactools" macro to list the registers in the system.

"mactools.mac" provided in the website http://www.pharmalabauto.com and also on the Chemstation software CD in the folder named "11". After typing "macro mactools.mac, go" at the command line, a menu will appear named "tools" under the method and run control view. Choose Tools->Registers->List Registers and the menu illustrated in Figure 7.5 will appear.

The names of all the registers are listed along with the number of objects they contain. The ChromReg register, the working register for all chromatograms, is displayed here as register

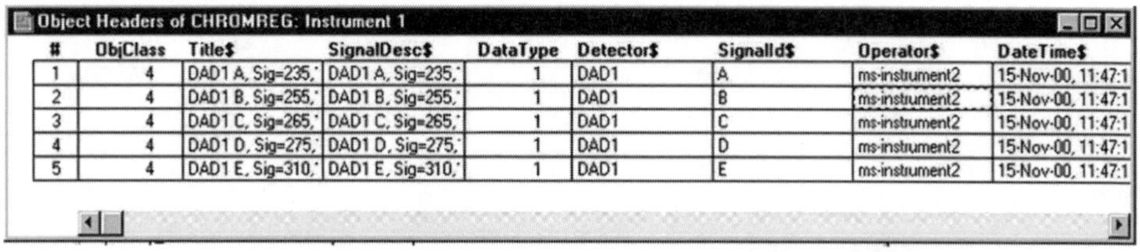

#	ObjClass	Title$	SignalDesc$	DataType	Detector$	SignalId$	Operator$	DateTime$
1	4	DAD1 A, Sig=235,	DAD1 A, Sig=235,	1	DAD1	A	ms-instrument2	15-Nov-00, 11:47:1
2	4	DAD1 B, Sig=255,	DAD1 B, Sig=255,	1	DAD1	B	ms-instrument2	15-Nov-00, 11:47:1
3	4	DAD1 C, Sig=265,	DAD1 C, Sig=265,	1	DAD1	C	ms-instrument2	15-Nov-00, 11:47:1
4	4	DAD1 D, Sig=275,	DAD1 D, Sig=275,	1	DAD1	D	ms-instrument2	15-Nov-00, 11:47:1
5	4	DAD1 E, Sig=310,	DAD1 E, Sig=310,	1	DAD1	E	ms-instrument2	15-Nov-00, 11:47:1

Object Headers of CHROMREG: Instrument 1

FIGURE 7.6 Displaying the corresponding headers for a register.

number 7. To look at the headers for all five objects in this register, click on CHROMREG and press the button List Hdrs. Figure 7.6 shows what will appear.

The SignalDesc$ header contains lots of information on the chromatogram separated by commas. If the user wanted to extract the wavelength at which each chromatogram was measured, the following macro should be used:

```
name reguse
local regno, SigDesc$, wave
Print "*** EXTRACTING WAVELENGTHS FROM EVERY CHROMATOGRAM ***"
for regno = 1 to RegSize(chromreg)
  Print "REGISTER No.",regno
  !Extract Wavelength from Signal Description
  SigDesc$ = ObjHdrText$(Chromreg[regno],"SignalDesc")
  Print "Signal Description: ", SigDesc$
  SigDesc$ = Substring$(SigDesc$, 2,, ",")
  Print "Token or Substring #2: ", SigDesc$
  wave = Val(SigDesc$[(Len(SigDesc$)-2):Len(SigDesc$)])
  Print "Wavelength: ", wave
  Print " "
next regno
endmacro
```

When this program is run, it will print the following output to the debugger window:

```
Loading Macros from D:\HPCHEM\CORE\LAB.MAC...
*** EXTRACTING WAVELENGTHS FROM EVERY CHROMATOGRAM ***
REGISTER No. 1
Signal Description: DAD1 A, Sig=235,16 Ref=600,100
Token or Substring #2: Sig=235
Wavelength: 235
REGISTER No. 2
Signal Description: DAD1 B, Sig=255,16 Ref=600,100
```

```
Token or Substring #2: Sig=255

Wavelength: 255

REGISTER No. 3

Signal Description: DAD1 C, Sig=265,16 Ref=600,100

Token or Substring #2: Sig=265

Wavelength: 265

REGISTER No. 4

Signal Description: DAD1 D, Sig=275,16 Ref=600,100

Token or Substring #2: Sig=275

Wavelength: 275

REGISTER No. 5

Signal Description: DAD1 E, Sig=310,16 Ref=600,100

Token or Substring #2: Sig=310

Wavelength: 310
```

Here the signal description string is extracted using the function ObjHdrText$(Chromreg[],"SignalDesc"). An index number must be supplied to Chromreg (i.e., Chromreg[3]) to indicate which chromatogram's signal description should be extracted.

Similarly, numerical data can be extracted from register objects by using the function ObjHdrVal(register,header).

```
name regval

! Demo Using ObjHdrVal

local endtime, SigDesc$, wave

Print "*** EXTRACTING ENDING TIME FROM EVERY CHROMATOGRAM ***"

for regno = 1 to RegSize(chromreg)

Print "REGISTER No.",regno

!Extract Wavelength from Signal Description

endtime = ObjHdrVal(Chromreg[regno],"end")

Print "Ending Time: ", endtime

Print " "

next regno

endmacro
```

When this program is run, it will print the following output to the debugger window:

```
Loading Macros from D:\HPCHEM\CORE\LAB.MAC...

*** EXTRACTING ENDING TIME FROM EVERY CHROMATOGRAM ***

REGISTER No. 1

Ending Time: 1.96233333
```

```
REGISTER No. 2

Ending Time: 1.96233333

REGISTER No. 3

Ending Time: 1.96233333

REGISTER No. 4

Ending Time: 1.96233333

REGISTER No. 5

Ending Time: 1.96233333
```

7.6   LOADING SEQUENCES INTO AN HPLC

The single most frustrating thing about working with an Agilent Chemstation HPLC is the difficulty involved in getting data into the sequence table. The sequence table holds all the information required to run a series of compounds, such as sample location, sample name, method name, for each individual sample to be run. A user can type the information in cell by cell, which works acceptably for a limited number of samples. The difficulty arises when the user wants to run 100 samples at a time, a task meant to be done by the HPLC with its autosampler.

Most researchers receive the data on the compounds they need to run in some type of spreadsheet format. The most intuitive thing to do would be to select a column of data in a spreadsheet and copy it into the sequence table. Unfortunately, the Chemstation does not support pasting multiple rows or columns into a sequence table and there is no method to import any type of file into the sequence table from the file menu. This leaves the user with two choices: type everything in by hand (not really an option for large numbers of samples), or cut and paste individual cells of data (also not really an option). To operate in a high throughput capacity, neither of these options is practical. However, it is possible to construct a macro to load a delimited file directly into a sequence table. To construct such a macro, it is necessary to know the type and names of the columns in the sequence table. Table 7.1 provides this information.

The following series of macros will import a comma-delimited file into the Agilent Chemstation sequence table (Figure 7.7). Order is important when the comma-delimited file is created, but the macro can be adjusted to reflect any ordering the user desires.

```
!!! file name: Lab_ls.mac

!!! macro name: LOADSEQ

!!! Generic Algorithm to Load Comma Delimited

!!! Sequence into Agilent Chemstation HPLC

name LOADSEQ

local cur_import_file$, sel_path$

local line_ctr, cur_line$, tot_len_cur_line

local msg$

!!! Extracted Parameters from Comma Delimited File

local cur_vial$, cur_sample$, cur_method$, cur_injpervial$,
cur_samptype$

local cur_istd$, cur_comment$, line_no$, hdrtxt$, cur_injvol$
```

TABLE 7.1
Column Information for Sequence Table

Predefined Columns in Sequence Table

Column Name	Type/Range	Meaning
Vial	Integer	The vial number of the run. −1 means blank run
Method	String 8 characters	Method to use
CalLine	String 1 characters	Sequence recalibration table line
InjVial	Integer	Number of injections for this vial
InjVolume	String 10 characters	Injection volume (not used for CE)
SampleName	String 16 characters	Sample name
Amount	String 12 characters	Sample amount
Multiplier	String 12 characters	Multiplier
ISTDAmt	String 12 characters	ISTD amount
Dilution	String 12 characters	Dilution factor
SampleInfo	String 4095 characters	Sample information (multiline text field, stored as a single text field, CR/LF separates lines)
SampleType	Integer	Identifies what type of sample is being defined on this sequence table line 1 Sample run 3 Calibration sample
CalLevel	Integer	Calibration level to be updated; valid only for SampleType = 3
UpdateRF	Integer	Defines how response factors will be updated in the calibration table when this calibration run occurs; valid only for SampleType = 3 0 No update 1 Average 2 Replace 3 Bracket
UpdateRT	Integer	Defines how response (or migration) times will be updated in the calibration table when this calibration run occurs; valid only for SampleType = 3. 0 No update 1 Average 2 Replace 3 Bracket
Interval	Integer	Defines the number of sample runs that will occur before this calibration is rerun. Leave blank for explicit (noncyclic) calibrations. Valid only for SampleType = 3
DataFileName	String 8 characters	Overrides the automatically generated data filename with the filename entered here

Source: HP Chemstation Macro Programming Guide, Hewlett-Packard, Palo Alto, CA, 1995. With permission.

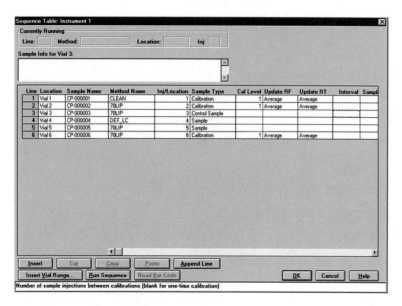

FIGURE 7.7 An HPLC sequence table.

```
local ErrorMessage$

echo = 1

On Error return

!!! confirm sequence deletion

if 0 = alert("OK to delete current sequence?",4,"Confirm
sequence deletion")

   message "User cancelled CSV importation."

   return

endif

!!! delete current sequence table

if 0 <> TABHDRVAL(_sequence[1],seqtable1,"numberofrows") then

   deltabrow _sequence[1],seqtable1,1:-1

endif

cur_import_file$=""

!!! get previously selected dir stored in win.ini

sel_path$ = ProfileString$("LCMS_UCL", "SEQ7LDDIR", "C:\")

!!! select CSV file

if 0 = selectfile(5,"SEQ7LD File
Import","*.csv",sel_path$,cur_import_file$) then

   message "No CSV sequence file selected!"

   return
```

```
endif
!!! save new dir in win.ini
SetProfileString "LCMS_UCL", "SEQ7LDDIR", sel_path$
cur_import_file$ = sel_path$ + cur_import_file$
line_ctr = 1 !!! counter for lines read
!!! open file for line-by-line read
open cur_import_file$ for input as #6
message cur_import_file$
On Error GoTo EndOfFile
!!! input #6,cur_line$ !!! read, discard header line
!!! read line-by-line from file, at eof goto EndOfFile
while line_ctr < 999 !!! max lines for sequence table
    ErrorMessage$ = "Format error in line: "
    On Error GoTo EndOfFile !!! On Error for file reads
    input #6, cur_line$
    msg$ = "Reading line: " + Val$(line_ctr)
    message msg$
    msg$ = "Contents: " + cur_line$
    message msg$
    On Error GoTo TheErroneousEnd !!! On Error for parsing data
    !!! Parse out substrings from comma delimited data
    cur_vial$ = Substring$(cur_line$, 1,, ",")
    message cur_vial$
    cur_sample$ = Substring$(cur_line$, 2,, ",")
    message cur_sample$
    cur_method$ = Substring$(cur_line$, 3,, ",")
    message cur_method$
    cur_injpervial$ = Substring$(cur_line$, 4,, ",")
    message cur_injpervial$
    cur_samptype$ = Substring$(cur_line$, 5,, ",")
    message cur_samptype$
    cur_istd$ = Substring$(cur_line$, 6,, ",")
    message cur_istd$
    cur_comment$ = Substring$(cur_line$, 7,, ",")
```

```
        message cur_comment$
        cur_injvol$ = Substring$(cur_line$, 8,, ",")
        message cur_injvol$
        !!! set sequence table values
        ErrorMessage$ = "Value error in line: "
        ! Insert a new Row in the Sequence Table
        InsTabRow _sequence[1], seqtable1
        ! Write Comma Delimited Information Extracted into Sequence
        Table
        SetTabval _sequence[1], seqtable1, line_ctr, vial,
        val(cur_vial$)
        SetTabtext _sequence[1], seqtable1, line_ctr, samplename,
        cur_sample$
        SetTabtext _sequence[1], seqtable1, line_ctr, method,
        cur_method$
        SetTabval _sequence[1], seqtable1, line_ctr, injvial,
        val(cur_injpervial$)
        SetTabval _sequence[1], seqtable1, line_ctr, sampletype,
        val(cur_samptype$)
        SetTabtext _sequence[1], seqtable1, line_ctr, istdamt,
        cur_istd$
        SetTabtext _sequence[1], seqtable1, line_ctr, sampleinfo,
        cur_comment$
        SetTabtext _sequence[1], seqtable1, line_ctr, injvolume,
        cur_injvol$
        ! Increment the line counter
        line_ctr=line_ctr+1
    endwhile
    EndOfFile:
    close #6
    print "Sequence imported."
    return
    TheErroneousEnd:
    print ErrorMessage$, line_ctr
    close #6
    return
    endmacro
```

```
name message
!!! If echo = 1 then print messages to screen
parameter msg$
If echo = 1 then
    Print msg$
endif
endmacro
name SLOADmenu
!!! adds menu item to Method and RunControl view menu.
if "TOP" = Uppercase$(TabHdrText$ (_CONFIG[1],Window,
"CurrentView"))
    menuadd "&LoadSeq"
    menuadd "&LoadSeq","Import a &CSV File",LOADSEQ,,"Import a
    Comma Delimited File"
EndIf
endmacro
!!! the following is automatically executed if the macro is
manually
!!! loaded with the ",go" parameter; or if the macro is
automatically
!!! loaded from the user.mac file
SetHook "PreViewMenu","SLOADmenu"
```

Notice the macro "message". When in the process of modifying or debugging a macro, it is often convenient to have different parameters printed out to the debug window so that the programmer can keep track of what is going on with all the different variables at different points in the macro's execution. However, when the macro is run on a day-to-day basis, it is undesirable to have all these messages cluttering up the screen. Macros are usually in a constant state of modification, so it is often counterproductive to remove print statements that programmers put in when debugging because they will no doubt be debugging the program again soon.

The message macro gives the programmer the best of both worlds. The variable echo can be set = 0 before the program is distributed, and no debugging comments will be printed out. When the macro is to be modified, only one line needs to be changed to allow the debugging comments to print out again. With echo = 1, and using the test file labls.csv, the following comments will be printed to the debug window:

```
C:\LABLS.CSV
Reading line: 1
Contents: 1,DMSO,clean,5,1,2,No sample just DMSO,1.2
1
DMSO
```

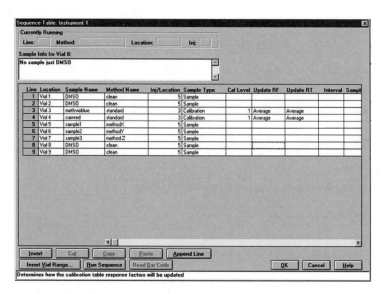

FIGURE 7.8 Filling the sequence table with data via a macro.

```
clean

5

1

2

No sample just DMSO

1.2
```

This sequence will continue to be printed to the debug window for as many lines (or samples) are imported. When the macro is finished, the sequence table will be filled with the data shown in Figure 7.8, which illustrates a typical sequence. The first and last two samples are solvent run under a cleaning method to ensure that the column is clean. The third and fourth samples are standards to check that the instrument is calibrated properly before the actual sample compounds are run, which consists of lines 5, 6, and 7. By clicking on any line, the comments loaded will appear for that sample at the "sample info for vial X" textbox at the top of the form.

Some additional commands need to be explained in order to understand how this macro works, and they specifically deal with accessing and setting data in tables. Just as specific commands were used earlier to access information in objects, specific commands are required to set and access information in tables as well. The following commands are used to populate table entries in the Chemstation software:

- InsTabRow inserts a new row in the sequence table for each line of comma-delimited data imported
- SetTabtext sets a single text element in the table
- SetTabval sets a single numeric element in a table

Keep in mind that the parameters necessary to use each of these commands can be accessed using the show command at the command line, for instance:

```
InsTabRow RegObj, TabName, RowRange

SetTabtext RegObj, TabName, RowIndex, ColName, Text

SetTabval RegObj, TabName, RowIndex, ColName, Value
```

The following commands are used to extract existing table values in the Chemstation software. (They are analogous to the ObjHdrText$ and ObjHdrVal used with objects.)

- TabVal extracts a single numeric object from a table

  ```
  TabVal RegObj, TabName, RowIndex, ColName
  ```

- TabText$ extracts a single string element from a table

  ```
  TabText RegObj, TabName, RowIndex, ColName
  ```

7.7   LOADING SEQUENCES THAT CONTAIN BOTH PLATES AND VIALS

The preceding macro works fine if the samples to be run are stored in vials. However, if the samples are stored in microplates, this macro will not work. Unlike when a vial number is used, a plate has three identification parameters that determine the position of the compound in the autosampler. A plate number, (typically P1 or P2), tells the Chemstation which plate the sample resides in. A row identifier (A ~ H) tells the Chemstation which row number in the plate the sample resides in (A = row 1, B = row 2, etc.). A column identifier tells the Chemstation which column number in the plate the sample resides in. For example:

Vial Ident	Plate	Row	Col
P1-A-01	1	1	1
P1-E-09	1	5	9
P2-H-12	2	8	12

A revised macro capable of dealing with vial positions in plate type format is illustrated next:

```
!!! file name: Lab_ls2.mac

!!! macro name: LoadVorP

!!! Specific Algorithm to Load Comma Delimited

!!! Sequence into Agilent Chemstation HPLC

!!!

!!! Handles the Agilent 1100 Well Plate Sampler (WPS)

!!! which encodes the plate, row and column

!!! designators as an integer "vial" number.

!!!

!!! The vial number is now labeled "Location" in

!!! ChemStation to indicate that both vials and

!!! well plates are supported.

!!!
```

```
!!! The vial or location is from 1-10 for vials
!!! in the WPS dual-plate, 10 vial tray. For well
!!! plates, the vial or location is coded as
!!! (PlateNum * 4096) + (Row - 1) * 64 + (Col -1)
Name GetLocNum
!!! Function to calculate sample position in plates !!!
Parameter cur_vial$
Local sPlate$, sCol$, sRow$, sEndChr$
Local nLocNum
!!! If current location in vial format then exit function !!!
If Len(cur_vial$) < 4 and InStr(cur_vial$,"-") = 0 then
   return Val(cur_vial$)
endif
!!! The Current Location is in plate format. Calculate Position
!!!
! Convert to upper case
cur_vial$ = convertText$(upper, cur_vial$)
! get column designator from end of string, i.e., "1"..."99"
sCol$ = ""
sEndChr$ = cur_vial$[Len(cur_vial$):Len(cur_vial$)]
While Instr("0123456789", sEndChr$) > 0
   sCol$ = sEndChr$ + sCol$
   cur_vial$ = cur_vial$[1:Len(cur_vial$) - 1]
   sEndChr$ = cur_vial$[Len(cur_vial$):Len(cur_vial$)]
EndWhile
! eliminate miscellaneous separatorsWhile asc(sEndChr$) < 65
or asc(sEndChr$) > 90
   cur_vial$ = cur_vial$[1:Len(cur_vial$) - 1]
   sEndChr$ = cur_vial$[Len(cur_vial$):Len(cur_vial$)]
EndWhile
! get row designator from end of string, i.e., "A"..."Z"
sRow$ = ""
While Asc(sEndChr$) > 64 And Asc(sEndChr$) < 91
   sRow$ = sEndChr$ + sRow$
   cur_vial$ = cur_vial$[1:Len(cur_vial$) - 1]
```

```
    sEndChr$ = cur_vial$[Len(cur_vial$):Len(cur_vial$)]
EndWhile
! eliminate miscellaneous separators
While (Asc(sEndChr$) < 65 Or Asc(sEndChr$) > 90) \
    and (Instr("0123456789", sEndChr$) = 0)
    cur_vial$ = cur_vial$[1:Len(cur_vial$) - 1]
    sEndChr$ = cur_vial$[Len(cur_vial$):Len(cur_vial$)]
EndWhile
! get plate designator, i.e., "1" or "2"
sPlate$ = ""
While Instr("0123456789", sEndChr$) > 0
    sPlate$ = sEndChr$ + sPlate$
    cur_vial$ = cur_vial$[1:Len(cur_vial$) - 1]
    sEndChr$ = cur_vial$[Len(cur_vial$):Len(cur_vial$)]
EndWhile
nLocNum = val(sPlate$)*4096
nLocNum = nLocNum + (asc(sRow$) - 65) * 64
nLocNum = nLocNum + val(scol$) - 1
print "plate - col - row - endchar - nLocNum"
print sPlate$, sCol$, sRow$, sEndChr$, nLocNum
return nLocNum
EndMacro
name LoadVorP
local cur_import_file$, sel_path$
local line_ctr, cur_line$, tot_len_cur_line
local msg$
!!! Extracted Parameters from Comma Delimited File
local cur_vial$, cur_sample$, cur_method$, cur_injpervial$,
cur_samptype$
local cur_istd$, cur_comment$, line_no$, hdrtxt$, cur_injvol$
local ErrorMessage$
echo = 1
On Error return
!!! confirm sequence deletion
```

```
if 0 = alert("OK to delete current sequence?",4,"Confirm
sequence deletion")

    message "User cancelled CSV importation."

    return

endif

!!! delete current sequence table

if 0 <> TABHDRVAL(_sequence[1],seqtable1,"numberofrows") then

    deltabrow _sequence[1],seqtable1,1:-1

endif

cur_import_file$=""

!!! get previously selected dir stored in win.ini

sel_path$ = ProfileString$("LCMS_UCL", "SEQ7LDDIR", "C:\")

!!! select CSV file

if 0 = selectfile(5,"SEQ7LD File
Import","*.csv",sel_path$,cur_import_file$) then

    message "No CSV sequence file selected!"

    return

endif

!!! save new dir in win.ini

SetProfileString "LCMS_UCL", "SEQ7LDDIR", sel_path$

cur_import_file$ = sel_path$ + cur_import_file$

line_ctr = 1 !!! counter for lines read

!!! open file for line-by-line read

open cur_import_file$ for input as #6

message cur_import_file$

On Error GoTo EndOfFile

!!! input #6,cur_line$ !!! read, discard header line

!!! read line-by-line from file, at eof goto EndOfFile

while line_ctr < 999 !!! max lines for sequence table

    ErrorMessage$ = "Format error in line: "

    On Error GoTo EndOfFile !!! On Error for file reads

    input #6, cur_line$

    msg$ = "Reading line: " + Val$(line_ctr)

    message msg$

    msg$ = "Contents: " + cur_line$
```

```
message msg$

On Error GoTo TheErroneousEnd !!! On Error for parsing data

!!! Parse out substrings from comma delimited data

cur_vial$ = Substring$(cur_line$, 1,, ",")

! Call Function getLocNum to determine vial position

cur_vial$ = val$(getLocNum(cur_vial$))

message cur_vial$

cur_sample$ = Substring$(cur_line$, 2,, ",")

message cur_sample$

cur_method$ = Substring$(cur_line$, 3,, ",")

message cur_method$

cur_injpervial$ = Substring$(cur_line$, 4,, ",")

message cur_injpervial$

cur_samptype$ = Substring$(cur_line$, 5,, ",")

message cur_samptype$

cur_istd$ = Substring$(cur_line$, 6,, ",")

message cur_istd$

cur_comment$ = Substring$(cur_line$, 7,, ",")

message cur_comment$

cur_injvol$ = Substring$(cur_line$, 8,, ",")

message cur_injvol$

!!! set sequence table values

ErrorMessage$ = "Value error in line: "

! Insert a new Row in the Sequence Table

InsTabRow _sequence[1], seqtable1

! Write Comma Delimited Information Extracted into Sequence
Table

SetTabval _sequence[1], seqtable1, line_ctr, vial,
val(cur_vial$)

SetTabtext _sequence[1], seqtable1, line_ctr, samplename,
cur_sample$

SetTabtext _sequence[1], seqtable1, line_ctr, method,
cur_method$

SetTabval _sequence[1], seqtable1, line_ctr, injvial,
val(cur_injpervial$)
```

```
        SetTabval _sequence[1], seqtable1, line_ctr, sampletype,
        val(cur_samptype$)

        SetTabtext _sequence[1], seqtable1, line_ctr, istdamt,
        cur_istd$

        SetTabtext _sequence[1], seqtable1, line_ctr, sampleinfo,
        cur_comment$

        SetTabtext _sequence[1], seqtable1, line_ctr, injvolume,
        cur_injvol$

        ! Increment the line counter
        line_ctr=line_ctr+1
    endwhile
    EndOfFile:
    close #6
    print "Sequence imported."
    return

    TheErroneousEnd:
    print ErrorMessage$, line_ctr
    close #6
    return
endmacro

name message
!!! If echo = 1 then print messages to screen
parameter msg$
If echo = 1 then
    Print msg$
endif
endmacro

name VOPmenu
!!! adds menu item to Method and RunControl view menu.
if "TOP" = Uppercase$(TabHdrText$ (_CONFIG[1],Window,
"CurrentView"))
    menuadd "&LoadVorP"
    menuadd "&LoadVorP","Import a &CSV File",LoadVorP,,"Import
    a Comma Delimited File"
EndIf
endmacro
```

```
!!! the following is automatically executed if the macro is
manually

!!! loaded with the ",go" parameter; or if the macro is
automatically

!!! loaded from the user.mac file

SetHook "PreViewMenu","VOPmenu"
```

In the preceding series of macros, the macro GetLocNum is used as a function to determine the location of the sample in the microplate. The location of the sample is found by means of the formula: = (PlateNum * 4096) + (Row - 1) * 64 + (Col -1). In the beginning of the GetLocNum macro, it checks the parameter passed to it to ensure that it is not in vial format, which would trigger an error in the macro function. The previous macro can be tested with the file "labls2.csv", which will yield the following output in the debug window:

```
Loading Macros from D:\HPCHEM\CORE\LAB_LS2.MAC...

C:\LABLS2.CSV

Reading line: 1

Contents: P1-A-01,DMSO,clean,5,1,2,No sample just DMSO,1.2

plate - col - row - endchar - nLocNum

1 01 A P 4096

4096

DMSO

clean

5

1

2

No sample just DMSO

1.2
```

This output sequence will be repeated for every sample (or line of data) loaded. Running the macro will produce the sequence table shown in Figure 7.9. As can be seen in the graphic, the location field in the spreadsheet stores locations of vials and of plates (although typically both are not used simultaneously in one run).

7.8 EXPORTING DATA AFTER EACH SAMPLE IS ANALYZED

Although macros can be run from the command line or by means of a menu, their true utility comes to light when they can automatically be included as part of a process or method. Macros can be selected to run at the beginning or the end of a process, and a special macro can even be utilized for the data analysis process. By selecting the menu Method->Run Time checklist, the GUI illustrated in Figure 7.10 will appear.

By checking the appropriate checkbox, and placing the name of the macro to be run in the corresponding textbox, a user-developed macro can be set to execute at various times during the

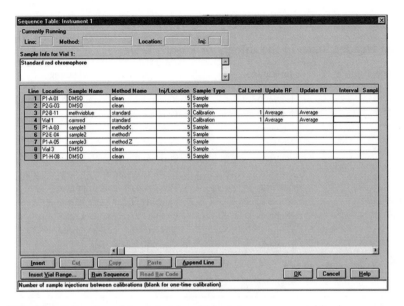

FIGURE 7.9 Loading plate and vial positions into a sequence table via a macro.

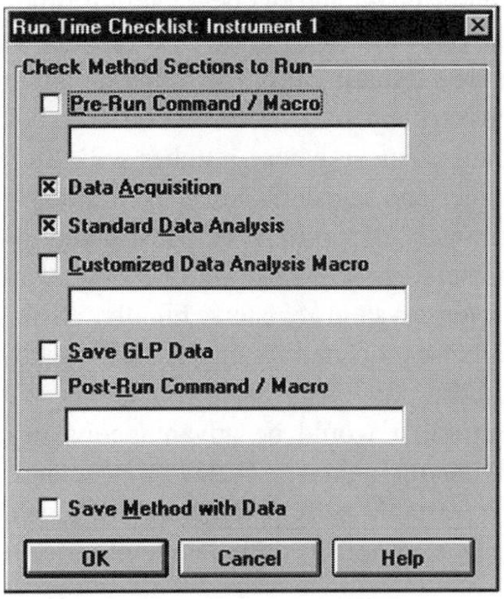

FIGURE 7.10 Using the run time checklist to execute macros.

method execution. The most lucrative of the three options in Figure 7.10 is to have a method execute postrun or after the analysis of each compound.

A postrun macro can be used to export any data from the run that will later be analyzed to determine a result. In this way, the crucial data required to produce the desired results will be stored after each compound is run. Should a run crash, the data for the compounds that have already undergone analysis will not be lost. In addition, if a comma-delimited file is used to export the important parameters, it can later be analyzed in any software package such as Excel, Origin, SAS, JMP, etc.

The results generated by the Chemstation's analysis are stored in the intresults table. Table 7.2 shows the column names of the results stored in the intresults table.

TABLE 7.2
Predefined Header Items in the Integrator Results Table

Column Name	Type/Range	Meaning
RetTime	Numeric	Retention/migration time (top of peak) in minutes
Area	Numeric	Peak area
Height	Numeric	Peak height
Width	Numeric	Peak width
Symmetry	Numeric	Peak symmetry
Baseline	Numeric	Baseline height at retention/migration time
TimeStart	Numeric	Time of peak start in minutes
LevelStart	Numeric	Signal height at peak start time
BaselineStart	Numeric	Baseline height at peak start time
TimeEnd	Numeric	Time of peak end in minutes
LevelEnd	Numeric	Signal height at peak end time
BaselineEnd	Numeric	Baseline height at peak end time

Source: HP Chemstation Macro Programming Guide, Hewlett-Packard, Palo Alto, CA, 1995. With permission.

It might be advantageous to export the following information about every peak observed in a comma-delimited format after each compound is successfully analyzed: (1) peak starting time, (2) peak retention time, (3) area of peak, and (4) peak ending time.

To extract this information and write it to a file in a manner such that it can be analyzed by means of another program can be a daunting task for someone unfamiliar with Chemstation macro programming. Many things need to happen in the correct order to be able to extract this information to a file. First, the fixed parameters (those whose number will not vary for each compound) must be extracted from system variables and registers. Next, the number of wavelengths used to analyze each compound must be determined. A loop must be constructed to loop through each wavelength and extract the variable parameters (those whose number will vary for each compound, such as peak information) from the various system registers. Finally, all this information must be written to a comma-delimited file that can be opened by any software analysis package.

It does little good to export the previously listed four items without any means of identifying where they came from. Therefore, it would be advantageous to create a header for each line consisting of: (1) filename, (2) samp_name$, (3) date_time$, and (4) method. After the header information is written, the four variable peak parameters are repeated for each peak in the chromatogram. The following macro accomplishes this task quite nicely:

```
!

!

!Programmer: Brian Bissett

!Function: Macro appends Compound information followed by
retention time

! and area to excel compatable *.csv file for later analysis.

!

!

!System requirements: Chemstation ver. a.01.00 or higher;

! Windows NT 4.0
```

```
!
!
!******************* Master Function Below *****************
Name PostRun
! Macro creates a comma delimited line of selected data to
be exported
! After the analysis of each individual compound in a
Chemstation run.
! Data is written to c:\temp\datafile.csv
! Declare Local Variables
local file_name$, samp_name$, date_time$
local totalreg, regno
local fixedpams$, num_peaks, SigDesc$, wave
local ret_time, ret_time$, area, area$
local start_time, start_time$, end_time, end_time$
local c$, variablepams$, dataline$
local loop_ctr,peak_get
local path$, fullpath$, filename$
! Trap all Errors
on Error GotoErrorTrap
! Print Debug Messages, set echo = 1
echo = 1 ! Print Debug Messages
! echo = 0 ! Suppress Debug Messages
! Comma Delimiter for File - c$ = ","
c$ = Chr$(44)
!*************************************************************
!* DATA ACQUISTION FROM SYSTEM VARIABLES AND REGISTERS OCCURS
HERE *
!*************************************************************
! obtain chrom info from chromreg
file_name$=objhdrtext$(chromreg[1],"rawdatafile")
samp_name$=objhdrtext$(chromreg[1],"samplename")
date_time$=objhdrtext$(chromreg[1],"datetime")
! Print Info Obtained from chromreg
messages "File Name: ",file_name$
```

```
messages "Sample Name: ",samp_name$

messages "Date/Time: ", date_time$

! Determine the Number of Wavelengths used by Counting the
Objects in chromreg registers

totalreg = RegSize(chromreg)

messages "Total Registers: ", Val$(totalreg)

! The METHOD can be extracted using the string system variable
_METHFILE$

messages "Method: ", _METHFILE$

!The First Four Parameters (Columns) of the ouput file are
Fixed (Always Same Number)

fixedpams$ = file_name$ + c$ + samp_name$ + c$ + date_time$
+ c$ + _METHFILE$ + c$

messages "Fixed Parameters: ", fixedpams$

! Loop through Each Register (Wavelength) and acquire both

! its retention times and areas

for regno = 1 to totalreg

    loop_ctr = 0 ! Reset the Loop Counter used in while loop below

    ! The Number of Peaks in the Intresults Table is determined
    by

    ! the number of rows the Intresults Table has. They are
    Equivilant.

    num_peaks=tabhdrval(chromreg[regno],"intresults
    ","numberofrows")

    !Extract Wavelength from Signal Description

    SigDesc$ = ObjHdrText$(Chromreg[regno],"SignalDesc")

    messages "Signal Description: ", SigDesc$

    SigDesc$ = Substring$(SigDesc$, 2,, ",")

    wave = Val(SigDesc$[(Len(SigDesc$)-2):Len(SigDesc$)])

    messages "Wavelength: ", Val$(wave)

            ! main loop gets peak information from the
                intresults table in chromreg

        ! While loop writes retention times and areas to
                    variablepams$

        ! Use tvpams$ as temporary storage to acquire
            all RT & Areas for a given Register

                while loop_ctr<num_peaks do
```

```
        peak_get = loop_ctr+1

        ! Extract Starting Time from chromreg

        start_time=tabval(chromreg[regno],"intresults",
        peak_get,"timestart")

        start_time$=val$(start_time)

        ! Extract Retention Time from chromreg

           ret_time=tabval(chromreg[regno],"intresults",
                      peak_get,"rettime")

                    ret_time$=val$(ret_time)

        ! Extract Area from chromreg

            area=tabval(chromreg[regno],"intresults",
                      peak_get,"area")

        area$=val$(area)

        ! Extract Ending Time from chromreg

        end_time=tabval(chromreg[regno],"intresults",
        peak_get,"timeend")

        end_time$=val$(end_time)

        tvpams$=tvpams$ + start_time$ + c$ + ret_time$ + c$ +
        area$ + c$ + end_time$ + c$

        loop_ctr=loop_ctr+1

                          endwhile

           ! Store the data from this register in
                      variablepams$

            variablepams$ = variablepams$ + tvpams$

      messages "Register No. ", Val$(regno)

      messages "Variable Parameters: ", variablepams$

         tvpams$ = "" ! Reset tvpams$ for next register
    next regno
    ! Create Line of Data to Append
    dataline$ = fixedpams$ + variablepams$
    messages "Line of Data: ", dataline$
    !**********************************************************
    !* DATA EXPORT OCCURS HERE *
    !**********************************************************
    ! Append this line of Data to the File
    ! Change Export Path here
```

```
path$ = "c:\Temp\"

filename$ = "DataFile.csv"

! Create full path for export

fullpath$ = path$ + filename$

! Declare Global Variable for Write Error

Write_Error$ = "False"

messages "FullPath: ",fullpath$

Open fullpath$ For Append as #3

   print #3, dataline$

close #3

If Write_Error$ = "False" then

     Print samp_name$, " Data Exported to: ",fullpath$

                              else

   Print "*** ERROR WRITING ", samp_name$, " Data !!! ***"

endif

EndMacro

!**************** Daughter Functions Follow
*************************

Name GotoErrorTrap

!!! This macro executes whenever an Error is encountered !!!

!!! Print to Debug Window the Occurrence of any Error !!!

Print "Error No. ", _Error, " has Occurred when executing
Command ", _ERRCMD$

!!! PostRun should only Respond To File Writing Errors - IGNORE
OTHERS !!!

If _Error = 41308 and _ErrCmd$ = "CLOSE" then

                 Write_Error$ = "True"

endif

Return

EndMacro

name messages

!!! If echo = 1 then print messages to screen

parameter msg1$, msg2$

If echo = 1 then

Print msg1$, msg2$
```

```
endif

endmacro
```

If the user sets the global variable echo = 1, comments such as these will be printed to the debugging window as the program is run:

```
Loading Macros from D:\HPCHEM\CORE\LAB_POST.MAC...

File Name: D:\MACROS\MARINA\VB\HPLCPLOT\TEST\111500\1CK-1401.D

Sample Name: CP-650903

Date/Time: 15-Nov-00, 11:47:10

Total Registers: 5

Method: DEF_LC.M

Fixed Parameters:
D:\MACROS\MARINA\VB\HPLCPLOT\TEST\111500\1CK-1401.D,CP-
650903,15-Nov-00, 11:47:10,DEF_LC.M,

Signal Description: DAD1 A, Sig=235,16 Ref=600,100

Wavelength: 235

Register No. 1

Variable Parameters:
0.242333,0.301206,1150.622925,0.372189,0.372189,0.394567,20.97
9219,0.481114

Signal Description: DAD1 B, Sig=255,16 Ref=600,100

Wavelength: 255

Register No. 2

Variable Parameters:
0.242333,0.301206,1150.622925,0.372189,0.372189,0.394567,20.97
9219,0.481114

Signal Description: DAD1 C, Sig=265,16 Ref=600,100

Wavelength: 265

Register No. 3

Variable Parameters:
0.242333,0.301206,1150.622925,0.372189,0.372189,0.394567,20.97
9219,0.481114

Signal Description: DAD1 D, Sig=275,16 Ref=600,100

Wavelength: 275

Register No. 4

Variable Parameters:
0.242333,0.301206,1150.622925,0.372189,0.372189,0.394567,20.97
9219,0.481114

Signal Description: DAD1 E, Sig=310,16 Ref=600,100
```

```
Wavelength: 310

Register No. 5

Variable Parameters:
0.242333,0.301206,1150.622925,0.372189,0.372189,0.394567,20.97
9219,0.481114

Line of Data: D:\MACROS\MARINA\VB\HPLCPLOT\TEST\111500\1CK-
1401.D,CP-650903,15-
Nov00,11:47:10,DEF_LC.M,0.242333,0.301206,1150.622925,0.372189
,0.372189,0.394567,20.979219,0.481114

FullPath: c:\Temp\DataFile.csv

CP-650903 Data Exported to: c:\Temp\DataFile.csv
```

As the output shows, this compound was analyzed at five wavelengths: 235, 255, 265, 275, and 310 nm. Hence, a total of five registers are used to store the data. The only shortcoming of this macro is that it exports the following information for every peak. This is a shortcoming because some peaks are obviously impurities, so including them in the output file will lead to unnecessary time allocated to filtering them out.

A very common criterion in analysis is to look at the size and shape of the peak in the chromatogram. For instance, a peak short (in height) and broad (in time) is usually an impurity (sometimes a carryover in the column from the previous run), which should not be included when exporting data. In most instances, a desirable peak is tall (in height) and short (in time) as shown in Figure 7.11.

To filter out impurities it would be desirable to omit from the data file peaks that are broad (having a total time start to finish in excess of a t_{max}) and short (having a height smaller than h_{min}). To accomplish this, data from the Height, TimeStart, and TimeEnd columns in the integrator results table would be used to qualify whether a peak fits a specific criterion for export. The previous

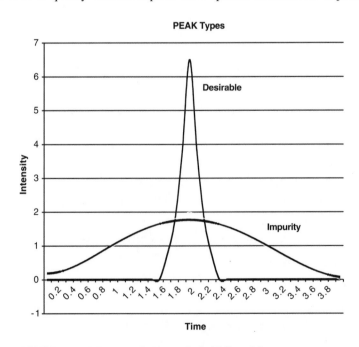

FIGURE 7.11 What constitutes a "desirable" peak?

macro can be modified to export only data that meet these requirements. The following macro is stored under "lab_pst2.mac" on the website http://www.pharmalabauto.com.

```
!

!

!Programmer: Brian Bissett

!Function: Macro appends Compound information followed by
retention time

! and area to excel compatible *.csv file for later analysis.

!

!

!System requirements: Chemstation ver. a.01.00 or higher;

! Windows NT 4.0

!

!

!************************************************************

!* MODIFIED TO ONLY EXPORT DATA WITH HEIGHT > hmin and Time
< tmax *

!************************************************************

!***************** Master Function Below
*************************

Name PostRun

! Macro creates a comma delimited line of selected data to
be exported

! After the analysis of each individual compound in a
Chemstation run.

! Data is written to c:\temp\datafile.csv

! Declare Local Variables

local file_name$, samp_name$, date_time$

local totalreg, regno

local fixedpams$, num_peaks, SigDesc$, wave

local ret_time, ret_time$, area, area$

local start_time, start_time$, end_time, end_time$

local c$, variablepams$, dataline$

local loop_ctr,peak_get

local path$, fullpath$, filename$

! New Parameters to determine Export Eligibility
```

```
local hmin ! minimum height of peak for export
local hght ! height of current peak
local tmax ! maximum time interval for export
local deltat ! duration of time interval for current peak
! Set Export Parameters
hmin = 100
tmax = 0.3
! Trap all Errors
on Error GotoErrorTrap
! Print Debug Messages, set echo = 1
echo = 1 ! Print Debug Messages
! echo = 0 ! Suppress Debug Messages
! Comma Delimiter for File - c$ = ","
c$ = Chr$(44)
!*************************************************************
!* DATA ACQUISTION FROM SYSTEM VARIABLES AND REGISTERS OCCURS
HERE *
!*************************************************************
! obtain chrom info from chromreg
file_name$=objhdrtext$(chromreg[1],"rawdatafile")
samp_name$=objhdrtext$(chromreg[1],"samplename")
date_time$=objhdrtext$(chromreg[1],"datetime")
! Print Info Obtained from chromreg
messages "File Name: ",file_name$
messages "Sample Name: ",samp_name$
messages "Date/Time: ", date_time$
! Determine the Number of Wavelengths used by Counting the
Registers in chromreg
totalreg = RegSize(chromreg)
messages "Total Registers: ", Val$(totalreg)
! The METHOD can be extracted using the string system variable
_METHFILE$
messages "Method: ", _METHFILE$
!The First Four Parameters (Columns) of the output file are
Fixed (Always Same Number)
```

```
fixedpams$ = file_name$ + c$ + samp_name$ + c$ + date_time$
+ c$ + _METHFILE$ + c$

messages "Fixed Parameters: ", fixedpams$

! Loop through Each Register (Wavelength) and acquire both

! its retention times and areas

for regno = 1 to totalreg

loop_ctr=0 ! Reset the Loop Counter used in while loop below

! The Number of Peaks in the Intresults Table is determined
by

! the number of rows the Intresults Table has. They are
Equivalent.

num_peaks= tabhdrval(chromreg[regno],"intresults","numberofrows")

!Extract Wavelength from Signal Description

SigDesc$ = ObjHdrText$(Chromreg[regno],"SignalDesc")

messages "Signal Description: ", SigDesc$

SigDesc$ = Substring$(SigDesc$, 2,, ",")

wave = Val(SigDesc$[(Len(SigDesc$)-2):Len(SigDesc$)])

messages "Wavelength: ", Val$(wave)

          ! main loop gets peak information from the
                    intresults table in chromreg

     ! While loop writes retention times and areas
                    to variablepams$

     ! Use tvpams$ as temporary storage to acquire
          all RT & Areas for a given Register

                while loop_ctr<num_peaks do

  peak_get=loop_ctr+1

  ! Extract Starting Time from chromreg

  start_time=tabval(chromreg[regno],"intresults",peak_get,
  "timestart")

  start_time$=val$(start_time)

  ! Extract Retention Time from chromreg

              ret_time=tabval(chromreg[regno],"
                intresults",peak_get,"rettime")

                  ret_time$=val$(ret_time)

  ! Extract Area from chromreg

          area=tabval(chromreg[regno],"intresults",
                    peak_get,"area")
```

```
    area$=val$(area)

    ! Extract Ending Time from chromreg

    end_time=tabval(chromreg[regno],"intresults",peak_get,
    "timeend")

    end_time$=val$(end_time)

    ! Determine Height of Peak

    hght =
    tabval(chromreg[regno],"intresults",peak_get,"height")

    ! Determine Duration of Interval

    deltat = end_time-start_time

    ! Iff the peak fits export criteria append to tvpams$ string

    If deltat<tmax and hght>hmin then

        tvpams$ = tvpams$ + start_time$ + c$ + ret_time$ + c$
        + area$ + c$ + end_time$ + c$

        else

        messages "Peak Rejected - Peak Height: ", Val$(hght)

        messages "Peak Duration: ", Val$(deltat)

        endif

        loop_ctr=loop_ctr+1

                              endwhile

            ! Store the data from this register in
                          variablepams$

            variablepams$ = variablepams$ + tvpams$
messages "Register No. ", Val$(regno)
messages "Variable Parameters: ", variablepams$
        tvpams$ = "" ! Reset tvpams$ for next register
next regno
! Create Line of Data to Append
dataline$ = fixedpams$ + variablepams$
messages "Line of Data: ", dataline$
!***********************************************************
!* DATA EXPORT OCCURS HERE *
!***********************************************************
! Append this line of Data to the File
! Change Export Path here
```

```
path$ = "c:\Temp\"

filename$ = "DataFile.csv"

! Create full path for export

fullpath$ = path$ + filename$

! Declare Global Variable for Write Error

Write_Error$ = "False"

messages "FullPath: ",fullpath$

Open fullpath$ For Append as #3

   print #3, dataline$

close #3

If Write_Error$ = "False" then

     Print samp_name$, " Data Exported to: ",fullpath$

                         else

         Print "*** ERROR WRITING ", samp_name$,
                     " Data !!! ***"

endif

EndMacro

!**************** Daughter Functions Follow
***********************

Name GotoErrorTrap

!!! This macro executes whenever an Error is encountered !!!

!!! Print to Debug Window the Occurrance of any Error !!!

Print "Error No. ", _Error, " has Occurred when executing
Command ", _ERRCMD$

!!! PostRun should only Respond To File Writing Errors - IGNORE
OTHERS !!!

If _Error = 41308 and _ErrCmd$ = "CLOSE" then

                 Write_Error$ = "True"

endif

Return

EndMacro

name messages

!!! If echo = 1 then print messages to screen

parameter msg1$, msg2$

If echo = 1 then
```

```
Print msg1$, msg2$

endif

endmacro
```

In the preceding macro, any peak that does not have a height >100 or is broader than 0.3 seconds will not be exported to the datafile "C:\temp\DataFile.csv". The height and time requirements for exporting can be changed by modifying the following two lines of code:

```
! Set Export Parameters

hmin = 100

tmax = 0.3
```

When a peak is rejected, its height and time duration are printed to the debug window so that the user can verify that the peak should have been omitted. Running the macro on a sample chromatogram yielded the following output to the debug window:

```
Loading Macros from D:\HPCHEM\CORE\LAB_PST2.MAC...

File Name: D:\MACROS\MARINA\VB\HPLCPLOT\TEST\111500\1CK-1401.D

Sample Name: CP-650903

Date/Time: 15-Nov-00, 11:47:10

Total Registers: 5

Method: DEF_LC.M

Fixed Parameters:
D:\MACROS\MARINA\VB\HPLCPLOT\TEST\111500\1CK-1401.D,CP-
650903,15-Nov-00, 11:47:10,DEF_LC.M,

Signal Description: DAD1 A, Sig=235,16 Ref=600,100

Wavelength: 235

Peak Rejected - Peak Height: 5.43684

Peak Duration: 0.108925

Peak Rejected - Peak Height: 3.395009

Peak Duration: 0.144085

Peak Rejected - Peak Height: 6.556548

Peak Duration: 0.350468

Peak Rejected - Peak Height: 1.916681

Peak Duration: 0.272524

Register No. 1

Variable Parameters: 0.242333,0.301206,1150.622925,0.372189,

Signal Description: DAD1 B, Sig=255,16 Ref=600,100

Wavelength: 255

Peak Rejected - Peak Height: 1.444806
```

```
Peak Duration: 0.153333

Peak Rejected - Peak Height: 15.096416

Peak Duration: 0.089919

Peak Rejected - Peak Height: 6.993433

Peak Duration: 0.123414

Peak Rejected - Peak Height: 1.565032

Peak Duration: 0.245598

Peak Rejected - Peak Height: 1.321166

Peak Duration: 0.31994

Register No. 2

Variable Parameters: 0.242333,0.301206,1150.622925,0.372189,

Signal Description: DAD1 C, Sig=265,16 Ref=600,100

Wavelength: 265

Peak Rejected - Peak Height: 1.423383

Peak Duration: 0.153333

Peak Rejected - Peak Height: 13.187468

Peak Duration: 0.092514

Peak Rejected - Peak Height: 6.268519

Peak Duration: 0.120819

Peak Rejected - Peak Height: 1.126864

Peak Duration: 0.343563

Register No. 3

Variable Parameters: 0.242333,0.301206,1150.622925,0.372189,

Signal Description: DAD1 D, Sig=275,16 Ref=600,100

Wavelength: 275

Peak Rejected - Peak Height: 1.339915

Peak Duration: 0.153333

Peak Rejected - Peak Height: 12.301945

Peak Duration: 0.089285

Peak Rejected - Peak Height: 6.980828

Peak Duration: 0.125507

Peak Rejected - Peak Height: 4.414949

Peak Duration: 0.28

Register No. 4
```

Variable Parameters: 0.242333,0.301206,1150.622925,0.372189,

Signal Description: DAD1 E, Sig=310,16 Ref=600,100

Wavelength: 310

Peak Rejected - Peak Height: 1.080081

Peak Duration: 0.146667

Peak Rejected - Peak Height: 8.386474

Peak Duration: 0.169245

Peak Rejected - Peak Height: 2.078013

Peak Duration: 0.28

Peak Rejected - Peak Height: 2.643066

Peak Duration: 0.325327

Register No. 5

Variable Parameters: 0.242333,0.301206,1150.622925,0.372189,

Line of Data: D:\MACROS\MARINA\VB\HPLCPLOT\TEST\111500\1CK-
1401.D,CP-650903,15-Nov-00,
11:47:10,DEF_LC.M,0.242333,0.301206,1150.622925,0.372189,

FullPath: c:\Temp\DataFile.csv

CP-650903 Data Exported to: c:\Temp\DataFile.csv

7.9 EXPORTING ACTUAL XY DATA FROM THE CHROMATOGRAMS

For most applications, the predefined header items in the integrator results table will be sufficient to produce the results desired. Some users, however, may want to have access to the actual data that produce the chromatogram plots in order to perform more sophisticated analyses such as regression analysis or statistical comparisons. The next macro will actually export the raw data from the chromatograms in a CSV format. The first column will be the X data (time) and the subsequent columns will be the Y data (absorbance) for every chromatogram. The first row in the spreadsheet is a header with the label "time" for the X column and the wavelength for all the subsequent Y data columns.

   The filename under which the data are stored is made up of the detector name followed by the signal ID of each chromatogram measured. Thus a sample filename could be "DAD1" + "ABCDE" + ".CSV" = "DAD1ABCDE.CSV". This file is then stored to the same path with all the individual chromatograms to avoid confusion. The macro code follows:

```
!******************************************************************

!* MACRO TO EXPORT RAW DATA FROM ALL CHROMATOGRAMS IN CSV
FORMAT *

!* BY BRIAN BISSETT *

!* JULY 24 2001 *

!******************************************************************
```

```
name RawData
Parameter Interval_Counter default 0
! Declare local variables
Local n,ydat,xdat,File_Name$,a$,button
Local obj_counter,num_objs
Local c$
Local temp$, samp_name$,ystream$
n=1
! Print Messages to Debug Window
echo = 1
! Comma Delimiter for File - c$ = ","
c$ = Chr$(44)
! Warn User if no Signals are Loaded
If RegSize(Chromreg)=-1 Then
        button = Alert("No signal loaded!",3,"Error")
                        Return
EndIf
! This is the number of wavelengths used or number of graphs
num_objs = RegSize(ChromReg)
! Construct an intelligent file name
File_Name$=DADataPath$+DADataFile$
CSV_File$ = ObjHdrText$(Chromreg[1],"Detector")
For obj_counter = 1 to num_objs
   CSV_File$ = CSV_File$ +
   ObjHdrText$(Chromreg[obj_counter],"SignalID")
next obj_counter
! Print Message
a$="Converting File: "+File_Name$ + " Signal: " + CSV_File$
Print a$
File_Name$ = File_Name$+"\"+CSV_File$+".CSV"
messages "filename: ", File_Name$
Open File_Name$ for output as #3
! Create header for file with column names
header$ = "time" + c$
```

```
For obj_counter = 1 to num_objs

   !Extract Wavelength from Signal Description

SigDesc$ =
Substring$(ObjHdrText$(Chromreg[obj_counter],"SignalDesc"),
2,, ",")

wave$ = SigDesc$[(Len(SigDesc$)-2):Len(SigDesc$)]

If obj_counter = num_objs then

   header$ = header$ + wave$

   Else

   header$ = header$ + wave$ + c$

Endif

next obj_counter

print #3, header$

messages "header: ", header$

! Loop through every row which has data

For n = 1 to datacols(ChromReg[1])

   ystream$ = "" ! Reset ystream$

   ! Extract X axis data (time) for this Wavelength

   xdat=data(ChromReg[1],0,n)

   ! Loop through every column which has data

   For obj_counter = 1 to num_objs

   ! Extract Y axis data (absorbance) for this Wavelength

   ydat = data(ChromReg[obj_counter],1,n)

   If obj_counter = num_objs then

      ystream$ = ystream$ + Val$(ydat)

      Else

      ystream$ = ystream$ + Val$(ydat) + c$

   Endif

Next obj_counter

dataline$ = Val$(xdat) + c$ + ystream$

print #3,dataline$

temp$ = "line no." + Val$(n)

if n = 1 then

   messages temp$, dataline$

Endif
```

```
next n

close #3

print File_Name$+" Created"

!Return

endMacro ! RawData

name messages

!!! If echo = 1 then print messages to screen

parameter msg1$, msg2$

If echo = 1 then

   Print msg1$, msg2$

endif

endmacro
```

With echo = 1 the following output would be printed to the debug macro for a compound with five chromatograms:

```
Loading Macros from D:\HPCHEM\CORE\LAB_CHRT.MAC...

Converting File: D:\MACROS\MARINA\VB\HPLCPLOT\TEST\111500\2EK-
IS01.D

Signal: DAD1ABCDE

filename: D:\MACROS\MARINA\VB\HPLCPLOT\TEST\111500\2EK-
IS01.D\DAD1ABCDE.CSV

header: time,235,255,265,275,310

line no.1 -0.038667,0.015736,-0.004768,-0.000477,-0.001431,
-0.002861

D:\MACROS\MARINA\VB\HPLCPLOT\TEST\111500\2EK-
IS01.D\DAD1ABCDE.CSV

Created
```

7.10 ADDING MACROS TO CHEMSTATION MENUS

So far macros have been run from the command line or been set to execute automatically before, during, or after an analysis. Another convenient way to run a macro is to have a menu item within the Chemstation software that will execute the macro when selected. This is possible by using the commands MenuAdd and SetHook.

A macro can be run from only two options for adding a menu item. A menu item can be added to an existing menu (at the bottom), or a brand new menu can be added (to the right of all the existing menu items). The following macro ("lab_hook.mac") demonstrates how to add menu items:

```
! Hook Demonstration Macro

name sillymac

! A silly macro to execute using a hook
```

```
button = Alert("Are You Senator McCarthy?!",4,"Senator McCarthy
Check")

endMacro ! sillymac

Name sillymac_hook

! Menu Items Can be Added to Existing Menu Items (at the bottom)

! or to new Menu Items at the end (on right)

! Adds Menu Item "Senator McCarthy Check" to the bottom of
the existing "View" Menu

If TabHdrText$(_Config,"Window","CurrentView") = "DA" Then

   MenuAdd "&View","Senator McCarthy Check",\

   "sillymac",,"Generate CSV (X,Y) File From Currently Loaded
   Signal"

   Print "silly.mac loaded - Menu Choice Senator McCarthy Check
   added to View menu"

EndIf

! Adds Menu Item "Senator McCarthy Check" to the bottom of
the existing "Graphics" Menu

If TabHdrText$(_Config,"Window","CurrentView") = "DA" Then

   MenuAdd "&Graphics","Senator McCarthy Check",\

   "sillymac",,"Generate CSV (X,Y) File From Currently Loaded
   Signal"

   Print "silly.mac loaded - Menu Choice Senator McCarthy Check
   added to GRAPHICS menu"

EndIf

! Creates New Menu Item - But It will always be added on the End.

If TabHdrText$(_Config,"Window","CurrentView") = "DA" Then

   MenuAdd "&Senator McCarthy","Senator McCarthy Check",\

   "silly",,"Generate CSV (X,Y) File From Currently Loaded
   Signal"

   Print "sillymac.mac loaded - Independent Menu Senator
   McCarthy added"

EndIf

EndMacro ! sillymac_hook

name erasestupid

! Deletes all the silly Menu Items

menudelete "&Senator McCarthy","Senator McCarthy Check"

menudelete "&Graphics","Senator McCarthy Check"
```

```
menudelete "&View","Senator McCarthy Check"

EndMacro !erasesilly

!! To Automatically Add the menus place this command in the
user.mac file

! SETHOOK "PreViewMenu","sillymac_hook"
```

The "sillymac_hook" macro is what sets up the menu item and assigns the macro to be run when that menu item is selected. Three examples of how to assign the macro "sillymac" to existing and to new menu items are shown. The macro "erasesilly" is provided to show how the user can erase any given menu item created by a developer from code.

To have the menu items loaded automatically upon startup and ready to go requires the use of the SetHook command. To have these three menus appear every time the Chemstation software is started, place the following line of code in the user.mac file:

```
SETHOOK "PreViewMenu","sillymac_hook"
```

The SetHook Command contains two parameters:

- Parameter 1 is the HOOK location (actually a table in the HOOK register) that dictates where and when in the Chemstation software the macro/command will be executed. In the previous example, a table in the HOOK register called "PreViewMenu" is where the command "sillymac_hook" is appended. This macro is then executed when a new view menu is created. In this case the macro "sillymac_hook" is added prior (PRE) to the addition of the VIEW menu. This is the only menu addition HOOK available.
- Parameter 2 is the command or macro that is to be executed when the HOOK location is processed within the Chemstation software.

The macro hook information in Appendix H describes all the possible values for parameter 1. SetHook can be used to initiate the execution of a macro in many trigger points within the Chemstation software.

7.11 SOME GENERIC MACROS AND TECHNIQUES

Many different modules are available for the Chemstation software system. This system can be used to control modules for liquid chromatography (LC), gas chromatography (GC), capillary electrophoresis (CE), and mass spectrometry (MS) — one or all of which a laboratory may be running. It is nice to have a way of loading only those macros applicable to a particular machine or method. The _FEATURES$ system string variable can be harnessed to allow the developer to do that.

For example, suppose a laboratory was running a capillary electrophoresis machine and a liquid chromatography machine. Some macros might be used in CE, some in LC, and others might be utilized in both. The following is some sample code that will add one menu item if the machine to be controlled is an LC and a different menu if it is a CE:

```
if _FEATURES$ = "LC"

   menuAdd "&ChemLab|&Export","&LC Export",lcrpt,,"Export
   *.csv LC\ELOGD information"

endif

if _FEATURES$ = "CE"
```

```
menuAdd "&ChemLab|&Export","&CE Export",cerpt,,"Export
*.csv CE\pKa information"
```
endif

In this example, the menu item is "LC export" for a liquid chromatography machine and "CE export" for a capillary electrophoresis machine. In other circumstances it might be similarly advantageous to load or suppress the use of certain macros based on the method that is currently loaded. The string system variable _METHFILE$ stores the name of the currently loaded method and can be utilized for just such a purpose. Other quite useful string system variables are _DATAFILE$, _ERRMSG$, _ERRCMD$, _INSTNAME$, _LICENSE$, _METHODINFO$, and _VERSION$.

The current value of any of these values can be obtained by typing Print _VariableName$ at the command line. The value of the variable will then be printed in the gray area above the command line, and to the debugging window if it is open.

Whenever a file is to be exported, it is prudent to check that the export path does exist. The macro shown next accomplishes this quite nicely:

```
name validpath

! if path$ does not exist this macro will create it!

Parameter path$

On Error Goto Error_Trap

! Make sure "\" not @ end of path$

If path$[Len(path$):Len(path$)] = "\" then

    path$ = path$[1:(Len(path$)-1)]

endif

! Try to change the working directory to path$

chdir path$

!Print path$," is valid and exists."

return

Error_Trap:

If _errmsg$[1:12] = "Invalid path" then

    ! If you can't change the working directory to the path
    it doesn't exist

    ! Create path$ on this machine

    MkDir path$

    !Print "creating ",path$

    return

endif

EndMacro
```

This macro tries to access the path by changing the working directory to the path passed to it by the parameter path$. If an error results, which would only happen if the path did not exist, the macro then creates the given path with the MkDir command.

The following macro extracts the last directory from a full file path. To modify the extracted directory, change the locations of the stopflg and startflg variables within the macro.

```
Name ExtractLastPath$

! EXTRACT A FILENAME FOR THE DATA FROM THE PATH

Parameter thestring$

local ii, trueindex

local startflg, stopflg

startflg = 0

!Print "The String$: ", thestring$

for ii = 1 to Len(thestring$)

trueindex = Len(thestring$) - (ii-1)

!Print "Loop: ",trueindex

currchar$ = thestring$[trueindex:trueindex]

if currchar$ = "\" then

   if startflg > 0 then

      stopflg = trueindex + 1

      !Print "Flags: ", startflg,
      stopflg,thestring$[stopflg:startflg]

      Return thestring$[stopflg:startflg]

      else

      startflg = trueindex - 1

   endif

endif

next ii

! In the event no path could be extracted use this default:
"cedata"

Return "cedata"

EndMacro
```

Example: Print ExtractLastPath$("c:\first\second\third\abc123.dat"). Yields: third.

Time is a very useful way to name files for archival purposes. The Chemstation software has two time functions. TIME() returns the current time in seconds since January 1, 1970 (the day leisure suits went out of style). For most archival applications, this function is worthless. Time$([Mode]) is quite useful, however. Mode is a value between 0 and 2.

```
Time$(0) will return: Thu Nov 07 12:43:18 2002
```

```
Time$(1) will return: 12:43:51 PM

Time$(2) will return: 12:44:19 PM
```

For some reason, modes 1 and 2 produce the same format output. It would have been nice if one of the modes could have produced a military format of time. However, the following function will return the current time in military format, which is often useful for archival purposes:

```
Name MilitaryTime$

! Returns Current Time in 24 hour format

local ctime$, AMorPM$, temp$

local hour

ctime$ = Time$(1)

!Oddly Enough Causes Error -> AMorPM$ = Time$(1)[Len(Time$(1))-
2:Len(Time$(1))]

AMorPM$ = ctime$[Len(Time$(1))-1:Len(Time$(1))]

hour = Val(Substring$(ctime$,1,,":"))

if AMorPM$ = "AM" then

   !Midnight Correction

   if hour = 12 then

      hour = 0

   endif

   ! Add Leading Zeros for AM Times

   if Len(Val$(hour)) = 1 then

      ctime$ = "0" + Val$(hour) + ":" +
      Substring$(ctime$,2,,":") + ":" +
      Substring$(ctime$,3,,":")

   else

      ctime$ = Val$(hour) + ":" + Substring$(ctime$,2,,":") +
      ":" + Substring$(ctime$,3,,":")

   endif

   return ctime$

else

   ! Must Add 12 to PM Times

   if hour <> 12 then

      hour = hour + 12

   endif

   ctime$ = Val$(hour) + ":" + Substring$(ctime$,2,,":") + ":"
   + Substring$(ctime$,3,,":")
```

```
        return ctime$

  endif
```

The Chemstation macro programming language does not have a function like Format$ in Visual Basic, which allows the programmer to specify the precision of the number he wishes to use. However, the following function will accomplish this task if it is passed the number and the number of decimal places to which to truncate:

```
name adjust_precision

       parameter number

       parameter precision

       local remainder, cmd$

  !! Multiple number ^10 based upon precision !!

       cmd$ = "1e" + Val$(precision)

       number = number * val(cmd$)

  !! Get digits to the right of the decimal point!!

  !! for rounding !!

       remainder = Mod(number,1)

  !! If digits to the right of the decimal point !!

  !! <0.5 then round down, if > = 0.5 then round !!

  !! up !!

  If remainder > = 0.5 then

       number = ceil(number)

  Else

       number = floor(number)

  EndIf

  !! Divide truncated number by ^10 based upon !!

  !! precision !!

       number = number/val(cmd$)

  !! Return control to system and adjusted !!

  !! number to calling macro !!

       return number

  EndMacro
```

Example: Print adjust_precision(1.234,2). Yields: 1.23.

It is often useful to adjust the precision of a parameter when it is to be included in a report or uploaded to a database.

7.12 SUMMARY

The Chemstation software is an ideal environment in which to operate because it offers control over a variety of instruments. More importantly, the methods for every instrument can be automated and customized via a standard scripting language. This feature gives the user an unprecedented amount of control over experiments to be run.

Appendix A

Website Contents and Organization

The website http://www.pharmalabauto.com associated with this text has nine directories that are defined as follows:

AgilentMacros — all sample macros included in the text for the Agilent HPLC Chemstation. These files have a suffix of .mac and should be placed in the "\Core" subdirectory of the HP Chemstation software (typically "C:\Program Files\HPChem\Core").

ApplicationNotes — all application notes in *.pdf format that are contained in Appendix D and Appendix H. Some additional application notes, which are extremely useful, but too large to include within the appendix, are also present.

Datasheets — *.pdf format datasheets for all devices used in Chapter 5. The most useful portions of the sheets are included in Appendix D and Appendix E, but this area contains the full version of the text.

Macrodebug — Agilent's self-extracting shareware utility that is used in debugging Chemstation macros.

Macropad — Agilent's shareware utility that provides a nice development environment (a glorified editor) for developers of macros to work in.

Origin — sample scripts and worksheets included in the text for use with Origin Lab's Origin.

PortIO — shareware program from Scientific Software Tools, Inc. that is used for performing hardware port I/O under Windows 95 and Windows NT. This product allows any 32-bit C/C++ or Visual Basic application (or any language that can call a DLL) direct access to the I/O ports over the range 0100H to FFFFH. This program is used extensively in Chapter 4 to accomplish low-level hardware interfacing.

VB — standalone Visual Basic project files for the Keithley Metrabyte hardware and the Kloehn syringe device driver, covered in Chapter 3 and Chapter 4.

VBA —Microsoft Excel's Visual Basic for Applications sample macros used in Chapter 1 and Chapter 2.

The companion software can be accessed from the website http://www.pharmalabauto.com in two different ways. The software for each section can be accessed and downloaded via the software page shown in Figure A.1.

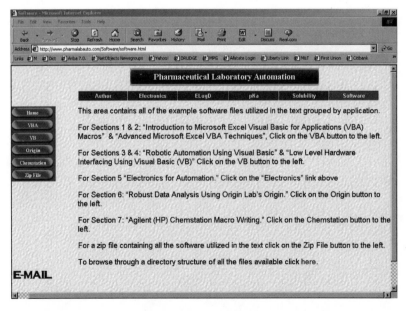

FIGURE A.1 The Software Page at http://www.pharmalabauto.com

The software can also be accessed just like a CD-ROM (see Figure A.2) by clicking on the link at the bottom of the page. This allows the user to browse a directory structure of the software and copy just the files of interest.

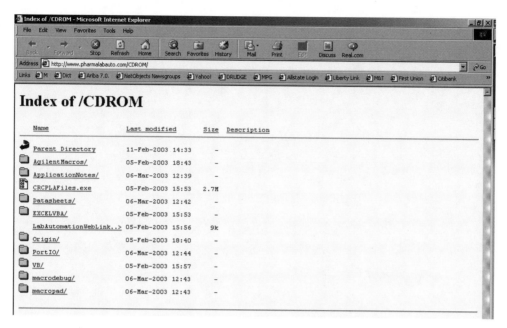

FIGURE A.2 directory structure access of the software available for the text.

Appendix B

Setting Up the Excel VBA Macro Examples

From the website http://www.pharmalabauto.com, copy all the files in "\VBA\XLStart" to Microsoft Office's XLStart directory ("C:\Program Files\Microsoft Office\Office\XLStart").

Next, copy all the files in "VBA\Templates" to Microsoft Office's templates directory ("C:\Program Files\Microsoft Office\Templates").

The files located in "VBA\TestFiles" are those used to test the sample macros in the text. They are subdivided into two directories labeled Section1 and Section2, each of which contains the sample files used in the respective section. These files can be accessed from the website and should be copied over to the user's hard drive.

Keep in mind that copied files from the website will have the "read only" attribute set, which will prevent any modification to these files unless this attribute is disabled. This attribute can be modified by selecting the filename, left clicking and selecting properties, and, from the general tab unchecking the "read-only" attribute. This attribute can also be removed from many files in "batch mode" by using the MS-DOS command at the command prompt "attrib -r path/s". For example, to remove the read-only attribute from all the files residing in and under the C:\TestFiles directory, type "attrib -r c:\testfiles/s" at the command prompt in DOS. In Windows 2000 and XP, the read only attribute can be removed from all files contained with a folder by right clicking the folder, unchecking the read only attribute, and when prompted selecting "apply changes to this folder, subfolders, and files".

Appendix C

Kloehn Syringe Protocol

(From *50300 Hardware User's Manual*, Appendices A – C, Kloehn Ltd., Las Vegas, NV, 1996, 50509 Rev. L, 03-20-02. With permission.)

APPENDIX A. MACHINE CODE COMMAND LIST

Notes:

1. Commands which have a (p) can be embedded within a program or executed immediately if followed by the "R" command.
2. Commands which have a (po) can be used only within a program.
3. Commands which have a (e) are executed when they are recognized and are not stored within a program.
4. Commands which have a (s) are normally setup commands, and the results of these commands are stored in nonvolatile memory by the "!" command.
5. After a Reset or power-up, the first command should be a "Q" and the reply, if any, should be ignored. A "W4", "Y4", or "Z4" must be sent before syringe move commands can be executed.
6. Numerical ranges in parenthesis are for the 24K model. The notation [48,000] means to replace the listed value with the bracketed value for the 48K model.
7. Commands which have a [@n] may use "@n" for the parameter value of "n", and the parameter value is determined as a normalized fraction of its full-scale range. See Section 3.15.4 for details.
8. Commands which have a [@a] may use "@n" for the parameter value of "n", and the parameter value is the actual input number that is read. See Section 3.15.4 for details.
9. Commands which do not have any [@_] brackets cannot use "@n" for the parameter value of "n". See Section 3.15.4 for a list of commands which can use the "@n" value.

A.1   SYRINGE COMMANDS

An   (p) [@n] Absolute syringe move. Move the syringe to position "n" with the "busy bit" in the status byte set to "busy" ("0"). (n: 0...24,000 [48,000])

an   (p) [@n] Absolute syringe move. Move the syringe to position "n" with the "busy bit" in the status byte set to "not busy" ("1"). (n: 0...24,000 [48,000])

cn   (p,s) [@n] Set the stopping speed. The factory default is 743. (n: 40...10,000)

Dn   (p) [@n] Relative dispense. Move the syringe "n" steps up from the current position and set the "busy bit" in the status byte to "busy" ("0"). (n: 0...24,000 [48,000])

dn   (p) [@n] Relative dispense. Move the syringe "n" steps up from the current position and set the "busy bit" in the status byte to "not busy" ("1"). (n: 0...24,000 [48,000])

hn   (p) Do a "handshake" dispense, using User Input #n and User Output #n for the handshake signals. See Section 3.15.8 for handshake operation. (n: 1...3)

kn   (p,s) [@n] Set the number of backlash compensation steps to "n". The factory default is 200. (n: 0...1000)

Ln   (p,s) [@a] Set the acceleration and deceleration slope. The acceleration/deceleration equals 2500 × "n" in Hz/sec. The factory default is 7. (n: 1...20)

ln   (p,s) [@a] Set the deceleration slope. The deceleration equals 2500 × "n" in Hz/sec. The factory default is 7. (n: 1...20)

Pn   (p) [@n] Relative pickup. Move the syringe "n" steps down (aspirate or air gap) from the current position and set the "busy bit" in the status byte to "busy" ("0"). (n: 0...24,000 [48,000])

pn (p) [@n] Relative pickup. Move the syringe "n" steps down (aspirate or air gap) from the current position and set the "busy bit" in the status byte to "ready" ("1"). (n: 0...24,000 [48,000])

Sn (p,s) [@n] Set the top speed. Note: the preferred method of setting top speed is to use the "V" command. (n: 0...34)

n	steps/sec	n	steps/sec	n	steps/sec
0	6400	12	1200	24	130
1	5600	13	1000	25	120
2	5000	14	800	26	110
3	4400	15	600	27	100
4	3800	16	400	28	90
5	3200	17	200	29	80
6	2600	18	190	30	70
7	2200	19	180	31	60
8	2000	20	170	32	50
9	1800	21	160	33	40
10	1600	22	150	34	30
11	1400	23	140		

Vn (p,s) [@n] Set the top speed to "n" steps per second. The factory default is 5000. (n: 40...10,000)

vn (p,s) [@n] Set the Start Speed to "n" steps per second. The factory default is 743. (n: 40...1000)

Wn (p) Initialize the syringe on systems with or without a valve. (n: 4 or 5)

n=4 Move the syringe to the "soft stop" position after moving the valve to port A (must be used after a Reset or a power-up, before syringe move commands can be recognized). This is the same as pressing the "Initialize" (lower) switch on the front panel.

n=5 Set the current syringe position as the "home" position (normally done only when syringes are changed). The result is automatically stored in nonvolatile memory. This is the same as pressing the "Home" switch on the front panel.

Yn (p) Initialize the syringe after moving the valve to the port corresponding to the number stored for the "~Yn" configuration command. The default value of "n" is "1". (n: 4...5, refer to command "Wn" for explanations of "n" values)

Zn (p) Initialize the syringe after moving the valve to the port corresponding to the number stored for the "~Zn" configuration command. The default value of "n" is "1". (n: 4...5, refer to command "Wn" for explanations of "n" values)

? (e) Query the absolute syringe position. Send the result to the host as an ASCII number.

?1 (e) Query the starting speed. Send the result to the host as an ASCII number.

?2 (e) Query the top speed. Send the result to the host as an ASCII number.

?3 (e) Query the stopping (cutoff) speed. Send the result to the host as an ASCII number.

A.2   Valve Commands

B   (p) Move a three way standard valve to the "bypass" position (port A-to-port B).

I   (p) Move a three way standard valve to the "input" position (port A-to-syringe).

O   (p) Move a three way standard valve to "output" position (port B-to-syringe).

on   (p) [@a] Move the valve to the position selected by "n". The values for n must be consistent with the valve type for which the pump is configured (see Configuration Commands). Negative numbers cause counterclockwise rotation as viewed from the front. (n: -8…8, exclusive of 0, where 1=A, 2=B, etc.)

?8   (e) Query the current valve position. Return the ASCII numerical value. (1=A, 2=B, etc.)

$ (e) Query the number of lost valve steps. Send the result to the host as an ASCII number. "0" = no error. "1" = a stall occurred and the move was completed on the first recovery attempt. "2" = 2 stalls occurred and 2 recovery attempts were made. If the second attempt fails, a valve overload error will be generated.

%   (e) Query the number of valve movements. Send the result to the host as an ASCII number.

A.3   I/O Commands

sn   (p) [@n] Send a serial byte from the User Serial expansion Port, MSB first. The value of the ASCII number "n" is the base 10 representation of the value of a binary byte. For transmitted bytes, positive logic applies ("1" = high logic level). In the 2-byte serial mode, "n" represents the second byte sent. The first byte is the same as the most recently specified first transmit byte from a "sn,m" instruction, or "0" if the second byte has not been specified by an "sn,m" instruction. See Section 3.15.4 for an explanation of "@n" usage. (n: 0…255)

sn,   (p) Send two bytes from the User Serial expansion Port, byte "m" first and byte "n" second.
m   Both bytes are sent MSB first. The values of "n" and "m" are expressed as the ASCII base 10 representation of the binary bytes. Positive logic applies. (n: 0…255, m: 0…255)

Un   (p) Turn the user parallel output "n" ON (shorted to ground) or turn off a serial I/O port output bit. (n:1…3 = parallel 1…3; 11…18 = serial byte 1, bit 1…8 where 18 [byte 1,bit 8] is the first byte MSB; 21…28 = serial byte 2, bit 1…8, where 28 [byte 2, bit 8] is the second byte MSB)

un   (p) Turn the user parallel output "n" OFF (open-circuited) or turn off a serial I/O port output bit. (n:1…3 = parallel 1…3; 11…18 = serial byte 1, bit 1…8 where 18 [byte 1,bit 8] is the byte 1 MSB; 21…28 = serial byte 2, bit 1…8, where 28 [byte 2, bit 8] is the byte 2 MSB)

?4   (e) Query the "digital input 1" status; send result to host as an ASCII "1" if "true" (low logic level) or an ASCII "0" if "false" (high logic level).

?5   (e) Query the "digital input 2" status; send result to host (see "?4").

?6   (e) Query the "digital input 3" status; send result to host (see "?4").

?7   (e) Query the "analog input" reading; send result to host as ASCII base 10 number. Voltage = number × 0.02 volts.

?10   (e) Query the byte value of the serial I/O input. An input byte is shifted into Din (MSB first), and the numerical value of the input byte is reported in a base 10 ASCII format. The value is derived from a negative logic convention (low level = 1, high level = 0). In 1-byte mode, this is the only byte. In 2-byte mode, this is the first of two bytes.

?n (e) Query the serial I/O port input bit designated by "n". A serial byte is shifted into Din (MSB first), and the state of the designated bit is reported as an ASCII "0" if the bit is "false" (high input logic level) or an ASCII "1" if "true" (low input logic level). (n: 11...18 = serial byte 1, bit 1...8 where 18 [byte 1,bit 8] is the first byte MSB; 21...28 = serial byte 2, bit 1...8, where 28 [byte 2, bit 8] is the byte 2 MSB) See also ?4, ?5, ?6.

?20 (e) Query the second byte value of the serial I/O port input in 2-byte mode. An input byte is shifted into Din (MSB first), and the numerical value of the input byte is reported in a base 10 ASCII format. The value is derived from a negative logic convention (low level = 1, high level = 0). This instruction is not valid in 1-byte modes.

A.4   PROGRAMMING COMMANDS

:p (po) Define a program label "p" at the current program location. (p: a...z, A...Z)

En (e) Enter the command string currently in temporary memory into nonvolatile memory as user program "n". Return the total number of write cycles to nonvolatile memory since manufacture. (n: 1...10, 1...99 if 8K expansion installed)

en (e) Erase or delete the user program #n from nonvolatile memory. Return the total number of write cycles to nonvolatile memory since manufacture. (n: 1...10, 1...99 if 8K expansion installed)

fn+ (po) Set the flag #n to the "on" state. (n: 1...8)

fn- (po) Clear the flag #n to the "off" state. (n: 1...8)

fnp (po) If flag #n is set, clear the flag and jump to label "p". (n: 1...8; p: a...z, A...Z)

f-np (po) If flag #n is clear, jump to label "p" and do not change the flag. (n: 1...8; p: a...z, A...Z)

Gn (po) [@a] Repeat "n" times the command sequence which lies between "g" (loop start) and "G". A value of "0" results in an infinitely repeating loop. (n: 0...30,000)

g (po) Loop start. Mark the start of a repeat sequence. (see also "Gn")

H (po) Halt the command string execution. Use in a command string to stop execution. A subsequent "R" command will cause execution to resume at the command following the Halt command. This can be used for a breakpoint.

i<np (po) [@n] If the analog input is less than the value "n", then execute the program instruction following label "p"; if not, execute the instruction following this "i" instruction. n=50 × (input voltage) (n: 0...255, p: a...z, A...Z)

i>np (po) [@n] If the analog input is greater than the value "n", then execute the program instruction following label "p"; if not, execute the instruction following this "i" instruction. n=50 × (input voltage) (n: 0...255, p: a...z, A...Z)

inp (po) [@a] If digital input "n" is "On" or if serial bit "n" is True (at a low level), then execute the program instruction following label "p"; if not, execute the instruction following this "i" instruction. (n:1...3 = parallel inputs; 11...18 = serial port input byte 1, bit 1...8 where 18 [byte 1,bit 8] is the first byte MSB; 21...28 = serial port input byte 2, bit 1...8, where 28 [byte 2, bit 8] is the second byte MSB; p: a...z, A...Z)

Jp (po) Jump to program label "p". Label "p" is within the current program. Note that label "Z" is used for a user-defined error handling routine. (p: a...z, A...Z)

J n (po) [@a] Jump to program "n", do program "n", and then resume execution of the calling program at the instruction following this jump instruction. This allows another program in memory to be "called" as a part of the current program. A program which has been "jumped to" cannot jump to yet another program.

k<np   (po) [@a] Jump to label "p" if the value of the software counter is less than "n". (n: 0...65,535)

k=np   (po) [@a] Jump to label "p" if the value of the software counter is equal to "n". (n: 0...65,535)

k>np   (po) [@a] Jump to label "p" if the value of the software counter is greater than "n". (n: 0...65,535)

R   (e) Run the command or command sequence in the temporary memory. If a program has been stopped by a "Halt" command within a program, resume execution at the command which follows the Halt. (See also "r n")

r n   (e) Run the user program "n" stored in nonvolatile memory. Begin with the first instruction of program "n", even if the program has been stopped by a "halt" command within the program. (n: 1...10)

s<np   (po) [@a] Jump to label "p" if the binary value of the S serial input byte is less than "n". If not, do the instruction following this one. In 1-byte mode, "n" is the single byte. In 2-byte mode, "n" is the first byte received. (n: 0...255, p: a-z, A-Z)

s>np   (po) [@a] Jump to label "p" if the binary value of the SIO serial input byte is greater than "n". If not, do the instruction following this one. In 1-byte mode, "n" is the single byte. In 2-byte mode, "n" is the first byte received. (n: 0...255, p: a-z, A-Z)

?9   (e) Query the amount of unused user nonvolatile program memory (NVM) in the basic and expanded NVM. Return the number of unused bytes as two ASCII numbers: the first is the basic NVM and the second is the expansion NVM.

?19   (e) Query the program numbers in NVM. A list of the program numbers stored in user program nonvolatile memory, separated by spaces, is returned.

qn   (e) Query the program "n". A copy of the program #n is sent to the host. (n: 1...10)

A.5   PUMP CONFIGURATION COMMANDS

Configuration commands all begin with a configuration mark "~" (tilde). These commands may be concatenated into a single configuration string using a single "~" at the beginning of the string; such a string cannot be combined with non-configuration mode commands.

~?   (e) Query the pump mode. If a "-1" is retuned, the unit is in the configuration mode. If any other value is returned, the unit is in the command (operational) mode and the number returned is the absolute syringe position.

~An   (e,s) [@a] Select the program to auto-start when the power is turned on. A selection of "0" means no program is started. If no parameter "n" is entered, the current value of "n" is returned. The factory default is 0. (n: 0...10)

~Bn   (e,s) [@a] Select the communications baud rate. If no parameter is entered, the current value of "n" is returned. The factory default is 9600. (See next page for rates.) (n: 0...7)

n	Baud rate	n	Baud rate
1	38,400	5	2,400
2	19,200	6	1,200
3	9,600	7	600
4	4,800	8	300

~Ln   (e,s) Set User Input #3 operating mode. If "n" is omitted, the current value of the operating mode will be returned. (n: 0 or 1) See also Section 3.5.6.

~Pn   (e,s) Select the communication protocol. If no parameter "n" is entered, the current value of "n" is returned. The Factory default is DT. (n: 1=DT, 2=OEM)

~Sn   (e,s) Select the serial I/O expansion port mode. If no parameter "n" is entered, the current value is returned. The factory default is 1-byte. (n: 1=1-byte transfers, 2=2-byte transfers)

~Vn   (e,s) Select the valve type. If no parameter "n" is entered, the current value is returned. The factory default is 0 for units ordered without valves. (n: 0...10)

The valve type numbers are:

n	Valve Types
0	no value
1	three way standard valve
2	three way distribution valve
3	four way standard valve
4	four way distribution valve
5	five way standard valve (Not applicable)
6	five way distribution valve
7	six way standard valve (Not applicable)
8	six way distribution valve
9	eight way standard valve (Not applicable)
10	eight way distribution valve

~Yn   (e,s) Select the position to which the valve will go just prior to moving the syringe to the soft limit using the "Y4" command. The "~Vn" value is checked for a valid entry before accepting the value of "n". The factory default is 1. (n: 1...8)

n	Port	n	Port
1	A	5	E
2	B	6	F
3	C	7	G
4	D	8	H

~Zn   (e,s) Select the position to which the valve will go just prior to moving the syringe to the soft limit using the "Z4" command. The "~Vn" value is checked for a valid entry before accepting the value of "n". The factory default is 1.

n	Port	n	Port
1	A	5	E
2	B	6	F
3	C	7	G
4	D	8	H

A.6   MISCELLANEOUS COMMANDS

!   (e) Store the current values of speeds and backlash into nonvolatile memory. The number returned is the total number "!" commands sent since manufacture.

fn+   (p) set the flag #n to "on". Note the special functions of flags #7 and #8 as explained in Section 3.5.6. (n: 1…8)

fn-   (p) clear the flag #n to "off". Note the special functions of flags #7 and #8 as explained in Section 3.5.6. (n: 1…8)

kn   (p) [@a] set the software counter value equal to "n". (n: 0…65,535)

k+n   (p) [@a] add "n" to the current value of the software counter. (n: 0…65,535)

k-n   (p) [@a] subtract "n" from the current value of the software counter. (n: 0…65535)

k   (e) return the current value of the software counter as an ASCII number to the host.

Mn   (p) [@n] Delay "n" milliseconds. (n: 1…60,000)

mn   (p) Turn the motor power ON or OFF according to the following table:

n	Meaning
0	syringe motor OFF
1	syringe motor ON
2	valve motor OFF
3	valve motor ON

Note: The default conditions are that the syringe motor is normally ON at half power, and the valve motor is normally OFF. Any valve move will turn the valve motor OFF at the end of the Move. Also, any syringe movement will end with the syringe motor in the ON (half power) condition.

T   (e) Terminate the current command, command sequence, or user program. This command is used to stop a program or command in progress. See also the "H" command.

X   (e) Repeat the last executed command or program string. This command is valid only if the preceding command string used "Program Mode" ("p" and "po") commands. Queries cannot be repeated with the "X" command.

A.7   QUERY COMMANDS

All query commands are immediate mode (e) and cannot be including within a stored program. Query commands cannot use the "@n" value.

Query commands can be concatenated (strung together). For example, the values of the Base Speed, the Top Speed, and the Stop Speed could be queried with the string "?1?2?3". The response might be "743 5000 743".

?   (e) Query the absolute syringe position. Return the position as an ASCII number of motor steps measured from the "home" position.

?1   (e) Query the syringe motor starting speed. Return the speed in steps per second as an ASCII number.

?2   (e) Query the syringe motor top speed. Return the speed in steps per second as an ASCII number.

?3   (e) Query the syringe motor stop speed. Return the speed in steps per second as an ASCII number.

?4   (e) Query the "User Input 1" status. Return an ASCII "1" if the input is "true" (at a logic low level) and return an ASCII "0" if the input is "false" (at a logic high level).

?5   (e) Query the "User Input 2" status. Return an ASCII "1" if the input is "true" (at a logic low level) and return an ASCII "0" if the input is "false" (at a logic high level).

?6   (e) Query the "User Input 3" status. Return an ASCII "1" if the input is "true" (at a logic low level) and return an ASCII "0" if the input is "false" (at a logic high level).

?7   (e) Query the voltage at the "User Analog Input". Return an ASCII number corresponding to the voltage between the analog input pin and the analog ground pin. The voltage is related to the number by: input voltage = number × 0.02 volts.

?8   (e) Query the current valve position. Return an ASCII number (1=port A or AS, 2=port B or AB, 3=port C or CS, 4=port D, etc.).

?9   (e) Query the number of unused user-program nonvolatile program memory (NVM) bytes. Returns two ASCII numbers separated by a space. The first is the number of unused bytes in the basic NVM. The second is the number of unused bytes in the expansion NVM. If the second number is "-1", then the expansion NVM was not detected as being installed.

?10   (e) Query the byte value of the serial I/O input. An input byte is shifted into Din (MSB first), and the numerical value of the input byte is reported in a base 10 ASCII format. The value is derived from a negative logic convention (low level = 1, high level = 0). In 1-byte mode, this is the only byte. In 2-byte mode, this is the first of two bytes.

?19   (e) Query the program numbers in NVM. A list of the program numbers stored in user program nonvolatile memory, separated by spaces, is returned.

?n   (e) Query the state of the serial I/O port input bit designated by "n". A serial byte is shifted into Din (MSB first), and the state of the designated bit is reported as an ASCII "0" if the bit is "false" (high input logic level) or an ASCII "1" if "true" (low input logic level). (n: 11...18 = serial byte 1, bit 1...8 where 18 [byte 1,bit 8] is the byte 1 MSB; 21...28 = serial byte 2, bit 1...8, where 28 [byte 2, bit 8] is the byte 2 MSB) See also ?4, ?5, ?6.

?20   (e) Query the second byte value of the serial I/O port input in 2-byte mode. An input byte is shifted into Din (MSB first), and the numerical value of the input byte is reported in a base 10 ASCII format. The value is derived from a negative logic convention (low level = 1, high level = 0). This instruction is not valid in 1-byte modes.

?29   (e) Query the contents of the syringe position capture FIFO. Return the two ASCII numbers which are the syringe positions at the time of the user input signal negative transitions. If one of the position values has not been captured, return a value of "-1" for that number. When the values have been read, set them back to "-1" and enable the capture function for two more captures.

?30   (e) Query the acceleration and deceleration values. Return two numbers separated by a space. The first number is the value of the acceleration parameter. The second number is the value of the deceleration parameter.

$ (e) Query the number of "lost valve steps". Return an ASCII number. "0" = good valve move. "1" = motor stalled, but the automatic error recovery routine completed the move "2" = move could not be completed and a valve motor stall condition exists. Consecutive errors result in an accumulation of the number of lost steps. A good move clears the number to zero.

%   (e) Query the number of valve movements since the pump was turned on. Return an ASCII number.

&   (e) Query the revision number of the pump firmware. Return an ASCII number preceded by a "P".

*   (e) Query the value of the pump supply voltage. Return an ASCII number corresponding to the supply voltage. Voltage = number × 0.2 volts.

fn?   (e) Query the status of the flag #n. (n: "0" = "clear", "1" = "set")

F   (e) Query the buffer status. Return an ASCII "0" if the buffer is empty.

k   (e) Return the current value of the software counter as an ASCII number.

Q   (e) Query the status and error byte. Return the bit values as an ASCII string of 1's and 0's, MSB (bit 7) first.

qn   Query the user program #n stored in nonvolatile memory. Return the complete program as an ASCII command string. This does not work for temporary memory.

~?   (e) Query the pump operating mode. Return an ASCII "-1" if in the configuration mode. If any other characters are returned, the ASCII number returned is the syringe absolute position.

~A   (e) Query the "auto-start" program number. Return the ASCII number of any program configured for auto-sharing a value of "0" indicates no auto-start.

~B   (e) Query the communications baud rate. Return an ASCII number corresponding to the rate as given in the Appendix A.5.

~L   (e) Query the User Input #3 operating mode. Return an ASCII "0" if in the "Logic" mode, or a "1" if in the "Limit" mode. (See Section 3.5.6, Dispense Limit Input.)

~P   (e) Query the communication protocol. Return an ASCII "1" if in DT mode, and a "2" if in OEM mode. (See Appendix B, Communication Protocols)

~S   (e) Query the serial I/O expansion port mode. Return an ASCII "1" for 1-byte transfers, or return an ASCII "2" for 2-byte transfers.

~V   (e) Query the type of valve. Return an ASCII number corresponding to the type of valve as given in Appendix A.5.

~Y   (e) Query the port to which the "Yn" initialization command will move the valve. See Appendix A.5 for details of valve port numbers.

~Z   (e) Query the port to which the "Zn" initialization command will move the valve. See Appendix A.5 for details of valve port numbers.

APPENDIX B. COMMUNICATION PROTOCOLS

There are two communication protocols. One is Data Terminal (DT) protocol; the other one is Original Equipment Manufacturer (OEM) protocol. The DT protocol is the default protocol. The data communication format of both protocols is defined as 8 bits, no parity, 1 stop bit, and half duplex at a baud rate between 300 and 39,400.

When addressing a single pump, the device always returns a response package. When addressing a group of pumps, none of the addressed devices will send a response package. Those commands which always require a reply from the pump do not work in the group addressing mode. Those commands are query commands, read input commands, and configuration query commands. Note that the configuration commands still work, but without a reply from the pump. All responses are ASCII character strings.

B.1   Data Terminal (DT) Protocol (Default Protocol)

The DT protocol is supported by nearly all serial communications software. It is supported by all PCs.

B1.1   Host Command Format

A command package is a sequence of bytes sent by a host computer from that host to a device. The package consists of a starting character, a device address, a command or command sequence, and an ending character.

Byte #	Description	ASCII	Hex
1	Start Character	/	2F
2	Address Character	(see Section 3.4.3)	
3 to N	Command Characters	(see Appendix A)	
N+1	End (Carriage Return)	<CR>	0D

Byte 1:   The starting character signals the beginning of a new package. It is the front slash character "/" on the computer keyboard, 2F hex.

Byte 2:   The device address is a address number for a device or for a group of devices. It can address a total of 15 devices in the network mode.

Byte 3:   The command or a sequence of commands starts with byte 3. A command or a command sequence with length n bytes, uses from byte 3 to byte 3+n-1.

Byte 3+n: The ending character indicates the end of a package. It is 0D hex, the carriage return on the keyboard.

B1.2   Device Response Package Format

This section describes the device response package format of the DT protocol. The device response package is a sequence of bytes sent by a device from that device to a host computer after receiving a command package. The format of the package is described as follows:

Byte #	Description	ASCII	Hex
1	Start Character	/	2F
2	Controller Address	0	30
3	Status Byte	(see Appendix C)	
4 to N	Response (if required)	(see Appendix B.1)	
N+1	End of Text	<ETX>	03
N+2	Carriage Return	<CR>	0D
N+3	Line Feed	<LF>	0A
N+4	End (Blank)	<Blank>	FF

Byte 1:	The starting character, 2F hex, which signals the beginning of a new package, is the front slash character "/" on a computer keyboard.
Byte 2:	The host address, 30 hex (ASCII 0), is the address number for the host computer.
Byte 3:	The status and error byte describes the device status. Please refer to Appendix C for the definitions of the status and errors.
Byte 4:	There may or may not be response byte(s) for a command. In general, all query commands, read input commands, and configuration query commands (~A, ~B, ~P, ~V, etc.) cause response bytes. Other commands do not cause a response.
Byte 4+n:	The end-of-response mark is 03 hex.
Byte 4+n+1:	Carriage return is 0D hex.
Byte 4+n+2:	End of package character is the line feed character, 0A hex.
Byte 4+n+3:	The extra ending character, FF hex, is an extra character to ensure the package is properly sent. This character might not be displayed by the host terminal.

B.2 ORIGINAL EQUIPMENT MANUFACTURER (OEM) PROTOCOL

The OEM is a special protocol which includes sequence numbers and checksums. It requires custom software to support it.

B2.1 Host Command Format of OEM Protocol

This section describes the command package format of the OEM protocol.

Byte #	Description	ASCII	Hex
1	Line synchronization character	<blank>	FF
2	Start Transmit character	<STX>	02
3	Device address	(see Section 3.4.3)	31 - 5F
4	Sequence number	(See following text)	
5	Command(s) (n bytes)	(See Appendix A)	
5+n	End of command(s)	<ETX>	03
5+n+1	Check sum	(See following text)	

Byte 1:	The line synchronization character, FF hex, indicates a command package is coming.
Byte 2:	The start transmit character, 02 hex, signals the beginning of a new package.
Byte 3:	The device address is a address number for a device or for a group of devices. Up to 15 devices can be addressed.
Byte 4:	Sequence number indicates the package sequence. If an error occurs during the communication, the host sends the last package again to the device with a new sequence number. The sequence number starts with 31 hex (ASCII 1). When repeating a command, the host sets bit 3 of the sequence number byte to 1 and increases the sequence number by 1. The valid sequence numbers are hexadecimal 31 for the first package, hexadecimal 3A for the second package (the first repeated package), 3B for the third package, and etc. The maximum number of repeat is 7 with a sequence number of 3F.
Byte 5:	The command or a sequence of commands starts with byte 5. A command or a command sequence with length n bytes uses byte 5 to byte 5+n-1.
Byte 5+n:	The end-of-command(s) character, 03 hex, indicates the end of a command or command sequence.
Byte 5+n+1:	The check sum is calculated by an exclusive-or operation on all bytes except line synchronization byte and check sum byte.

B.2.2   Device Response Data Format of OEM Protocol

This section describes the response package format in the OEM protocol.

Byte #	Description	ASCII	Hex
1	Line synchronization character	<blank>	FF
2	Starting character	<STX>	02
3	Host address	0	
4	Status and error byte	(See Appendix C)	
5	Response, if any (n bytes)	(See Appendix A)	
5+n	End-of-response mark	<ETX>	03
5+n+1	Check sum	(See following text)	
5+n+2	Extra ending character	<blank>	FF

Byte 1:	The line synchronization character, FF hex, indicates a command package is coming.
Byte 2:	The start transmit character, 02 hex, signals the beginning of a new package.
Byte 3:	The host address, 30 hex, is the address number for a the host computer.
Byte 4:	The status and error byte describes the device status. Please refer to Appendix C for the definitions of the status and errors.
Byte 5:	There may or may not be response byte(s) for a command. In general, all query commands, read input commands, and configuration query commands (~A, ~B, ~P, ~V, etc.) produce response bytes. Other commands do not produce a response.
Byte 5+n:	The end-of-response mark, 03 hex, indicates the end of the response byte(s).
Byte 5+n+1:	The check sum is calculated by an exclusive-or operation on all bytes except the line synchronization byte and check sum byte.
Byte 5+n+2:	The extra ending character, FF hex, is an extra character to ensure the package is properly sent. This character might not be displayed by the host terminal.

APPENDIX C. STATUS AND ERROR CODES

C.1 STATUS BYTE CODES

The status byte indicates the device status and any existing error conditions. The status and error code are defined as follows:

ASCII	Binary	Hexadecimal	Decimal	Code	Description
bsy or rdy	7 6 5 4 3 2 1	bsy or rdy	bsy or rdy		
@ or '	0 1 X 0 0 0 0 0	40 or 60	64 or 96	0	No error
A or a	0 1 x 0 0 0 0 1	41 or 61	65 or 97	1	Syringe not initialized
B or b	0 1 X 0 0 0 1 0	42 or 62	66 or 98	2	Invalid command
C or c	0 1 x 0 0 0 1 1	43 or 63	67 or 99	3	Invalid operand
D or d	0 1 x 0 0 1 0 0	44 or 64	68 or 100	4	Communication error
E or e	0 1 x 0 0 1 0 1	45 or 65	69 or 101	5	Invalid R command
F or f	0 1 x 0 0 1 1 0	46 or 66	70 or 102	6	Low voltage
G or g	0 1 x 0 0 1 1 1	47 or 67	71 or 103	7	Device not initialized

ASCII	Binary	Hexadecimal	Decimal	Code	Description
H or h	0 1 x 0 1 0 0 0	48 or 68	72 or 104	8	Program in progress
I or I	0 1 x 0 1 0 0 1	49 or 69	73 or 105	9	Syringe overload
J or j	0 1 x 0 1 0 1 0	4A or 6A	74 or 106	10	Valve overload
K or k	0 1 x 0 1 0 1 1	4B or 6B	75 or 107	11	Syringe move not allowed in valve bypass position
L or l	0 1 x 0 1 1 0 0	4C or 6C	76 or 108	12	No move against limit
M or m	0 1 x 0 1 1 0 1	4D or 6D	77 or 109	13	NVM memory failure
N or n	0 1 x 0 1 1 1 0	4E or 6E	78 or 110	14	Reserved
O or o	0 1 x 0 1 1 1 1	4F or 6F	79 or 111	15	Command buffer full
P or p	0 1 x 1 0 0 0 0	50 or 70	80 or 112	16	For 3-way valve only
Q or q	0 1 x 1 0 0 0 1	51 or 71	81 or 113	17	Loops nested too deep
R or r	0 1 x 1 0 0 1 0	52 or 72	82 or 114	18	Label not found
S or s	0 1 x 1 0 0 1 1	53 or 73	83 or 115	19	No end of program
T or t	0 1 x 1 0 1 0 0	54 or 74	84 or 116	20	Out of program space
U or u	0 1 x 1 0 1 0 1	55 or 75	85 or 117	21	Home limit not set
V or v	0 1 x 1 0 1 1 0	56 or 76	86 or 118	22	Call stack overflow
W or w	0 1 x 1 0 1 1 1	57 or 77	87 or 119	23	Program not present
X or x	0 1 x 1 1 0 0 0	58 or 78	88 or 120	24	Valve postion error
Y or y	0 1 x 1 1 0 0 1	59 or 79	89 or 121	25	Syringe position error
Z or z	0 1 x 1 1 0 1 0	5A or 7A	90 or 122	26	Syringe may crash

Bit #5 is the busy/ready bit which is logic "0" if busy executing commands and logic "1" if ready for new commands. The "busy" version of the syringe move commands should be used so that the true status of the device will be reported.

C.2 ERROR CODE DEFINITIONS

A	Syringe not initialized	The "W4" command was not issued or the "Initialize" panel button was not pressed prior to any syringe move commands. See Section 3.9.1.
B	Invalid command	The command syntax (letter or structure) does not exist or was inappropriate.
C	Invalid operand	The number sent with the command was out of range or is not a valid selection. See Appendix A.
D	Communication error	In the OEM protocol, an error occurred in the sequence number or the checksum.
E	Invalid R command	An attempt was made to run a nonexistent program or an illegal command was found in the program string.
F	Low voltage	The pump supply voltage went below the minimum allowed for syringe or valve moves (20Vdc). See Section 3.3.
G	Device not initialized	Not implemented.
H	Program in progress	The command string cannot be executed because a program is currently running and cannot be overwritten. See Section 3.14.
I	Syringe Overload	The syringe stalled or lost steps. Position is no longer accurately known. The syringe must be reinitialized. See Section 3.9.1.
J	Valve overload	The valve did not reach the commanded position. The valve motor is stalled due to a mechanical problem with the valve or the quiescent valve position has been disturbed. See Section 3.16.8.
K	Syringe move not allowed	The valve is in the bypass position (nondistribution valves only). A syringe move is not allowed.
L	Cannot move against limit	The external dispense limit input (User Input #3) is active (low level) and the limit mode is enabled. A dispense direction move is not allowed. See Section 3.5.6.
M	Progrm Memory Failure	A write to the nonvolatile expansion memory returned false data. The expansion memory has failed or has been corrupted. See Section 3.12.4.
O	Command buffer full	The string (program or command sequence) is too long to fit within the remaining communications buffer space. A previous program or command string may be taking too much space. See Section 3.4.7 and Section 3.12.1.
P	Standard 3-way valve command	Commands "B". "I", and "O" can be used only with a 3-way standard type of valve. Pump is not configured for a 3-way standard valve. See Appendix A.5.
Q	Loops nested too deep	A program loop is illegally contained within another loop.
R	Label not found	A program branch (If...then, or Jump To) is referencing a nonexistent program label.
S	End of program not found	Nonvolatile memory program space has been corrupted. Programs should be deleted and then stored again.
T	Out of program space	Insufficient memory remains for the program to be stored. See Section 3.12.

U	**Home limit not established**	A syringe move cannot be performed until the location of home (zero position) has been established. Also, the home position must be above the "soft limit" position. See Section 3.9.2.
V	**Call stack overflow**	A program has illegally called another program. A program which has been called by another program cannot call yet another program. Only one program return at a time may be pending.
W	**Selected program not present**	A numbered user program in nonvolatile memory has been called which cannot be found in the memory.
X	**Valve position error**	The valve position has been displaced and is no longer positioned on an encoder wheel slot. The valve cannot be reliably positioned anywhere but "home" (Port A).
Y	**Syringe position value out of bounds**	The internal syringe position value has exceeded the allowable range of values and is no longer valid.
Z	**Syringe may go past home**	The internal syringe location is greater than the known distance to home from the soft limit. The position value is no longer valid.

Appendix D

Communications Datasheets

Robust DataComm Inc.
St. Paul, Minnesota USA & Singapore
info@robustdc.com   www.robustdc.com

Grounding and RS-422/485 (No Free Lunch!)

RobustDC Application Note #5

Quick Index:

Grounding non-isolated RS-422/485
Grounding partially isolated RS-422/485
Grounding fully isolated RS-422/485
"I don't want (or can no longer afford) to run an extra wire ..."

RS-422 and the more robust RS-485 are the most common standard data communication inter-faces found in industrial equipment today. Yet it is common to hear users complain that these products communicate unreliably -- sometimes they work, sometimes they do not. But rather than the product being at fault, it is often a poor understanding of the basic principles behind these standards which causes this. Poor ground design is one of the most common reasons users have trouble with RS-422 and RS-485. This application note attempts to explain the design options avail-able for grounding RS-422 and RS-485 data communication systems. It starts with the older non-isolated designs and progresses to the more modern designs which require complete galvanic isola-tion between devices.

Grounding non-isolated RS-422/485

The primary source of confusion over RS-422/485 grounding is the "magic" nature of the ground in the common non-isolated RS-422/485 system. Most users incorrectly assume that the two wires within each twisted pair consist of a signal wire and a return wire. While this may be true for current loop systems, this is completely incorrect for RS-422/485. Both transmit

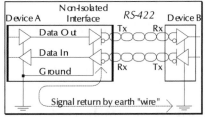

wires at each device are supplying current to maintain a voltage level relative to an external refer-ence. Quoting the EIA/RS-485 standard:

*"Proper operation of the generator and receiver circuits requires the presence of a signal return path between the circuit grounds of the equipment at each end of the interconnection. The circuit reference may be established by a third conductor connecting the common leads of devices, **or** it may be estab-lished by connections in each using equipment to an earth reference."*

Even without turning a screw or pulling a wire, you still have a magic ground "wire" through the de-vice's normal earth reference or physical earth (PE). So a non-isolated RS-422/485 connection works as expected without a specific ground wire. In fact, adding such a wire just leads to destructive ground loop problems such as suffered by RS-232.

While using the "earth wire" is free and easy, it is one of the noisiest "wires" known. Every surge, system fault, and lightning strike on the planet dumps current into this "wire", plus earth materials with variable resistance's causes unpredictable movements of charges around plant sites. Besides minor back-ground noise, a more serious problem exists with this simple earth return. *Common*

Making Industrial Data Flow Like Water ... Safely, Silently, & Sanely

Robust DataComm Inc.
St. Paul, Minnesota USA & Singapore
info@robustdc.com   www.robustdc.com

mode surges will be caused by momentary shifts in the ground potential between device A and device B. For minor day-to-day fluctuations of ground potential this is not a problem -- RS-485 was specifically designed to both function normally with a ±7v ground potential difference and survive ± 25v surges.

However ±25v is not a big voltage when compared to the potentials in the surrounding industrial environment. Power system faults or equipment malfunctions can easily cause brief but powerful ground potential differences as large as the voltage of the power system -- for example 110, 230, or 480 volts. What about lightning discharges? These can cause momentary ground potential difference of hundreds or even thousands of volts. This refers not to direct strikes -- even the best "surge protection" device in world will not save a CMOS device struck by lightning. This refers to the common discharges on a local site which are caught and safely directed to earth by regulation lightning protection equipment. Even these good discharges elevate local ground potentials until they can fully dissipate into the earth.

Of all the fatal surge damage suffered by data communications equipment, this is the most common culprit. Non-isolated devices often completely fail when damaged by surges due to an unpredictable number of failures and/or protection diodes shorting power rails to ground. For example, a surge may burn out both the serial port and a hard disk or even a video display. The actual damage is somewhat unpredictable. A common fix to this problem is to install external surge protection devices which offer a lower impedance path for surges to ground. These attempt to keep the surge energy out of the protected device. Another solution is the current trend for galvanic isolation of data communication in systems.

In Summary: A *non-isolated RS-422/485* link must never have a direct ground wire between the two ends. It must rely on each device having a good physical earth connection, plus have external surge protection to divert the inevitable common mode surges to ground.

Grounding & Partially isolated RS-422/485

Robustness can be greatly improved at low cost by partial isolation of the RS-422/485 system. The interface is isolated from the local device, but often grounded with an external power supply to the same earth ground as the device. This can also be referred to as 2-port galvanic isolation. Partial isolation is more robust than no isolation because it

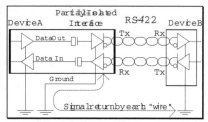

virtually guarantees that surges will by-pass the delicate, high-speed CMOS circuits used in most intelligent devices today. From a surge stand-point, RS-422/485 interface chips are an order of magnitude more robust than - for example - a Pentium CPU or ultra low-power hard disk controller.

Yet partial isolation is still susceptible to the large common mode surges that affect non-isolated systems. The main advantage over no isolation is that surge damage in partially isolated systems is predictably limited to the interface circuit alone. Many PLC devices use partial isolation for their communications ports and isolated field I/O. It is a good compromise between keeping costs low

---

Robust DataComm Inc.
St. Paul, Minnesota USA & Singapore
info@robustdc.com   www.robustdc.com

while increasing system robustness. As testimony to it's effectiveness, I remember a site where a low-cost PLC was connected by non-isolated modems to a DCS 2Km away. A lightning discharge caused a common mode surge which destroyed both modems and the PLC's partially isolated communication port -- destroyed meaning black, charred components. Yet the PLC itself was un-affected and continued it's control function without interruption.

> **In Summary:** A **partially isolated RS-422/485** link generally is locally grounded and would not have a direct ground wire between the two ends. It still relies on external surge protection to divert the inevitable common mode surges to ground.

Grounding & Fully isolated RS-422/485

The ultimate form of isolation with metal wires is full galvanic isolation.   The interface is both isolated from the device and the local ground. This can also be referred to as 3-port galvanic isolation. Returning to the EIA/RS-485 standard:

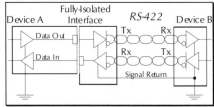

"Proper operation of the generator and receiver

*circuits requires the presence of a signal return path between the circuit grounds of the equipment at each end of the interconnection. The circuit reference may be established by a third conductor con-necting the common leads of devices, **or** it may be established by connections in each using equipment to an earth reference."*

Since we no longer have a signal return path through the earth "wire", full galvanic isolation requires a physical wire to act as a circuit ground. While from an installation stand-point this can be an un-expected cost and even a nuisance, one must consider the trade-offs involved. A properly isolated RS-422/485 system will be almost as effective as fiber optics with none of the technical difficulties of managing fiber cables and connectors.

One great characteristic of full galvanic isolation is that to a common mode surge the data communication link now looks like an electrical dead-end. Ground potential differences up to the rated voltage of the isolation barrier will cause no common mode surge. Ratings of fully isolated RS-422/485 interfaces are often in the range of 250-500 or 1500-2500 volts.

Virtually all modern, system-oriented data communications standards require full galvanic isola-tion. For example, thick Ethernet requires at least 1500v isolation, while even the thin Ethernet requires 500v. The new Fieldbus standard requires full galvanic isolation between each device and the bus. All successful, big-name proprietary control bus systems include either optical or trans-former isolation at each node. And of course, the current interest in fiber optics is due in part to the complete galvanic isolation it offers between all devices.

> **In Summary:** A **fully isolated RS-422/485** link generally requires a direct ground wire between the two ends. External surge protection is only required if large lightning related surges to ground are expected.

Robust DataComm Inc.
St. Paul, Minnesota USA & Singapore
info@robustdc.com   www.robustdc.com

"I don't want (or can no longer afford) to run an extra wire ..."

If running the extra ground wire for full galvanic isolation is "too costly", there are three alternatives. The
first is to connect the floating ground locally, converting the full galvanic isolation to partial isolation.
It solves the ground problem without extra cost and
still retains some advantages of isolation.

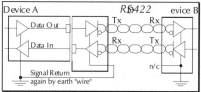

Fully Isolated Interface
Converted to Partial Isolation

The second alternative is slightly controversial.
You can run the floating ground through the cable shield. This practice is widely used and accepted for signals above 1MHz, such as for Ethernet and most other systems with coax cable.
However, your signal is likely less than 1MHz -- for example 9600 baud is only about 0.01MHz. This will
cause a current flow in your shield, which increases the noise in your data communication link. So
we now have a design trade-off : running a reference ground through your shield increases noise,
but so does grounding a circuit at both ends. Which solution is better or worse I have seen no good
answer to yet. But it is clear that grounding the circuit at both ends (ie: partial isolation) will cause
the common mode surge problems mentioned above, while running the ground through the shield
eliminates these surges.

The third alternative is to connect the floating earth of the isolated interface to the surge protection earth. Therefore the surge devices keep the signal voltages referenced to approximately
the correct level.

Specific recommendations

1) Both devices are situated in the same functional area and sharing the same power system.

     Need for isolation:.. Low
     Need for common mode surge protection: .... Low

Examples: a computer and control device in the same room and both plugged into power points in the same circuit of the power distribution system.

Since both of these device share an almost identical ground, the probability of a significant ground potential developing between them is slight.

2) Devices are situated in the different functional areas but sharing the same power system

     Need for isolation:.. Medium
     Need for common mode surge protection: .... Low

Examples: a computer in a control room and a control device on the plant floor. Each is on a different circuit of the same power distribution system.

The greater the electrical distance of the ground connection between the two devices, the greater the probability of a significant ground potential developing between them. While engineers like to quote the theorem "connected grounds are like grounds", *this only applies in normal, steady-state conditions*. Even if the two device are on very nearly the same power distribution circuit today, future changes to the power distribution panel may put them into completely different sub-systems. There is also another risk here. Since the functional areas are different, the person maintaining (or vendor supplying) each end of the link may be different. Adding isolation is a wise political move to prevent finger pointing in the event of a system failure.

3) Devices are in the same building but having different power systems

     Need for isolation:.. **High**
     Need for common mode surge protection: .... Medium

Examples: a computer powered by 230vac and a control device powered by a 110vac UPS or two control devices powered by different physical generators.

There is a very high probability of a significant ground potential developing between them during lightning storms or system faults. The small data communications wire may become a bridge or short circuit between two very large power systems. Even if two devices normally share the same UPS, if users can easily move the power plugs, someday they will not be sharing the same UPS. One will be connected to the USP and one to main supply.

4) Devices are in different buildings

     Need for isolation:.. **High**
     Need for common mode surge protection: .... **High**

Examples: a computer connected to a control device in an outside building, or a computer connected to a terminal device in a detached guard house.

Since the two "sites" are not connected by a metal structural ground, there is a very high probability of a significant ground potential developing between them during lightning storms.

Robust DataComm can truly make your data flow like water - safely, sanely, and silently.

For more information, contact:

Robust DataComm, Inc.
2142, 3rd St N.W.
Saint Paul, MN 55112, U.S.A.
Tel: (612)628-0533 Fax: (612)628-9642
email : info@robustdc.com

Robust DataComm Pte Ltd
Block 32, Defu Lane 10, #04-34
Singapore 539213
Tel: (65)487-5624 Fax: (65)487-5634/288-3306
web : http//www.robustdc.com

Ten Ways to Bulletproof RS-485 Interfaces

National Semiconductor
Application Note 1057
John Goldie
October 1996

Despite its widespread use, RS-485 is not as well under-stood as it should be. However, if you invest a little time on familiarizing yourself with the bus and pay attention to 10 as-pects of your application, you'll find that designing rock-solid implementations is easy.

Recommended Standard 485 (RS-485) has become the in-dustry's workhorse interface for multipoint, differential data transmission. RS-485 is unique in allowing multiple nodes to communicate bidirectionally over a single twisted pair. No other standard combines this capability with equivalent noise rejection, data rate, cable length, and general robustness. For these reasons, a variety of applications use RS-485 for data transmission. The list includes automotive radios, hard-disk drives, LANs, cellular base stations, industrial pro-grammable logic controllers (PLCs), and even slot ma-chines. The standard's widespread acceptance also results from its generic approach, which deals only with the inter-face's electrical parameters. RS-485 does not specify a con-nector, cable, or protocol. Higher level standards, such as the ANSI's SCSI standards and the Society of Automotive Engineers' (SAE's) J1708 automotive-communication stan-dard, govern these parameters and reference RS-485 for the electrical specifications.

Although RS-485 is extremely popular, many system design-ers must learn how to address its interface issues. You should review 10 areas before you design an RS-485 inter-face into a product. Understanding the issues during system design can lead to a trouble-free application and can reduce time to market.

RS-485 addresses a need beyond the scope of RS-422, which covers buses with a single driver and multiple receiv-ers. RS-485 provides a low-cost, bidirectional, multipoint in-terface that supports high noise rejection, fast data rates, long cable, and a wide common mode range. The standard specifies the electrical characteristics of drivers and receiv-ers for differential multipoint data transmission but does not specify the protocol, encoding, connector mechanical char-acteristics, or pinout. RS-485 networks include many sys-tems that the general public uses daily. These applications appear wherever a need exists for simple, economical com-munication among multiple nodes. Examples are gas-station pumps, traffic and railroad signals, point-of-sale equipment, and aircraft passenger seats. The Electronic Industries As-sociation (EIA) Technical Recommendation Committee, TR30, made RS-485 a standard in 1983. The Telecommunit-cations Industry Association (TIA) is now responsible for re-visions. RS-485 is currently being revised. After successful balloting, the revised standard will become "ANSI TIA/ EIA-485-A."

The 10 considerations that you should review early in a sys-tem design are:

- Mode and nodes,
- Configurations,
- Interconnect media,
- Data rate vs cable length,
- Termination and stubs,
- Unique differential and RS-485 parameters,
- Grounding and shielding,
- Contention protection,
- Special-function transceivers, and
- Fail-safe biasing.

MODE AND NODES

In its simplest form, RS-485 is a bidirectional half-duplex bus comprising a transceiver (driver and receiver) located at each end of a twisted-pair cable. Data can flow in either di-rection but can flow only in one direction at a time. A full-duplex bus, on the other hand, supports simultaneous data flow in both directions. RS-485 is mistakenly thought to be a full-duplex bus because it supports bidirectional data transfer. Simultaneous bidirectional transfers require not one but two data pairs, however.

RS-485 allows for connection of up to 32 unit loads (ULs) to the bus. The 32 ULs can include many devices but com-monly comprise 32 transceivers. *Figure 1* illustrates a multi-point bus. In this application, three transceivers—two re-ceivers and one driver—connect to the twisted pair. You must observe the 32-UL limitation, because the loads appear in parallel with each other and add to the load that the termination resistors present to the driver. Exceed-ing 32-UL loads excessively limits the drivers and attenuates the differential signal, thus reducing the differential noise margin.

RS-485 drivers are usually called "60 mA drivers." The name relates to the allowable loading. Developing 1.5V across the 60Ω termination load (120Ω at each end of the bus) requires 25 mA. The worst-case input current of a UL is 1 mA (at ex-treme common mode, explained later). *Figure 2* shows the loading curve of a full UL. The worst-case UL input resis-tance is 10.56 kΩ, although a frequently quoted incorrect value is 12 kΩ. Thus, 32 ULs require 32 mA drive capability. Adding this current to the 25 mA for the terminations yields 57 mA, which rounds up to an even 60 mA. A driver that can-not supply the full 60 mA violates the standard and reduces the bus's performance. The resulting problems include re-duced noise margin, reduction in the number of unit loads or allowable cable length, and limited common-mode voltage tolerance.

Designers frequently ask, "What is the maximum number of transceivers the bus allows?" The standard does not specify a maximum number of transceivers, but it does specify a maximum of 32 ULs. If a transceiver imposes one unit load, the maximum number of transceivers is also 32. You can now obtain transceivers with ½- and ¼-UL ratings, which al-low 64 and 128 transceivers. However, these fractional-UL devices, with their high-impedance input stages, typically op-erate much more slowly than do single-UL devices. The lower speed is acceptable for buses operating in the low hundreds of kilobits per second, but it may not be acceptable for a 10 Mbps bus.

Note:

EDN Design Feature Article, Reprinted with permission, Copyright 1996 Reed Elsevier Inc.

A solution exists for high-speed buses: You can use RS-485 repeaters to connect multiple buses end to end. In this setup, each bus must have no more than 32 loads. Directional control of the repeaters is complex, but hardware can handle it (Reference 1). Therefore, a conservative estimate is that, without using special transceivers, a bus can include 32 transceivers.

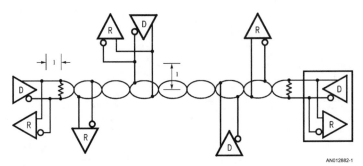

An RS-485 bus supports two-way data transfer over a single pair of wires. A typical bus includes multiple nodes. Each transceiver includes a differential driver, D, and a differential receiver, R. The stub length is I. The bus is terminated only at the ends — not at each node.

FIGURE 1.

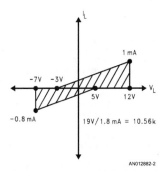

The loading of a tranceiver must remain within the shading region to be one unit load.

FIGURE 2.

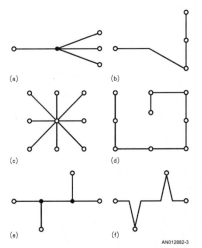

Although you can make RS-485 buses with all of these configurations work, you should avoid the ones in (a), (c), and (e). Those in (b), (d), and (f), offer superior transmission-line performance.

FIGURE 3.

A final word of caution: Do not connect too many transceivers to the bus or lump too many transceivers too close together. The AC loading, which results mainly from the devices' pin I/O capacitance (usually about 15 pF), can alter the interconnect-medium impedance and cause transmission-line problems.

CONFIGURATIONS

Because RS-485 allows connecting multiple transceivers, the bus configuration is not as straightforward as in a point-to-point bus (RS-232C, for example). In a point-to-point bus, a single driver connects to one receiver alone. The optimal configuration for the RS-485 bus is the daisy-chain connection from node 1 to node 2 to node 3 to node n. The bus must form a single continuous path, and the nodes in the middle of the bus must not be at the ends of long branches, spokes, or stubs. *Figure 3a*, *Figure 3c*, and *Figure 3e* illustrate three common but *improper* bus configurations. (If you mistakenly use one of these configurations, you can usually make it work but only through substantial effort and modification.) *Figure 3b*, *Figure 3d*, and *Figure 3f* show equivalent daisy-chained configurations.

Connecting a node to the cable creates a stub, and, therefore, every node has a stub. Minimizing the stub length minimizes transmission-line problems. For standard transceivers with transition times around 10 ns, stubs should be shorter than 6 in. A better rule is to make the stubs as short as possible. A "star" configuration (*Figure 3c*) is a special case and a cause for concern. This configuration usually does not provide a clean signaling environment even if the cable runs are all of equal length. The star configuration also presents a termination problem, because terminating every endpoint would overload the driver. Terminating only two endpoints solves the loading problem but creates transmission-line problems at the unterminated ends. A true daisy-chain connection avoids all these problems.

INTERCONNECT MEDIA

The standard specifies only the driver-output and receiver-input characteristics — not the interconnection medium. You can build RS-485 buses using twisted-pair cables, flat cable, and other media, even backplane pc traces. However, twisted-pair cable is the most common. You can use a range of wire gauges, but designers most frequently use 24 AWG. The characteristic impedance of the cable should be 100Ω to 120Ω. A common misconception is that the cable's chatacteristic impedance (Z_0) must be 120Ω, but 100Ω works equally well in most cases. Moreover, the 120Ω cable's higher Z_0 presents a lighter load, which can be helpful if the cable runs are extremely long.

Twisted pair offers noise benefits over flat or ribbon cables. In flat cable, a noise source (usually a conductor carrying an unrelated signal) can be closer to one member of the conductor pair than to the other over an entire wiring-run length. In such cases, more noise capacitively couples to the closer conductor than to the more distant one, producing a differential noise signal that can be large enough to corrupt the data. When you use the twisted pair, the noise source is closest to each of the conductors for roughly half of the wiring-run length. Therefore, the two conductors pick up roughly equal noise voltages. The receiver rejects these voltages because they appear mainly as common mode.

A special ribbon cable that is useful for noise reduction intermixes relatively long twisted sections with short flat sections.

This cable provides the advantages of twisted pair between the flat sections and allows the use of insulation displacement connectors at the flat points.

DATA RATE VS CABLE LENGTH

You can transmit data over an RS-485 bus for 4000 ft (1200m), and you can also send data over the bus at 10 Mbps. But, you cannot send 10 Mbps data 4000 ft. At the maximum cable length, the maximum data rate is not obtainable: The longer the cable, the slower that data rate, and vice versa. *Figure 4a* shows a conservative curve of data rate vs cable length for RS-422 and RS-485. The two slopes result from different limitations. The maximum cable length is the result of the voltage divider that the cable's DC loop resistance and the termination resistance create. Remember, for differential buses, the loop resistance is twice as high as you might expect, because both conductors in the pair equally contribute.

The curve's sloped portion results from AC limitations of the drivers and the cable. *Figure 4b* shows four limits for a DS3695 transceiver that drives a common twisted-pair cable. Notice that the data rate vs cable length depends significantly on how you determine the necessary signal quality. This graph includes two types of criteria. The first is a simple ratio of the driver's transition time to the unit interval. A curve showing the results for the common 30% ratio defines the most conservative set of operating points.

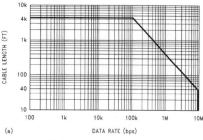

(a)

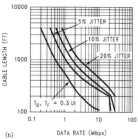

(b)

AN012882-4

As you increase the cable length, the maximum data rate decreases. The more jitter you can accept, the greater is the allowable data rate for a given length of cable.

FIGURE 4.

Special Transceivers Solve Special Problems		
DS3696/A — Thermal shutdown reporting pin: This device provides an open collector pin that reports the occurrence of a severe bus fault that has caused a thermal shutdown of the driver (>150˚C junction temperature).	**DS36276** — Fail-safe transceiver: Standard transceiver pinout with fail-safe detecting receiver, optimal for use with UARTs and asynchronous buses.	**DS36C279** — Ultra-low-power CMOS transceiver with automatic-sleep mode: Optimizes current with the automatic-sleep mode. With inactivity on the enable lines, I_{CC} drops to less than 10 µA.
DS3697 — Repeater pinout: Special pinout that internally connects a receiver port to a driver port. You need two of these devices for a bidirectional repeater.	**DS36277** — Fail-safe transceiver with active-low driver enable: Similar to the DS36276 but includes an active-low driver-enable pin. This feature allows a simplified connection to a UART and supports dominant-mode operation (use of the enable pin as the data pin).	**DS36C280** — Ultra-low-power CMOS transceiver with adjustable-slew-rate control: Adjustable driver slew rate allows tailoring for long stub lengths and reduced emissions.
DS3698 — Repeater pinout with thermal-shutdown pin: A repeater device that also provides the thermal-shutdown reporting pin.	**DS36C278** — Ultra-low-power CMOS transceiver: µA supply current and full RS-485 drive capability. One-fourth unit load allows up to 128 transceivers on the bus.	**DS36954** — Quad transceiver: Offers four independent transceivers in a single package. Useful for parallel buses.

A second method of determining the operating points uses eye-pattern (jitter) measurements. To make such measurements, you apply a pseudo random bit sequence (PRBS) to the driver's input and measure the resulting eye pattern at the far end of the cable. The amount of jitter at the receiver's threshold vs the unit interval yields the data point. Less jitter means better signal quality. Common operating curves use 5, 10, or 20% jitter. Above 50%, the eye pattern starts to close, and error-free data recovery becomes difficult (Reference 2). The key point is that you can't obtain the maximum data rate at the maximum cable length. But, if you operate the bus within the published, conservative curves, you can expect an error-free installation.

TERMINATION AND STUBS

Most RS-485 buses require termination because of fast transitions, high data rates, or long cables. The purpose of the termination is to prevent adverse transmission-line phenomena, such as reflections. Both ends of the main cable require termination. A common mistake is to connect a terminating resistor at each node — a practice that causes trouble on buses that have four or more nodes. The active driver sees the four termination resistors in parallel, a condition that excessively loads the driver. If each of the four nodes connects a 100Ω termination resistor across the bus, the active driver sees a load of 25Ω instead of the intended 50Ω. The problem becomes substantially worse with 32 nodes. If each node includes a 100Ω termination resistor, the load becomes 3.12Ω. You can include provisions for termination at every node, but you should activate the termination resistors only at the end nodes (by using jumpers, for example).

Stubs appear at two points. The first is between the termination and the device behind it. The second is between the main cable and a device at the middle of the cable. *Figure 1* shows both stubs. The symbol "l" denotes the stub length. Keep this distance as short as possible. Keeping a stub's electrical length below one-fourth of the signal's transition time ensures that the stub behaves as a lumped load and not as a separate transmission line. If the stub is long, a signal that travels down the stub reflects to the main line after hit-ting the input impedance of the device at the end of the stub. This impedance is high compared with that of the cable. The net effect is degradation of signal quality on the bus. Keeping the stubs as short as possible avoids this problem. Instead of adding a long branch stub, loop the main cable to the device you wish to connect. If you must use a long stub, drive it with a special transceiver designed for the purpose.

TERMINATION OPTIONS

You have several options for terminating an RS-485 bus. The first option is no termination. This option is feasible if the cable is short and if the data rate is low. Reflections occur, but they settle after about three round-trip delays. For a short-cable, the round-trip delay is short and, if the data rate is low, the unit interval is long. Under these conditions, the reflections settle out before sampling, which occurs at the middle of the bit interval.

The most popular termination option is to connect a single resistor across the conductor pair at each end. The resistor value matches the cable's differential-mode characteristic impedance. If you terminate the bus in this way, no reflections occur, and the signal fidelity is excellent. The problem with this termination option is the power dissipated in the termination resistors.

If you must minimize power dissipation, an RC termination may be the solution. In place of the single resistor, you use a resistor in seris with a capacitor. The capacitor appears as a short circuit during transitions, and the resistor terminates the line. Once the capacitor charges, it blocks the DC loop current and presents a light load to the driver. Lowpass effects limit use of the RC termination to lower data-rate applications, however (Reference 3).

Another popular option is a modified parallel termination that also provides a fail-safe bias. A detailed discussion of fail-safe biasing occurs later in the article. *Figure 5* compares the four popular termination methods. The main point to remember is that, if you use termination, you should locate the termination networks at the two extreme ends of the cable, not at every node.

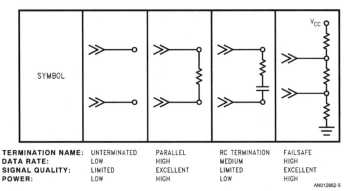

TERMINATION NAME:	UNTERMINATED	PARALLEL	RC TERMINATION	FAILSAFE
DATA RATE:	LOW	HIGH	MEDIUM	HIGH
SIGNAL QUALITY:	LIMITED	EXCELLENT	LIMITED	EXCELLENT
POWER:	LOW	HIGH	LOW	HIGH

AN012882-5

RS-485 buses can use four methods of termination. Achieving the best electrical performance requires accepting higher power dissipation in the termination resistors.

FIGURE 5.

UNIQUE DIFFERENTIAL AND RS-485 PARAMETERS

Four parameters that are important to differential data transmission and RS-485 are V_{OD}, V_{OS}, V_{GPD}, and V_{CM}. *Figure 6*, *Figure 7* and *Figure 8* illustrate these parameters, which are not common in the world of single-ended signaling and standard logic families.

V_{OD} represents the differential output voltage of the driver across the termination load. The RS-485 standard refers to this parameter as "termination voltage" (V_T), but V_{OD} is also commonly used. You measure V_{OD} differentially across the transmission line — not with respect to ground. On long cable runs, the DC resistance attenuates V_{OD}, but the receivers require only a 200 mV potential to assume the proper state. Attenuation, therefore, is not a problem. At the driver output, V_{OD} is 1.5V minimum. The IC manufacturer should guarantee this voltage under two test conditions: The first uses a simple differential load resistor. The second includes two 375Ω resistors connected to a common-mode supply. These resistors model the input impedance of 32 parallel ULs, all referenced to an extreme common-mode voltage. To make the 1.5V limit in this test, the driver must source or sink roughly 60 mA. This test is difficult and is important, because it essentially guarantees the system's differential-noise margin under worst-case loading and common-mode conditions.

Data sheets for RS-485 drivers usually do not include V_{OL} or V_{OH} specifications. The driver's V_{OL} is typically around 1V. Even for CMOS devices, V_{OH} is slighly above 3V, because both the source and sink paths of the output structure include a series-connected diode, which provides the common-mode tolerance for an Off driver. Because V_{OL} is usually greater than 0.8V, an RS-485 driver is not TTL-compatible.

V_{OS} represents the driver's offset voltage measured from the center point of the load with respect to the driver's ground reference. V_{OS} is also called "V_{OC}" for output common-mode voltage. This parameter is related to V_{CM}.

V_{CM} represents the common-mode voltage for which RS-485 is famous. The limit is −7V to +12V. Common-mode voltage is defined as the algebraic mean of the two local-ground-referenced voltages applied to the referenced terminals (receiver input pins, for example). The common-mode voltage represents the sum of three voltage sources. The first is the active driver's offset voltage. The second is coupled noise that shows up as common mode on both signal lines. The third is the ground-potential difference between the node and the active driver on the bus. Mathematically,

$$V_{CM} = V_{OS} + V_{NOISE} + V_{GPD}.$$

V_{GPD} represents the ground-potential difference that can exist between nodes in the system. RS-485 allows for a 7V shift in grounds. A shift of 7V below the negative (0V) power rail yields the −7V common-mode limit, whereas 7V above the 5V positive power rail yields the other common-mode limit of 12V. Understanding these parameters enables improved component selection, because some devices trade off certain parameters to gain others.

To further illustrate RS-485's common-mode noise-rejection capability, you can conduct the following test: Connect a driver to a receiver via an unshielded twisted-pair cable. Then, couple a noise signal onto the line, and, from the scope, plot the resulting waveform at the receiver input (*Figure 9*). The plot includes the receiver's output signal. Note that the receiver clearly detects the correct signal state, despite the common-mode noise. Differential transmission offers this high noise rejection; a single-ended system would erroneously switch states several times under these test conditions.

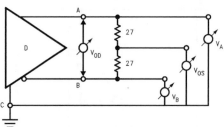

AN012882-6

Because RS-485 buses are differential, they involve parameters that have no counterparts in single-ended systems.

FIGURE 6.

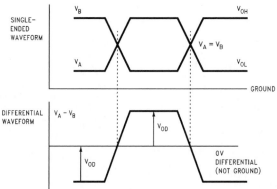

AN012882-7

Although the voltages on the bus's two conductors vary symmetrically about a value that is halfway between the driver's Low and High output-state values, the differential voltage between the conductors varies about zero. In addition, the differential voltage swing is double the swing on either of the bus conductors.

FIGURE 7.

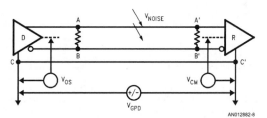

AN012882-8

The common-mode voltage, V_{CM}, is the average of the voltages at a driver's two outputs or a receiver's two inputs. You measure all voltages with respect to the driver's or receiver's common terminal.

FIGURE 8.

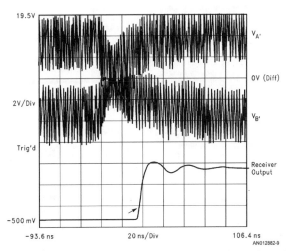

19.5V

$V_{A'}$

0V (Diff)

2V/Div

$V_{B'}$

Trig'd

Receiver
Output

−500 mV

−93.6 ns   20 ns/Div   106.4 ns
                                        AN012882-9

Because of the bus's differentail nature and the receiver's common-mode rejection, the receiver's output (lowest trace) is clean despite large common-mode noise voltages on both of the bus conductors.

FIGURE 9.

Standards Related to RS-485

EIA RS-485 — Originally published in 1983, the multipoint standard specifies the concept of the unit load along with electrical characteristics of the drivers and receivers. It was developed from the RS-422 standard, adding multipoint capability, extended common-mode range, increased drive capability, and contention protection. It is an "electrical-only" standard that does not specify the function of the bus or any connectors.	**TIA/EIA-485-A (PN-3498)** — The TIA expects completion of the first revision to RS-485 this year (Project Number 3498). The goals of the revision are to clean up vague text and to provide additional information to clarify certain technical topics. This work will become the "A" revision of RS-485 once the balloting (approval) process is complete. In addition to the revision work, an application bulletin (PN-3615) is also in process. This document will provide additional application details and system considerations to aid the designer.	**ISO/IEC 8482.1993** — The current revision of this international standard maps closely to RS-485. The original ISO standard specified different limits and conditions. However, the 1993 revision changed many of these differences, and RS-485 and ISO 8482 are now similar.

GROUNDING AND SHIELDING

Although the potential difference between the data-pair conductors determines the signal without officially involving ground, the bus needs a ground wire to provide a return path for induced common-mode noise and currents, such as the receivers' input current. A typical mistake is to connect two nodes with only two wires. If you do this, the system may radiate high levels of EMI, because the common-mode return current finds its way back to the source, regardless of where the loop takes it. An intentional ground provides a low-impedance path in a known location, thus reducing emissions.

Electromagnetic-compatibility and application requirements determine whether you need a shield. A shield both prevents the coupling of external noise to the bus and limits emissions from the bus. Generally, a shield connects to a solid ground (normally, the metal frame around the system or subsystem) with a low impedance at one end and a series RC network at the other. This arrangement prevents the flow of DC ground-loop currents in the shield.

CONTENTION PROTECTION

Because RS-485 allows for connecting multiple drivers to the bus, the standard addresses the topic of contention. When two or more drivers are in contention, the signal state on the bus is not guaranteed. If two drivers are on at the same time and if they are driving the same state, the bus state is valid. However, if the drivers are in opposite states, the bus state is undetermined, because the differential voltage on the bus drops to a low value within the receiver's threshold range. Because you do not know the driver states, you must assume the worst—namely, that the data on the bus is invalid.

Contention can also damage the ICs. If several drivers are in one state, a single driver in the opposite state sinks a high current (as much as 250 mA). This large current causes excessive power dissipation. A difference in ground potential between nodes only aggravates this dissipation. In this situation, the driver's junction temperature can increase beyond safe limits. The RS-485 standard recommends the use of special circuitry, such as a thermal shutdown circuit, to prevent such damage. Most RS-485 devices use this technique. The shutdown circuit disables the driver outputs when the junction temperature exceeds 150°C and automatically re-enables the outputs when the junction cools. If the fault is still present, the device cycles into and out of thermal shutdown until someone clears the fault.

Besides thermal shutdown, other current limiting is required to prevent accidental damage. If an active output is shorted to any voltage within the −7V to +12V range, the resulting current must not exceed 250 mA. In addition, the outputs of a driver must not sustain damage if they are shorted together indefinitely. (Entering thermal shutdown is allowed, of course.) Lastly, RS-485 drivers must source and sink large currents (60 mA). This situation requires outputs of rather large geometry, which provide robust ESD protection.

SPECIAL-FUNCTION TRANSCEIVERS

You can handle many of the above-mentioned issues by using special transceivers, of which there are several types, differing in pinout or functions supported. The most common device is a standard transceiver (DS3695/DS75176B), which provides a two-pin connection to the RS-485 bus and a four-pin TTL interface (driver input, driver enable, receiver output, and receiver enable). Among the problems you can solve with an appropriate transceiver are these:

For ultra-low-power applications, the DS36C279 provides an auto-sleep function. Inactivity on the two enable lines automatically triggers the sleep mode, dropping the power-supply current to less than 10 µA. This characteristic is extremely valuable in applications that provide an interface connection but that are connected to their cables for down-loading only a small percentage of the time. This is the case for package-tracking boxes carried by many overnight-delivery services. With this sleep feature, idle transceivers do not consume precious battery current.

For applications that are asynchronous and based on a standard UART, fail-safe biasing is an issue. UARTs look for a low or a high state, and, between characters, the line usually remains high. With RS-485, this condition is troublesome, because, when there are no active drivers on the bus, the bus state is undetermined. (See the following section for a detailed discussion of fail-safe biasing.) In this case, the DS36276 simplifies the hardware design. This unique transceiver's receiver detects a high state for a driven high and also for the nondriven (V_{ID} = 0) bus state, thus providing the UART with the high state between characters and only valid start bits.

Although the discussion of configurations and the section on stubs advises minimizing stub length to avoid transmission-line problems, the application may not permit minimizing stub length. Another approach is to increase the driver's transition time to permit longer stubs without transmission-line effects. If you use the DS36C280, long stubs can branch off the main cable. This arrangement keeps the main cable short, whereas looping the cable back and forth to reach inconveniently located nodes would greatly increase the main-cable length. Besides allowing longer stubs, the slower edge rates generate lower emissions. Thus, this transceiver is also useful for applications that severely limit emitted noise.

FAIL-SAFE BIASING

The need for fail-safe operation is both the principal application issue and most frequently encountered problem with RS-485. Fail-safe biasing provides a known state in which there are no active drivers on the bus. Other standards do not have to deal with this issue, because they typically define a point-to-point or multidrop bus with only one driver. The one driver either drives the line or is off. Because there is only one source on the bus, the bus is off when the driver is off. RS-485, on the other hand, allows for connection of multiple drivers to the bus. The bus is either active or idle. When it is idle with no drivers on, a question arises as to the state of the bus. Is it high, low, or in the state last driven? The answer is any of the above. With no active drivers and low-impedance termination resistors, the resulting differential voltage across the conductor pair is close to zero, which is in the middle of the receivers' thresholds. Thus, the state of the bus is truly undetermined and cannot be guaranteed.

Some of the functional protocols that many applications use aggravate this problem. In an asynchronous bus, the first transition indicates the start of a character. It is important for the bus to change states on this leading edge. Otherwise, the clocking inside the UART is out of sync with the character and creates a framing error. The idle bus can also randomly switch because of noise. In this case, the noise emulates a valid start bit, which the UART latches. The result is a framing error or, worse, an interrupt that distracts the CPU from other work.

The way to provide fail-safe operation requires only two additional resistors. At one end of the bus (the master node, for example), connect a pullup and pulldown resistor (*Figure 10*). This arrangement provides a simple voltage divider on the bus when there are no active drivers. Select the resistors so that at least 200 mV appears across the conductor pair. This voltage puts the receivers into a known state. Values that can provide this bias are 750Ω for the pullup and pulldown resistors, 130Ω across the conductor pair at the fail-safe point, and a 120Ω termination at the other end of the cable. For balance, use the same value for the pullup and pulldown resistors. Reference 4 provides extensive details on this issue.

Forethought into these 10 areas before production greatly reduces the likelihood of problems. RS-485 is unique in its capabilities and requirements. Fully understanding these 10 issues leads to a rock-solid, trouble-free, multipoint differential interface that maximizes the benefits of RS-485 and provides the application with robust, rugged, highly noise-tolerant data communication.

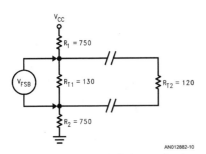

AN012882-10

Unless you do something to keep the situation from occurring, when no driver is driving the bus, the receivers cannot determine the bus state. Fail-safe biasing is a bus-termination method, which ensures that, even when no driver is active, a differential voltage large enough to unambiguously determine the bus state appears at the receiver inputs.

FIGURE 10.

REFERENCES

1. Murdock, G and J Goldie, "AN-702: Build a direction-sensing bidirectional repeater," *Interface Databook*, National Semiconductor Corp, 1996.

2. True, K, "AN-808: Long transmission lines and data-signal quality." *Interface Databook*, National Semiconductor Corp, 1996.

3. Vo, J. "AN-903: A comparison of differential-termination techniques," *Interface Databook*, National Semiconductor Corp, 1996.

4. Goldie, J, "AN-847: Fail-safe biasing of differential buses," *Interface Databook*, National Semiconductor Corp, 1996.

5. ANSI/TIA/EIA-422-B-1995, *Electrical characteristics of balanced-voltage digital-interface circuits.*

6. EIA RS-485-1983, *Electrical characteristics of generators and receivers for use in balanced digital-multipoint systems.*

7. Sivasothy, S, "AN-409: Transceivers and repeaters meeting the EIA RS-485 interface standard," *Interface Databook*, National Semiconductor Corp, 1996.

8. ISO/IEC 8482:1993, *Information technology — telecommunications and information exchange between systems — twisted-pair multipoint interconnections.*

LIFE SUPPORT POLICY

NATIONAL'S PRODUCTS ARE NOT AUTHORIZED FOR USE AS CRITICAL COMPONENTS IN LIFE SUPPORT DE-
VICES OR SYSTEMS WITHOUT THE EXPRESS WRITTEN APPROVAL OF THE PRESIDENT OF NATIONAL SEMI-
CONDUCTOR CORPORATION. As used herein:

1. Life support devices or systems are devices or systems which, (a) are intended for surgical implant into
   the body, or (b) support or sustain life, and whose failure to perform when properly used in accordance
   with instructions for use provided in the labeling, can
   be reasonably expected to result in a significant injury
   to the user.

2. A critical component in any component of a life support
   device or system whose failure to perform can be reasonably expected to cause the failure of the life support
   device or system, or to affect its safety or effectiveness.

National Semiconductor Corporation Americas	National Semiconductor Europe	National Semiconductor Asia Pacific Customer Response Group	National Semiconductor Japan Ltd.
Tel: 1-800-272-9959	Fax: +49 (0) 1 80-530 85 86	Tel: 65-2544466	Tel: 81-3-5620-6175
Fax: 1-800-737-7018	Email: europe.support@nsc.com	Fax: 65-2504466	Fax: 81-3-5620-6179
Email: support@nsc.com	Deutsch Tel: +49 (0) 1 80-530 85 85	Email: sea.support@nsc.com	
www.national.com	English Tel: +49 (0) 1 80-532 78 32		
	Français Tel: +49 (0) 1 80-532 93 58		
	Italiano Tel: +49 (0) 1 80-534 16 80		

AN-1057

The Practical Limits of RS-485

National Semiconductor
Application Note 979
Todd Nelson
March 1995

INTRODUCTION

This application note discusses the EIA-485 standard for differential multipoint data transmission and its practical limits. It is commonly called RS-485, however its official name is EIA-485 which reflects the name of the committee at the time it was released. It is expected to be revised soon and will then become TIA/EIA-485-A.

Differential data transmission is ideal for transmitting at high data rates, over long distances and through noisy environments. It nullifies the effects of ground shifts and noise signals which appear as common mode voltages on the transmission line. TIA/EIA-422-B is a standard that defines differential data transmission from a single driver to multiple receivers. RS-485 allows multiple drivers in operation, which makes multipoint (party line) configurations possible.

This application note will discuss the specifications as defined in the RS-485 document. Interpretations of the standard and device specifications can vary among manufacturers. However, there are some guarantees required to be completely compliant with the standard.

There are many possibilities and trade-offs associated with being partially compliant— or ™compatible. Some applications can tolerate the trade-offs in return for increased performance or added value. For that reason, this application note will discuss the practical application of the specifications.

A detailed explanation of each requirement of the standard will not be given as this is beyond the scope of this note. Also beyond the scope are advanced topics relating to new technology.

KEY RS-485 REQUIREMENTS

The key features are:

- Differential (Balanced) Interface
- Multipoint Operation
- Operation from a single +5V Supply
- -7V to +12V Bus Common Mode Range
- Up to 32 Unit Loads (Transceivers)
- 10 Mbps Maximum Data Rate (@40 feet)

4000 Foot Maximum Cable Length (@100 kbps)

A typical application is shown in *Figure 4.*

The key requirement of the driver is its guaranteed differential output voltage as measured: with no load; with a minimum configuration of two nodes; and with the full load of 32 nodes. The terms used in the specification are:

V_{OA} True output voltage with respect to ground

V_{OB} Complimentary output voltage with respect to ground

V_{OD} Differential output voltage ($V_{OA} \pm V_{OB}$)

V_{OS} Offset voltage, or center point of V_{OA} or V_{OA}, also called V_{OC}

V_{CM} Algebraic mean of V_{OA} and V_{QB}, including any ground potential difference or noise

The specifications are best represented by the following figures and table.

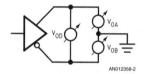

FIGURE 1. No Load Configuration

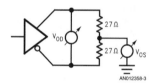

FIGURE 2. Termination Load Configuration

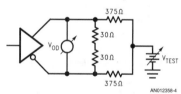

FIGURE 3. Full Load Configuration

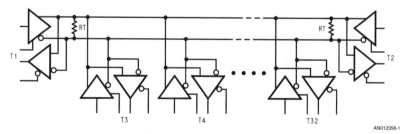

FIGURE 4. Typical RS-485 Application

TABLE 1. Driver Output Voltage Requirements

Configuration	Test	Min	Max	Units
No Load	V_{OD1}	1.5	6.0	V
Figure 1	V_{OA}	0	6.0	V
	V_{OB}	0	6.0	V
Termination	V_{OD2}	1.5	5.0	V
Figure 2	V_{OS}	-1.0	3.0	V
Full Load	V_{OD3}	1.5	5.0	V
Figure 3	with $-7V \leq V_{CM} \leq +12V$			

There is also a condition that the driver must not be damaged when the outputs are shorted to each other or any potential within the common mode range of ‚7V to +12V. The peak current under shorted conditions must be less than 250 mA. This point is key to multipoint operation, since contention may occur.

The data rate requirements have implications on the speed of the device. Switching characteristics must specify that the transition time (t_r, t_f) be ≤ 0.3 of the unit interval. The minimum unit interval for 10 Mbps at 40 feet is 100 ns so $t_r/t_f \leq 33$ ns; for 100 kbps at 4000 feet it is 10 s so $t_r/t_f \leq 3.3$ s.

A Unit Load is defined as a load on the bus, it is commonly a driver and a receiver. The result should be that the unit load does not load down the bus under power-on or power-off conditions. Driver leakage tends to be in micro-amps but receiver input current can be significant compared to driver leakage. Four points define the unit load, shown in *Figure 5*.

Rec. Input current (I_{IN}) at +12V ≤ 1 mA

I_{IN} between +5V and +12V ≥ 0 mA

I_{IN} at ‚7V $\geq$ ‚0.8 mA

I_{IN} between ‚3V and ‚7V ≤ 0 mA

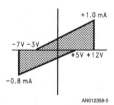

AN012358-5

FIGURE 5. V/I Relationship defining a Unit Load

The shaded area effectively defines the receiver input impedance (R_{IN}), ≥ 10.6 kΩ (19V/1.8 mA). The standard does not require a specific impedance, only that it falls within the shaded area.

The key receiver requirements are its threshold voltage levels and common mode range. The receiver output must be HIGH if the true input is more than 200 mV above the complimentary input; LOW if it is more than 200 mV below the complimentary input. This must be possible with the inputs varying from -7V to +12V. A graphic representation is shown

in *Figure 6*. In this diagram, the lightly shaded region represents the range of points where RI is more than 200 mV below RI*; therefore the output is LOW.

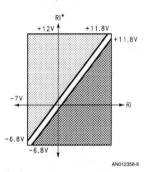

AN012358-6

FIGURE 6. Receiver Input Range

The 200 mV receiver threshold and the 1.5V minimum differential driver output voltage provide 1.3V of differential noise margin. Since the bus is typically a twisted pair, ground noise is canceled out by the differential operation. The result is a bus that is well suited for high data rates and noisy environments.

There are further requirements such as balance of terminated voltage, balance of offset voltage and timing which can be reviewed in the standard. Note that all of these requirements should be met over the full supply voltage and temperature range in which the device will operate.

Interpreting the standard and creating device specifications appears to be straight forward. However, the range of practices shows that there are differing opinions.

COMPATIBILITY TRADEOFFS

It is not always practical to meet all of the requirements. The devices may have limitations, the applications may not need full compliance or there may be a possible improvement in one area at the expense of another.

Commonly accepted minimum specifications for compatibility include V_{OD1}, V_{OD2}, I_{OS}, V_{CM}, V_{TH}. At times, these are specified at controlled conditions− not over the full operating range, as is required. Furthermore, ™Upto 32 unit loads . . ." implies V_{OD3} and R_{IN} or the V/I relationship discussed above. V_{OD3} can be traded off if the application is not expected to be fully loaded.

R_{IN} can be increased and thereby allowing more than 32 nodes to be connected without exceeding the 32 unit loads. For example, a R_{IN} of 24 kΩ implies that 64 nodes equates to 32 unit loads.

In many applications, I_{CC} is the differentiating factor. Optimizing a device for low power may slow switching speed. The end user may define the acceptable speed but switching speed is quite often defined by the choice of protocol.

Many low power technologies have lower breakdown voltages, which reduces the recommended maximum voltage range for the bus pins. The recommended voltage range for the bus pins defines how much protection the device has beyond the -7V to +12V common mode range. If the environment demands that the bus survive voltages up to †24V, then the device must guarantee this, otherwise external protection must be included.

External limitations may dictate controlled edge rates to allow greater stub lengths or reduced EMI, which may result in a device that does not meet the prescribed data rate. All of these trade-offs must be considered in the design of the system.

IMPLEMENTATION ISSUES

Topoloqy: RS-485 is defined as a multi-point bus, (*Figure 7*) therefore multiple drivers and receivers can be connected to the bus at the same time (see discussion regarding unit load).

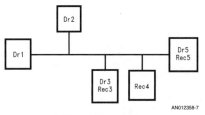

FIGURE 7. Bus Topology

In such a configuration, only one driver has control at a time and all the active receivers receive the same signals.

A ring which is created by connecting both ends of a bus together will not work. A traditional ring uses point-to-point links between the nodes. This can be implemented using RS-485, however, there are many other point-to-point technologies available.

Star configurations are also discouraged. In a star configuration (*Figure 8*) the device is effectively at the end of a very long stub, and this causes reflection and termination problems.

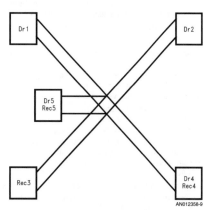

FIGURE 8. Star Topology

Stubs: RS-485 recommends keeping the stubs as short as practical. A stub is the distance from the device to the bus, or the termination resistor (in the case at the ends of the bus), see *Figure 9*. The maximum length is not defined by the standard, but longer stubs will have a negative impact on signal quality. This affect can be reduced by controlling the transition time of the driver.

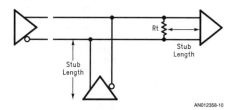

FIGURE 9. Stub Length

The driven signal encounters a reflection at the end of the stub, if this occurs within the rising edge of the signal then it can be neglected. A general rule is that stubs should be less than 1/3 of the transition time. Therefore, slowing the transition time can extend the practical stub length.

stub $\ell \leq$ 1/3 (transition time)/(velocity)

ℓft $\leq$ 1/3 (t_r or t_f)/(1.5 ns/ft)

Number of Nodes: The standard allows 32 unit loads, as defined by the driver leakage and receiver impedance− this was intended to mean 32 transceivers. If a number of the devices guarantee a greater impedance, then it is possible to add more than 32 transceivers to a bus.

Termination: RS-485 has defined the termination as 120Ω parallel termination at each end of the bus. This assumes a characteristic impedance in the range of Z_O = 100Ω to 120Ω for the cable. Other termination schemes could be implemented, but a thorough analysis must be done to assure adequate signal quality. For more information on termination, see AN-903.

Bus Faults: This bus is defined to be resistant to many of the faults associated with a cable environment such as noise and variations in device ground. It is built for party line applications so it can withstand driver contention. In most cases, there is enough noise margin to detect a valid HIGH or LOW. However, in the case where both lines are open or there is a short between the two lines, the state may be unknown. Such a case requires the designer to implement a "failsafe" scheme to bias the receiver to a known state. See AN-847 and AN-903 for a detailed discussion of failsafe techniques.

Data Rate: Earlier, the data rate vs distance guidelines were given as 10 Mbps at 40 feet and 100 kbps at 4000 feet. Advances in technology continue to push these limits. At long distances the practical limitation is dominated by the rise time degradation due to the cable. The approximate delay associated with 100Ω cable is 1.5 ns/foot. Therefore, 4000 feet of cable will cause 6 s of delay− which limits the data rate to 333 kbps (166 kHz) before device delays are involved. At 100 feet only 150 ns of delay are added by the cable, so an ideal driver/receiver could switch at 10 Mbps theoretically. Further complications are added by encoding schemes (PWM, RTZ, etc.) and protocol requirements (idle time, overhead, etc.). If an off-set bias is implemented for receiver failsafe, this may induce some signal distortion or cause slight duty cycle distortion which must be factored in to the data rate considerations.

RS-485 is defined as a half-duplex bus, though many applications use multiple channels in parallel or full-duplex. In parallel bus applications, channel-to-channel skew becomes a critical issue. These and possible protocol requirements would have to be considered in the evaluation of each device.

Supply Power: I_{CC} is not always the dominant indicator of power requirements. Low power CMOS devices require little quiescent current, typically less than 1 mA. However, when switching against a heavy load, the load current can be over 60 mA! And switching at higher frequencies also requires more current. Comparing bipolar and CMOS devices should include the case when switching a heavy load at high frequencies as well as the quiescent case. The total requirement will depend on the portion of time at idle versus switching.

Signal Quality: At the extremes of distance and data rate, the signal quality will be degraded. This is a qualitative parameter that is usually judged with eye patterns or in probabilities of errors.

Eye patterns show the effects of intersymbol interference, a hypothetical example is shown in *Figure 10*. A full discussion on signal quality is given in AN-808.

Interfacing to other standards: This bus is not intended to be inter-operable with other standards such as TIA/EIA-232-E or ECL. TIA/EIA-422-B buses can accept RS-485 devices, but the opposite case is not true for the drivers. For a full discussion on this topic, see AN-972. The international standard ISO 8482.1994 has recently become compatible with RS-485.

PRACTICAL LIMITS

Theoretical limits defined by the standard should not be exceeded without fully examining the trade-offs discussed above. However, there are some common practices which can extend RS-485 beyond its defined limits.

The maximum number of nodes can exceed 32. R_{IN} can be defined as 1/2 unit load or 1/4 unit load, thus extending the number of nodes that can be attached to a single bus to 64 or 128 respectively. The leakage specifications must also support the stated unit load. Note that a bus with 128 nodes requires that the average loading be 1/4 unit load− including any third-party nodes that may be attached. Not all devices need the same unit load rating, but the total cannot exceed 32 unit loads.

The common-mode voltage range requirements continue to be a factor that limits many other types of interfaces. TIA/EIA-422-B offers a common-mode range of †7V, but does not allow multiple drivers on the bus. The process technologies and design techniques required to meet the -7V to +12V range are somewhat unique. In fact, many applications see common-mode voltages beyond this range, such as †24V! Generally, reducing common-mode voltage in trade for any performance or integration gains has not been acceptable. Increasing common-mode beyond the RS-485 limits depends on the specific devices; wider common-mode may affect the thresholds and hysteresis of the receiver which reduces the noise margin.

Speed and power requirements are opposing trends: higher data rates tend to use more power, yet lower power (I_{CC}) technologies tend to be slower. Technologies that effectively combine both high speed and low power are becoming available, and will come down in cost. Optimizing for speed in excess of the RS-485 limits may require technologies that consume greater quiescent current. In applications that are not transmitting for extended periods, optimizing for low power is common. Such devices may not meet the 10 Mbps data rate referenced in RS-485, which is acceptable since many of these applications are specified between 9600 bps and 1 Mbps.

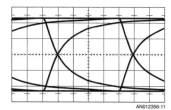

AN012358-11

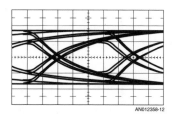

AN012358-12

FIGURE 10. Eye Patterns

CONCLUSION

RS-485 is a well-defined, multi-purpose electrical specification for multi-point data transmission. The standard allows manufacturers to optimize devices for speed and power. Despite the definition, there is still potential for compatibility issues if the devices are not fully specified.

Many of the limits imposed in RS-485 can be exceeded at some cost and with increased risk. But technology barriers are continuously being removed and this promises tremendous performance gains, perhaps eliminating those costs and risks.

RS-485 is a very rugged standard for multi-point applications. It has proven to be popular over a span of many years.

With the breadth of devices available and new technologies being applied, RS-485 will continue for many years to come.

REFERENCES

EIA RS-485 standard for differential multi-point data transmission

TIA/EIA-422-B standard for differential multi-drop data transmission

ISO 8482.1994 Information processing systems – Data communication Twisted pair multipoint inter-connections

The Practical Limits of RS-485

LIFE SUPPORT POLICY

NATIONAL'S PRODUCTS ARE NOT AUTHORIZED FOR USE AS CRITICAL COMPONENTS IN LIFE SUPPORT DE-
VICES OR SYSTEMS WITHOUT THE EXPRESS WRITTEN APPROVAL OF THE PRESIDENT OF NATIONAL SEMI-
CONDUCTOR CORPORATION. As used herein:

1. Life support devices or systems are devices or systems which, (a) are intended for surgical implant into
   the body, or (b) support or sustain life, and whose failure to perform when properly used in accordance
   with instructions for use provided in the labeling, can
   be reasonably expected to result in a significant injury
   to the user.

2. A critical component in any component of a life support
   device or system whose failure to perform can be reasonably expected to cause the failure of the life support
   device or system, or to affect its safety or effectiveness.

 National Semiconductor
Corporation
Americas
Tel: 1-800-272-9959
Fax: 1-800-737-7018
Email: support@nsc.com

www.national.com

National Semiconductor
Europe
          Fax: +49 (0) 1 80-530 85 86
          Email: europe.support@nsc.com
Deutsch Tel: +49 (0) 1 80-530 85 85
English Tel: +49 (0) 1 80-532 78 32
Fran'ais Tel: +49 (0) 1 80-532 93 58
Italiano Tel: +49 (0) 1 80-534 16 80

National Semiconductor
Asia Pacific Customer
Response Group
Tel: 65-2544466
Fax: 65-2504466
Email: sea.support@nsc.com

National Semiconductor
Japan Ltd.
Tel: 81-3-5620-6175
Fax: 81-3-5620-6179

AN-979

National does not assume any responsibility for use of any circuitry described, no circuit patent licenses are implied and National reserves the right at any time without notice to change said circuitry and specifications.

DALLAS
SEMICONDUCTOR

Application Note 83
Fundamentals of RS–232
Serial Communications

Due to it's relative simplicity and low hardware overhead (as compared to parallel interfacing), serial communications is used extensively within the electronics industry. Today, the most popular serial communications standard in use is certainly the EIA/TIA–232–E specification. This standard, which has been developed by the Electronic Industry Association and the Telecommunications Industry Association (EIA/TIA), is more popularly referred to simply as "RS–232" where "RS" stands for "recommended standard". In recent years, this suffix has been replaced with "EIA/TIA" to help identify the source of the standard. This paper will use the common notation of "RS–232" in its discussion of the topic.

The official name of the EIA/TIA–232–E standard is "Interface Between Data Terminal Equipment and Data Circuit–Termination Equipment Employing Serial Binary Data Interchange". Although the name may sound intimidating, the standard is simply concerned with serial data communication between a host system (Data Terminal Equipment, or "DTE") and a peripheral system (Data Circuit–Terminating Equipment, or "DCE").

The EIA/TIA–232–E standard which was introduced in 1962 has been updated four times since its introduction in order to better meet the needs of serial communication applications. The letter "E" in the standard's name indicates that this is the fifth revision of the standard.

RS–232 SPECIFICATIONS

RS–232 is a "complete" standard. This means that the standard sets out to ensure compatibility between the host and peripheral systems by specifying 1) common voltage and signal levels, 2) common pin wiring configurations, and 3) a minimal amount of control information between the host and peripheral systems. Unlike many standards which simply specify the electrical characteristics of a given interface, RS–232 specifies electrical, functional, and mechanical characteristics in order to meet the above three criteria. Each of these aspects of the RS–232 standard is discussed below.

ELECTRICAL CHARACTERISTICS

The electrical characteristics section of the RS–232 standard includes specifications on voltage levels, rate of change of signal levels, and line impedance.

The original RS–232 standard was defined in 1962. As this was before the days of TTL logic, it should not be surprising that the standard does not use 5 volt and ground logic levels. Instead, a high level for the driver output is defined as being +5 to +15 volts and a low level for the driver output is defined as being between –5 and –15 volts. The receiver logic levels were defined to provide a 2 volt noise margin. As such, a high level for the receiver is defined as +3 to +15 volts and a low level is –3 to –15 volts. Figure 1 illustrates the logic levels defined by the RS–232 standard. It is necessary to note that, for RS–232 communication, a low level (–3 to –15 volts) is defined as a logic 1 and is historically referred to as "marking". Likewise a high level (+3 to +15 volts) is defined as a logic 0 and is referred to as "spacing".

The RS–232 standard also limits the maximum slew rate at the driver output. This limitation was included to help reduce the likelihood of cross–talk between adjacent signals. The slower the rise and fall time, the smaller the chance of cross talk. With this in mind, the maximum slew rate allowed is 30 V/μs. Additionally, a maximum data rate of 20k bits/second has been defined by the standard. Again with the purpose of reducing the chance of cross talk.

The impedance of the interface between the driver and receiver has also been defined. The load seen by the driver is specified to be 3kΩ to 7kΩ. For the original RS–232 standard, the cable between the driver and the receiver was also specified to be a maximum of 15 meters in length. This part of the standard was changed in revision "D" (EIA/TIA–232–D). Instead of specifying the maximum length of cable, a maximum capacitive load of 2500 pF was specified which is clearly a more adequate specification. The maximum cable length is determined by the capacitance per unit length of the cable which is provided in the cable specifications.

RS–232 LOGIC LEVEL SPECIFICATIONS Figure 1

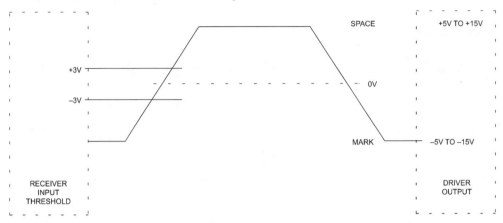

FUNCTIONAL CHARACTERISTICS

Since RS–232 is a "complete" standard, it includes more than just specifications on electrical characteristics. The second aspect of operation that is covered by the standard concerns the functional characteristics of the interface. This essentially means that RS–232 has defined the function of the different signals that are used in the interface. These signals are divided into four different categories: common, data, control, and timing. Table 1 illustrates the signals that are defined by the RS–232 standard. As can be seen from the table there is an overwhelming number of signals defined by the standard. The standard provides an abundance of control signals and supports a primary and secondary communications channel. Fortunately few applications, if any, require all of these defined signals. For example, only eight signals are used for a typical modem. Some simple applications may require only four signals (two for data and two for handshaking) while others may require only data signals with no handshaking. Examples of how the RS–232 standard is used in some "real world" applications are discussed later in this paper. The complete list of defined signals is included here as a reference, but it is beyond the scope of this paper to review the functionality of all of these signals.

MECHANICAL INTERFACE CHARACTERISTICS

The third area covered by RS–232 concerns the mechanical interface. In particular, RS–232 specifies a 25–pin connector. This is the minimum connector size that can accommodate all of the signals defined in the functional portion of the standard. The pin assignment for this connector is shown in Figure 2. The connector for DCE equipment is male for the connector housing and female for the connection pins. Likewise, the DTE connector is a female housing with male connection pins. Although RS–232 specifies a 25–position connector, it should be noted that often this connector is not used. This is due to the fact that most applications do not require all of the defined signals and therefore a 25–pin connector is larger than necessary. This being the case, it is very common for other connector types to be used. Perhaps the most popular is the 9–position DB9S connector which is also illustrated in Figure 2. This connector provides the means to transmit and receive the necessary signals for modem applications, for example. This will be discussed in more detail later.

RS–232 DEFINED SIGNALS Table 1

CIRCUIT MNEMONIC	CIRCUIT NAME*	CIRCUIT DIRECTION	CIRCUIT TYPE
AB	Signal Common	–	Common
BA	Transmitted Data (TD)	To DCE	Data
BB	Received Data (RD)	From DCE	
CA	Request to Send (RTS)	To DCE	
CB	Clear to Send (CTS)	From DCE	
CC	DCE Ready (DSR)	From DCE	
CD	DTE Ready (DTR)	To DCE	
CE	Ring Indicator (RI)	From DCE	
CF	Received Line Signal Detector** (DCD)	From DCE	Control
CG	Signal Quality Detector	From DCE	
CH	Data Signal Rate Detector from DTE	To DCE	
CI	Data Signal Rate Detector from DCE	From DCE	
CJ	Ready for Receiving	To DCE	
RL	Remote Loopback	To DCE	
LL	Local Loopback	To DCE	
TM	Test Mode	From DCE	
DA	Transmitter Signal Element Timing from DTE	To DCE	
DB	Transmitter Signal Element Timing from DCE	From DCE	Timing
DD	Receiver Signal Element Timing From DCE	From DCE	
SBA	Secondary Transmitted Data	To DCE	Data
SBB	Secondary Received Data	From DCE	
SCA	Secondary Request to Send	To DCE	
SCB	Secondary Clear to Send	From DCE	Control
SCF	Secondary Received Line Signal Detector	From DCE	

*Signals with abbreviations in parentheses are the eight most commonly used signals.
**This signal is more commonly referred to as Data Carrier Detect (DCD).

RS–232 CONNECTOR PIN ASSIGNMENTS Figure 2

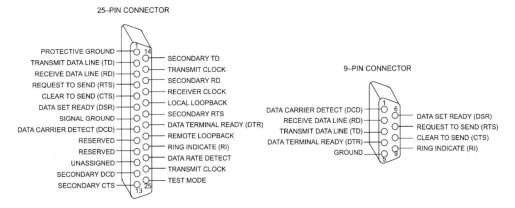

030998 3/9

PRACTICAL RS–232 IMPLEMENTATION

Most systems designed today do not operate using RS–232 voltage levels. Since this is the case, level conversion is necessary to implement RS–232 communication. Level conversion is performed by special RS–232 IC's. These IC's typically have line drivers that generate the voltage levels required by RS–232 and line receivers that can receive RS–232 voltage levels without being damaged. These line drivers and receivers typically invert the signal as well since a logic 1 is represented by a low voltage level for RS–232 communication and likewise a logic 0 is represented by a high logic level. Figure 3 illustrates the function of an RS–232 line driver/receiver in a typical modem application. In this particular example, the signals necessary for serial communication are generated and received by the Universal Asynchronous Receiver/Transmitter (UART). The RS–232 line driver/receiver IC performs the level translation necessary between the CMOS/TTL and RS–232 interface.

The UART just mentioned performs the "overhead" tasks necessary for asynchronous serial communication. For example, the asynchronous nature of this type of communication usually requires that start and stop bits be initiated by the host system to indicate to the peripheral system when communication will start and stop. Parity bits are also often employed to ensure that the data sent has not been corrupted. The UART usually generates the start, stop, and parity bits when transmitting data and can detect communication errors upon receiving data. The UART also functions as the intermediary between byte–wide (parallel) and bit–wide (serial) communication; it converts a byte of data into a serial bit stream for transmitting and converts a serial bit stream into a byte of data when receiving.

Now that an elementary explanation of the TTL/CMOS to RS–232 interface has been provided we can consider some "real world" RS–232 applications. It has already been noted that RS–232 applications rarely follow the RS–232 standard precisely. Perhaps the most significant reason this is true is due to the fact that many of the defined signals are not necessary for most applications. As such, the unnecessary signals are omitted. Many applications, such as a modem, require only nine signals (two data signals, six control signals, and ground). Other applications may require only five signals (two for data, two for handshaking, and ground), while others may require only data signals with no handshake control. We will begin our investigation of "real world" implementations by first considering the typical modem application.

RS–232 IN MODEM APPLICATIONS

Modem applications are one of the most popular uses for the RS–232 standard. Figure 4 illustrates a typical modem application utilizing the RS–232 interface standard. As can be seen in the diagram, the PC is the DTE and the modem is the DCE. Communication between each PC and its associated modem is accomplished using the RS–232 standard. Communication between the two modems is accomplished via telecommunication. It should be noted that although a microcomputer is usually the DTE in RS–232 applications, this is not mandatory according to a strict interpretation of the standard.

TYPICAL RS±232 MODEM APPLICATION Figure 3

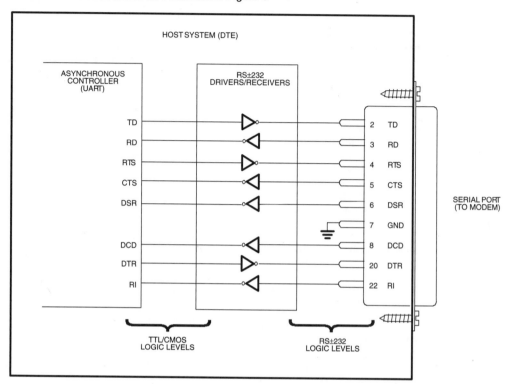

MODEM COMMUNICATION BETWEEN TWO PC'S Figure 4

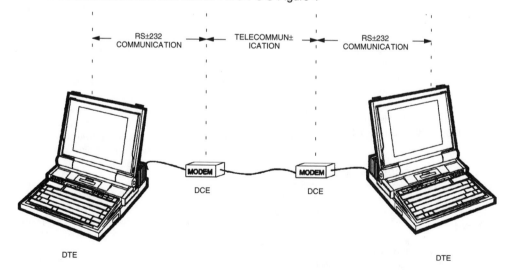

Many modem applications require only nine signals (including ground). Although some designers choose to use a 25–pin connector, it is not necessary since there are only nine interface signals between the DTE and DCE. With this in mind, many have chosen to use to use 9– or 15–pin connectors (see Figure 2 for 9–pin connector pin assignment). The "basic nine" signals used in modem communication are illustrated in Figure 3. Note that with respect to the DTE, three RS–232 drivers and five receivers are necessary. The functionality of these signals is described below. Note that for the following signal descriptions, "ON" refers to a high RS–232 voltage level (+5 t o +15 volts) and "OFF" refers to a low RS–232 voltage level (–5 to –15 volts). Keep in mind that a high RS–232 voltage level actually represents a logic 0 and a low RS–232 voltage level refers to a logic 1.

Transmitted Data (TD): One of two separate data signals. This signal is generated by the DTE and received by the DCE.

Received Data (RD): The second of two separate data signals. This signals is generated by the DCE and received by the DTE.

Request to Send (RTS): When the host system (DTE) is ready to transmit data to the peripheral system (DCE), RTS is turned ON. In simplex and duplex systems, this condition maintains the DCE in receive mode. In half–duplex systems, this condition maintains the DCE in receive mode and disables transmit mode. The OFF condition maintains the DCE in transmit mode. After RTS is asserted, the DCE must assert CTS before communicationcan commence.

Clear to Send (CTS): CTS is used along with RTS to provide handshaking between the DTE and the DCE. After the DCE sees an asserted RTS, it turns CTS ON when it is ready to begin communication.

Data Set Ready (DSR): This signal is turned on by the DCE to indicate that it is connected to the telecommunications line.

Data Carrier Detect (DCD): This signal is turned ON when the DCE is receiving a signal from a remote DCE which meets its suitable signal criteria. This signal remains ON as long as the a suitable carrier signal can be detected.

Data Terminal Ready (DTR): DTR indicates the readiness of the DTE. This signal is turned ON by the DTE when it is ready to transmit or receive data from the DCE. DTR must be ON before the DCE can assert DSR.

Ring Indicator (RI): RI, when asserted, indicates that a ringing signal is being received on the communications channel.

The signals described above form the basis for modem communication. Perhaps the best way to understand how these signals interact is to give a brief step by step example of a modem interfacing with a PC. The following step s describe a transaction in which a remote modem calls a local modem.

1. The local PC monitors the RI (Ring Indicate) signal via software.

2. When the remote modem wants to communicate with the local modem, it generates an RI signal. This signal is transferred by the local modem to the local PC.

3. The local PC responds to the RI signal by asserting the DTR (Data Terminal Ready) signal when it is ready to communicate.

4. After recognizing the asserted DTR signal, the modem responds by asserting DSR (Data Set Ready) after it is connected to the communications line. DSR indicates to the PC that the modem is ready to exchange further control signals with the DTE to commence communication. When DSR is asserted, the PC begins monitoring DCD for indication that data is being sent over the communication line.

5. The modem asserts DCD (Data Carrier Detect) after it has received a carrier signal from the remote modem that meets the suitable signal criteria.

6. At this point data transfer can began. If the local modem has full–duplex capability, the CTS (Clear to Send) and RTS (Request to Send) signals are held in the asserted state. If the modem has only half–duplex capability, CTS and RTS provide the handshaking necessary for controlling the direction of the data flow. Data is transferred over the RD and TD signals.

7. When the transfer of data has been completed, the PC disables the DTR signal. The modem follows by inhibiting the DSR and DCD signals. At this point the PC and modem are in the original state described in step number 1.

RS–232 IN MINIMAL HANDSHAKE APPLICATIONS

Even though the modem application discussed above is simplified from the RS–232 standard in terms of the number of signals needed, it is still more complex than the requirements of many systems. For many applications, two data lines and two handshake control lines are all that is necessary to establish and control communication between a host system and a peripheral system. For example, an environmental control system may need to interface with a thermostat using a half–duplex communication scheme. At times the control systems may desire to read the temperature from the thermostat and at other times may need to load temperature trip points to the thermostat. In this type of simple application, five signals may be all that is necessary (two for data, two for handshake control, and ground).

Figure 5 illustrates a simple half–duplex communication interface. As can be seen in this diagram, data is transferred over the TD (Transmit Data) and RD (Receive Data) pins and handshake control is provided by the RTS (Ready to Send) and CTS (Clear to Send) pins. RTS is driven by the DTE to control the direction of data. When it is asserted, the DTE is placed in transmit mode. When RTS is inhibited, the DTE is placed in receive mode. CTS, which is generated by the DCE, controls the flow of data. When asserted, data can flow. However, when CTS is inhibited, the transfer of data is interrupted. The transmission of data is halted until CTS is reasserted.

HALF–DUPLEX COMMUNICATION SCHEME Figure 5

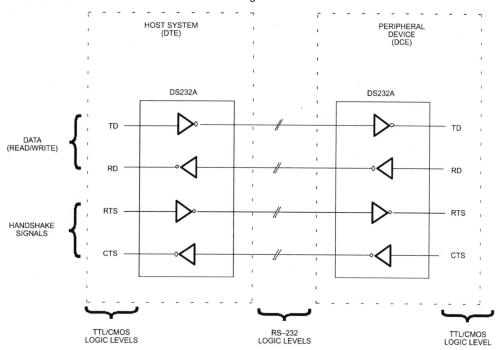

RS–232 APPLICATION LIMITATIONS

As mentioned earlier in this paper, the RS–232 standard was first introduced in 1962. In the more than three decades since, the electronics industry has changed immensely and therefore there are some limitations in the RS–232 standard. One limitation, the fact that over twenty signals have been defined by the standard, has already been addressed – simply do not use all of the signals or the 25–pin connector if they are not necessary. Other limitations in the standard are not necessarily as easy to correct, however.

GENERATION OF RS–232 VOLTAGE LEVELS

As we saw in the section on RS–232 electrical characteristics, RS–232 does not use the conventional 0 and 5 volt levels implemented in TTL and CMOS designs. Drivers have to supply +5 to +15 volts for a logic 0 and –5 to –15 volts for a logic 1. This means that extra power supplies are needed to drive the RS–232 voltage levels. Typically, a +12 volt and a –12 volt power supply are used to drive the RS–232 outputs. This is a great inconvenience for systems that have no other requirements for these power supplies. With this in mind, RS–232 products manufactured by Dallas Semiconductor have on–chip charge–pump circuits that generate the necessary voltage levels for RS–232 communication. The first charge pump essentially doubles the standard +5 volt power supply to provide the voltage level necessary for driving a logic 0. A second charge pump, inverts this voltage and provides the voltage level necessary for driving a logic 1. These two charge pumps allow the RS–232 interface products to operate from a single +5 volt supply.

MAXIMUM DATA RATE

Another limitation in the RS–232 standard is the maximum data rate. The standard defines a maximum data rate of 20k bits/second. This is unnecessarily slow for many of today's applications. RS–232 products manufactured by Dallas Semiconductor guarantee up to 250k bits/second and typically can communicate up to 350k bits/second. While providing a communication rate at this frequency, the devices still maintain a maximum 30V/μs maximum slew rate to reduce the likelihood of cross–talk between adjacent signals.

MAXIMUM CABLE LENGTH

A final limitation to discuss concerning RS–232 communication is cable length. As we have already seen, the cable length specification that was once included in the RS–232 standard has been replaced by a maximum load capacitance specification of 2500 pF. To determine the total length of cable allowed, one must determine the total line capacitance. Figure 6 shows a simple approximation for the total line capacitance of a conductor. As can be seen in the diagram, the total capacitance is approximated by the sum of the mutual capacitance between the signal conductors and the conductor to shield capacitance (or stray capacitance in the case of unshielded cable).

As an example, let's assume that the user has decided to use non–shielded cable when interconnecting the equipment. The cable mutual capacitance (Cm) of the cable is found in the cable's specifications to be 20 pF per foot. If we assume that the input capacitance of the receiver is 20 pF, this leaves the user with 2480 pF for the interconnecting cable. From the equation in Figure 6, the total capacitance per foot is found to be 30 pF. Dividing 2480 pF by 30 pF reveals that the maximum cable length is approximately 80 feet. If a longer cable length is required, the user would need to find a cable with a smaller mutual capacitance.

INTERFACE CABLE CAPACITIVE MODEL PER UNIT LENGTH Figure 6

Cm = Mutual capacitance between conductors.

Cs = Conductor to interface cable shield capacitance (if shielded cable is used) or stray capacitance to earth (if unshielded cable is used).

= 2(Cm) for shielded cable

= 0.5(Cm) for unshielded cable

Cc = Cm + Cs = Total line capacitance per unit length

Appendix E

Electronic Devices Datasheets

TECHNICAL

INFORMATION

DECOUPLING: BASICS

by Arch Martin
AVX Corporation
Myrtle Beach, S.C.

Abstract:

This paper discusses the characteristics of multilayer ceramic capacitors in decoupling applications and compares their performance with other types of decoupling capacitors. A special high-frequency test circuit is described and the results obtained using various types of capacitors are shown.

Introduction

The rapid changes occurring in the semiconductor industry are requiring new performance criteria of their supporting components. One of these components is the decoupling capacitor used in almost every circuit design. As the integrated circuits have become faster and more dense, the application design considerations have created a need to redefine the capacitor parameters and its performance in high-speed environments. Faster edge rates, larger currents, denser boards and spiraling costs have all served to focus upon the need for better and more efficient decoupling techniques.

As integrated circuits have grown, so has the demand for multilayer ceramic capacitors.

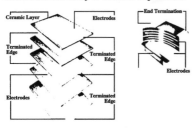

The phenomenal growth of multilayer ceramic capacitors over the last few years has been a result of their ability to satisfy these new requirements. We at AVX are continually studying these new requirements from the application view in order to better define what is required of the capacitor now and in the future, so that we can develop even better capacitor designs. Some results of these studies are the subject of this paper.

Background

A capacitor is an electrical device consisting of two metal conductors isolated by a nonconducting material capable of storing electrical charge for release at a controlled rate and at a specified time. Its usefulness is determined by its ability to store electrical energy.

An equivalent circuit for capacitors is shown in Fig. 2. This equivalent circuit of three series impedances can be represented by one lumped impedance which is used as a measure of the capacitance (Fig. 3). In other words, the amount of coulombs stored are not measured; what is measured is the lumped impedance and from this value an equivalent capacitance value is calculated.

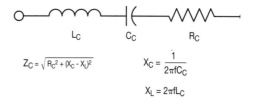

$$Z_C = \sqrt{R_C^2 + (X_C - X_L)^2} \qquad X_C = \frac{1}{2\pi f C_C}$$

$$X_L = 2\pi f L_C$$

Fig. 2. Total capacitor impedance

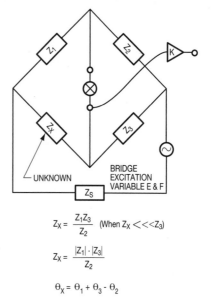

$$Z_X = \frac{Z_1 Z_3}{Z_2} \quad (\text{When } Z_X <<< Z_3)$$

$$Z_X = \frac{|Z_1| \cdot |Z_3|}{Z_2}$$

$$\theta_X = \theta_1 + \theta_3 - \theta_2$$

Fig. 3. Basic impedance bridge

Thus capacitance as measured is actually a combination of the capacitive reactance, the inductive reactance and the equivalent series resistance.

As shown in Fig. 4, all three of these series impedances vary differently with frequency. Since all three vary at different rates with frequency, the capacitance calculated from the resultant impedance is made up of different components at different frequencies. This can be seen by increasing the lead length of a capacitor while it is being measured at 1 MHz on an equivalent series capacitance bridge and watching the capacitance increase. Increased inductance (increased lead length) actually increases the capacitance value read by the capacitance bridge (Fig. 5).

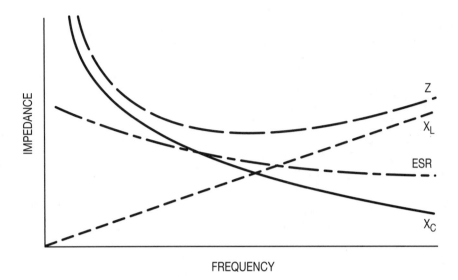

Fig. 4. Variation of impedance with frequency

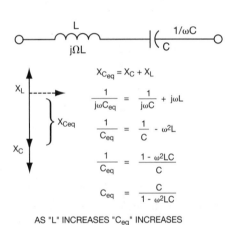

AS "L" INCREASES "C$_{eq}$" INCREASES

Fig. 5. Effects of inductance

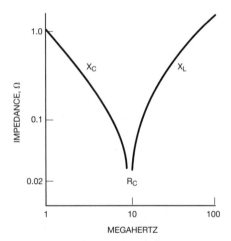

Fig. 6. Major impedance components

The major components of a typical impedance curve vs. frequency are shown in Fig. 6. Below resonance, the major component is capacitive reactance, at resonance it is equivalent series resistance, and above resonance it is inductive reactance. In decoupling today's high-speed digital circuits the capacitor is primarily being used to eliminate high-speed transient noise which is above its resonance point, an area where inductive reactance is the major impedance component. In these applications it is desirable to maintain as low an inductance or totol impedance as possible. For effective and economical designs it is important to define the performance of the capacitor under the circuit conditions in which it will be used.

High-Frequency Testing

The test schematic shown in Fig. 7 can be used to determine the performance of various styles of capacitors at high frequencies. Lead lengths are minimized through the use of the test fixture shown in Fig. 8. This test set-up is intended to duplicate the performance of the capacitor under actual use conditions. The Hewlett-Packard current generator supplies high-frequency pulses equivalent to an actual digital IC circuit. Edge rates of 200 mA/10 ns or 200 mA/5 ns are fed to the capacitor to determine its performance. Edge rates in the 5 to 10-ns range were chosen for testing since transients in this range are common to many designs such as shown in Fig. 9, where transients in the 5 to 10-ns range occur regularly on a DEC VAX 11/780 memory board using Mostek 4116 dynamic RAMs.

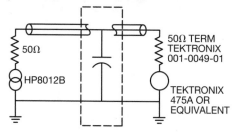

INDUCTANCE TEST FIXTURE

Fig. 7. High-frequency test schematic

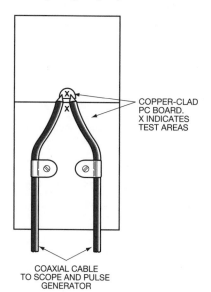

COPPER-CLAD PC BOARD. X INDICATES TEST AREAS

COAXIAL CABLE TO SCOPE AND PULSE GENERATOR

Fig. 8. Inductance test fixture

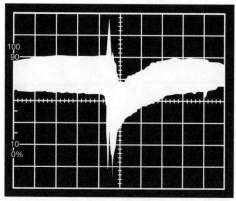

SCALE:
HORIZONTAL = 1 μS/DIV
VERTICAL   = 5MV/DIV

Fig. 9. Transients from V_{DD} to V_{SS} in 256K byte dynamic memory (Mostek 4116) in DEC VAX 11/780

All three components (capacitive, resistive, and inductive) are evident on the scope trace (Fig. 10) obtained by using the test circuit of Fig. 7. Thus the performance of a capacitor can be described under typical use conditions for capacitance ($C = I \frac{dt}{dv}$), equivalent series resistance ($ESR = V_A/I$), and inductance ($L = V_L \frac{dt}{di}$). This information can then be used to select a capacitor having the required characteristics for the application.

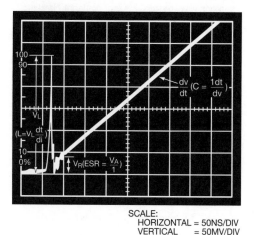

SCALE:
HORIZONTAL = 50NS/DIV
VERTICAL   = 50MV/DIV

Fig. 10. Scope trace of 0.22-μF film capacitor using test setup of Fig. 7 and 200 mA/10 ns input

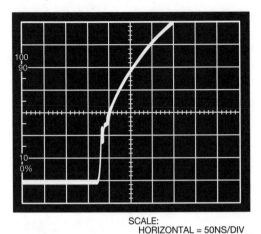

SCALE:
HORIZONTAL = 50NS/DIV
VERTICAL   = 100MV/DIV

Fig. 11 (A). Scope trace of 0.1-μF tantalum capacitor using test setup of Fig. 7 and 200 mA/10 ns input

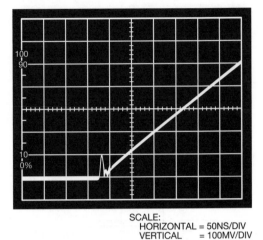

SCALE:
HORIZONTAL = 50NS/DIV
VERTICAL   = 100MV/DIV

Fig. 11 (B). Scope trace of 0.1-μF multilayer ceramic capacitor using same test setup and input

Test Results

Figures 11(A) and 11(B) compare the results of testing a 0.1-μF rated multilayer ceramic capacitor with a 0.1-μF rated tantalum capacitor. The slope of the curve, dv/dt, is a measure of the capacitance in the time domain of interest. For the ceramic this is approximately 0.1 μF but for the tantalum the value is 0.05 μF or approximately half the capacitance of the ceramic.

Another way to state this is at constant current:

$$C_{Tantalum} \frac{dv}{dt} = I = C_{Ceramic} \frac{dv}{dt}$$

$$C_T \frac{0.2}{50 \times 10^9} = C_C \frac{0.1}{50 \times 10^9}$$

$$C_T = \tfrac{1}{2} C_C$$

The induced voltage generated in response to the 200 mA/10 ns edge rate can be used to determine the inductance of the capacitor. Under these conditions the tantalum shows 2.5 times the inductance of the multilayer ceramic (MLC) capacitor:

$$L = V_L \frac{dt}{di}$$

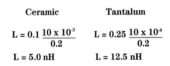

Ceramic	Tantalum
$L = 0.1 \dfrac{10 \times 10^{-9}}{0.2}$	$L = 0.25 \dfrac{10 \times 10^{-9}}{0.2}$
$L = 5.0 \text{ nH}$	$L = 12.5 \text{ nH}$

The ESR is obtained from the minimum voltage reached after the voltage spike. In the case of the ceramic the ESR is too small to measure on the scale used, i.e., less than 5 m|. The ESR for the tantalum can be determined with the V_R = 150-mV reading giving an ESR = 750 m|.

Both the MLC and the tantalum capacitors were tested with equal lead lengths. Removal of the leads reduces the inductance obtained for the MLC (now an MLC chip capacitor) to V_L = 20 mV for an inductance of 1 nH.

This test procedure can be used to compare the performance of various styles of capacitors. Fig. 12 compares the performance of MLC and a film capacitor. Both units show dv/dt equivalent to the capacitance values read at 1 kHz (0.22 μF for the film and 0.20 μF for the MLC). The MLC shows a voltage spike of 75 mV vs. the 275-mV spike for the film, i.e., the film had 3.7 times the inductance of the MLC. In addition, the MLC had half the ESR of the film capacitor.

The initial voltage spikes found from an input of 200 mA/5 ns edge rate are given in Table I for various capacitor values and styles. The equivalent inductances for these values are shown in Table II. These results are based on essentially zero lead lengths for all types. From a practical standpoint, the lead length inductances shown in Table III must be added to values of Table II in order to mate the capacitor with the PC board.

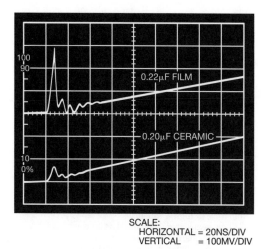

SCALE:
HORIZONTAL = 20NS/DIV
VERTICAL   = 100MV/DIV

Fig. 12. Comparison of 0.2-µF MLC with
0.22-µF film capacitor

CAPACITOR	.01µF	.1µF	.22µF	1.0µF	2.2µF
FILMS:					
BRAND A	170	240	—	470	—
BRAND B	110	120	120	140	—
BRAND C	180	300	—	350	—
BRAND D	300	—	230	—	—
TANTALUMS:					
BRAND E	—	380	—	380	—
ALUMINUMS:					
BRAND F	—	630	—	400	—
AVX MLC'S					
CONFORMAL					
RADIALS	—	70	—	70	90
MOLDED					
RADIALS	—	—	—	50	—
DIP'S	—	50	40	—	—
MOLDED					
AXIALS	80	130	140	210	—

Table I. Voltage spikes (mV) from 200 mA/5 ns edge rate
with essentially zero lead lengths

CAPACITOR	.01µF	.1µF	.22µF	1.0µF	2.2µF
FILMS:					
BRAND A	4.3	6.0	—	11.8	—
BRAND B	2.8	3.0	3.0	3.5	—
BRAND C	4.5	7.5	—	8.8	—
BRAND D	7.5	—	5.8	—	—
TANTALUMS:					
BRAND E	—	9.5	—	9.5	—
ALUMINUMS:					
BRAND F	—	15.8	—	10.0	—
AVX MLC'S					
CONFORMAL					
RADIALS	—	1.8	—	1.8	2.3
MOLDED					
RADIALS	—	—	—	1.3	—
DIP'S	—	1.3	1.0	—	—
MOLDED					
AXIALS	2.0	3.3	3.5	5.3	—

Table II. Inductances (nH) from 200 mA/5 ns edge rate
with essentially zero lead lengths

CAPACITOR TYPE	TYPICAL LEAD LENGTH (MM)	LEAD L (NH)	LEAD V_L (MV)	LEAD IMP. (Ω)
FILMS:				
BRAND A	3.81	1.5	60	0.9
BRAND B	2.54	1.0	40	0.6
BRAND C	3.81	1.5	60	0.9
BRAND D	2.54	1.0	40	0.6
TANTALUMS:				
BRAND E	5.08	2.0	80	1.3
ALUMINUMS:				
BRAND F	4.45	1.8	70	1.1
AVX MLC'S				
CONFORMAL				
RADIALS	2.54	1.0	40	0.6
MOLDED				
RADIALS	2.54	1.0	40	0.6
DIP'S	5.08	2.0	80	1.3
MOLDED				
AXIALS	7.62	3.0	120	1.9

Table III. Typical lead lengths required to connect capacitor to
PC board

Decoupling

The above capacitor models can be used to optimize the decoupling of integrated-circuit designs. As an example, the decoupling requirements of dynamic RAMs will be discussed. Dynamic RAMs have large transients which are generated during their refresh cycle. These large transients require careful attention to the decoupling techniques to avoid "V bump" or "soft" error problems.

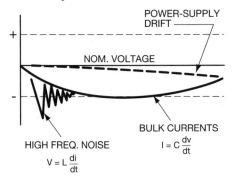

Fig. 13. Three factors causing voltage variations

The function of the capacitor can be illustrated by referring to Fig. 13. Through time, the three factors causing voltage variations are: dc drift, bulk variations, and switching transients. Dc drift is independent upon power-supply design and not on the board level decoupling. Bulk variations come about by the current demands of recharging the internal storage cells during the refresh cycle. Transient "noise" comes from switching currents internal to the IC chip. The total of these three voltage variations must be maintained within the allowed tolerance for the IC device. In other words, if the voltage drops below the operating margin of the IC, a "soft" error will occur.

The board low-frequency surge current to be supported by the bulk capacitance is effectively the IC's average refresh active current for the length of a refresh cycle. This can be determined by referring to the IC spec (and assuming worst-case that stand-by current is zero and active current is maximum). The traditional approach for calculating bulk capacitance requirements is to multiply the capacitance needed per package times the number of packages on the board. This arrives at the total capacitance required, which is then approximated with large-value capacitors around the periphery of the circuits.

It is more effective to distribute this capacitance throughout the design with smaller value capacitors whose combined total meets or exceeds the bulk-current requirement. Distribution of the bulk requirement to capacitors adjacent to the current requirements (the IC chip) is beneficial in reducing

inductance and resistive voltage drops. Slight increases in the distributed transient capacitor values can typically eliminate the need for large capacitors to supply bulk current.

Transient noise is commonly called in the industry "V bump," and reflects the supply transient induced by the chip itself on the decoupling capacitor when various clocks fire on-chip and drive on-chip capacitance associated with that event (such as address decoding). Since actual loads switched are small (20 pF), the size of the decoupling capacitor is less important than its inductance.

Using capacitors for reducing the line "noise" that comes from switching internal to the IC chip requires low inherent inductance within the decoupling capacitor and effective board design. Multilayer ceramic capacitors are available in values high enough to meet distributed bulk requirements while maintaining low inductance at high frequencies.

When a capacitor is mounted on a board, lead lengths and board lines (device to capacitor to ground) are a major source of inductance. This inductance must be minimized to obtain good decoupling performance under high-speed transient conditions. Minimum lead lengths, wiring, and gridding of power supplies and ground with alternate parallel paths are important as is the quality of the capacitor (Fig. 14). The use of multiple capacitors instead of a few large bulk capacitors can be used to decrease line lengths and to increase path numbers (gridding) for reduced inductance and more effective surge-current availability.

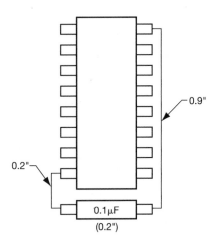

Fig. 14. Circuit board line and capacitor lead lengths

Myrtle Beach, SC / Tel: 803-448-9411 / FAX: 803-448-1943
Vancouver, WA / Tel: 206-696-2840 / FAX: 206-695-5836
Olean, NY / Tel: 716-372-6611 / FAX: 716-372-6316
Raleigh, NC / Tel: 919-878-6200 / FAX: 919-878-6462
Biddeford, ME / Tel: 207-282-5111 / FAX: 207-283-1941
AVX Limited, Fleet, Hants, England / Tel: (01252) 770000 / FAX: (01252) 770001
AVX S.A., France / Tel: (1) 6918 4600 / FAX: (1) 6928 7387
AVXGmbH, Germany / Tel: 08131 9004-0 / FAX: 08131 9004-44
AVX s.r.l., Milano, Italy / Tel: 02-665 00116 / FAX: 02-614 2576
AVX/Kyocera (HK) Ltd. / Tel: 852-363-3303 / FAX: 852-765-8185
AVX/Kyocera (Singapore) Pte. Ltd. / Tel: (65) 258-2833 / FAX: (65) 258-8221
AVX Israel Ltd. / Tel: 972-957-3873 / FAX: 972-957-3853
AVX/Kyocera Corp. / Tel: 75-593-4518 / FAX: 75-501-4936   S-DB2.5M1194-C

N *National Semiconductor*

August 2000

LM124/LM224/LM324/LM2902 Low Power Quad Operational Amplifiers

LM124/LM224/LM324/LM2902
Low Power Quad Operational Amplifiers

General Description

The LM124 series consists of four independent, high gain, internally frequency compensated operational amplifiers which were designed specifically to operate from a single power supply over a wide range of voltages. Operation from split power supplies is also possible and the low power supply current drain is independent of the magnitude of the power supply voltage.

Application areas include transducer amplifiers, DC gain blocks and all the conventional op amp circuits which now can be more easily implemented in single power supply systems. For example, the LM124 series can be directly operated off of the standard +5V power supply voltage which is used in digital systems and will easily provide the required interface electronics without requiring the additional †15V power supplies.

Unique Characteristics

n In the linear mode the input common-mode voltage range includes ground and the output voltage can also swing to ground, even though operated from only a single power supply voltage

n The unity gain cross frequency is temperature compensated

n The input bias current is also temperature compensated

Advantages

n Eliminates need for dual supplies

n Four internally compensated op amps in a single package

n Allows directly sensing near GND and V_{OUT} also goes to GND

n Compatible with all forms of logic

n Power drain suitable for battery operation

Features

n Internally frequency compensated for unity gain

n Large DC voltage gain   100 dB

n Wide bandwidth (unity gain)   1 MHz (temperature compensated)

n Wide power supply range:
    Single supply   3V to 32V
    or dual supplies   †1.5V to †16V

n Very low supply current drain (700 A) − essentially independent of supply voltage

n Low input biasing current   45 nA (temperature compensated)

n Low input offset voltage   2 mV and offset current: 5 nA

n Input common-mode voltage range includes ground

n Differential input voltage range equal to the power supply voltage

n Large output voltage swing   0V to V^+ ، 1.5V

Connection Diagram

Dual-In-Line Package

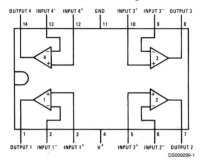

DS009299-1

Top View
Order Number LM124J, LM124AJ, LM124J/883 (Note 2), LM124AJ/883 (Note 1), LM224J,
LM224AJ, LM324J, LM324M, LM324MX, LM324AM, LM324AMX, LM2902M, LM2902MX, LM324N, LM324AN,
LM324MT, LM324MTX or LM2902N LM124AJRQML and LM124AJRQMLV(Note 3)
See NS Package Number J14A, M14A or N14A

Note 1: LM124A available per JM38510/11006
Note 2: LM124 available per JM38510/11005

Connection Diagram (Continued)

Note 3: See STD Mil DWG 5962R99504 for Radiation Tolerant Device

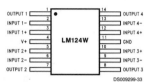

DS009299-33

Order Number LM124AW/883, LM124AWG/883, LM124W/883 or LM124WG/883
LM124AWRQML and LM124AWRQMLV(Note 3)
See NS Package Number W14B
LM124AWGRQML and LM124AWGRQMLV(Note 3)
See NS Package Number WG14A

Schematic Diagram (Each Amplifier)

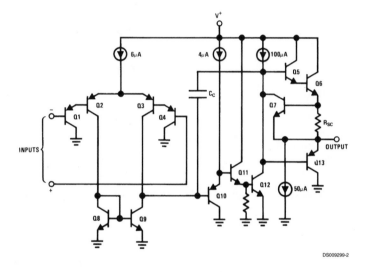

DS009299-2

Absolute Maximum Ratings (Note 12)

If Military/Aerospace specified devices are required, please contact the National Semiconductor Sales Office/ Distributors for availability and specifications.

	LM124/LM224/LM324 LM124A/LM224A/LM324A	LM2902
Supply Voltage, V^+	32V	26V
Differential Input Voltage	32V	26V
Input Voltage	−0.3V to +32V	−0.3V to +26V
Input Current		
(V_{IN} < −0.3V) (Note 6)	50 mA	50 mA
Power Dissipation (Note 4)		
Molded DIP	1130 mW	1130 mW
Cavity DIP	1260 mW	1260 mW
Small Outline Package	800 mW	800 mW
Output Short-Circuit to GND		
(One Amplifier) (Note 5)		
$V^+ \leq 15V$ and $T_A = 25°C$	Continuous	Continuous
Operating Temperature Range		−40°C to +85°C
LM324/LM324A	0°C to +70°C	
LM224/LM224A	−25°C to +85°C	
LM124/LM124A	−55°C to +125°C	
Storage Temperature Range	−65°C to +150°C	−65°C to +150°C
Lead Temperature (Soldering, 10 seconds)	260°C	260°C
Soldering Information		
Dual-In-Line Package		
Soldering (10 seconds)	260°C	260°C
Small Outline Package		
Vapor Phase (60 seconds)	215°C	215°C
Infrared (15 seconds)	220°C	220°C

See AN-450 "Surface Mounting Methods and Their Effect on Product Reliability" for other methods of soldering surface mount devices.

| ESD Tolerance (Note 13) | 250V | 250V |

Electrical Characteristics

V^+ = +5.0V, (Note 7), unless otherwise stated

Parameter	Conditions	LM124A			LM224A			LM324A			Units
		Min	Typ	Max	Min	Typ	Max	Min	Typ	Max	
Input Offset Voltage	(Note 8) $T_A = 25°C$		1	2		1	3		2	3	mV
Input Bias Current (Note 9)	$I_{IN(+)}$ or $I_{IN(-)}$, V_{CM} = 0V, $T_A = 25°C$		20	50		40	80		45	100	nA
Input Offset Current	$I_{IN(+)}$ or $I_{IN(-)}$, V_{CM} = 0V, $T_A = 25°C$		2	10		2	15		5	30	nA
Input Common-Mode Voltage Range (Note 10)	V^+ = 30V, (LM2902, V^+ = 26V), $T_A = 25°C$	0	V^+−1.5		0	V^+−1.5		0	V^+−1.5		V
Supply Current	Over Full Temperature Range $R_L = \infty$ On All Op Amps										mA
	V^+ = 30V (LM2902 V^+ = 26V)		1.5	3		1.5	3		1.5	3	
	V^+ = 5V		0.7	1.2		0.7	1.2		0.7	1.2	
Large Signal Voltage Gain	V^+ = 15V, $R_L \geq 2k\Omega$, (V_O = 1V to 11V), $T_A = 25°C$	50	100		50	100		25	100		V/mV
Common-Mode Rejection Ratio	DC, V_{CM} = 0V to V^+ − 1.5V, $T_A = 25°C$	70	85		70	85		65	85		dB

LM124/LM224/LM324/LM2902

Electrical Characteristics (Continued)

$V^+ = +5.0V$, (Note 7), unless otherwise stated

Parameter		Conditions		LM124A			LM224A			LM324A			Units
				Min	Typ	Max	Min	Typ	Max	Min	Typ	Max	
Power Supply Rejection Ratio		V^+ = 5V to 30V (LM2902, V^+ = 5V to 26V), T_A = 25°C		65	100		65	100		65	100		dB
Amplifier-to-Amplifier Coupling (Note 11)		f = 1 kHz to 20 kHz, T_A = 25°C (Input Referred)			-120			-120			-120		dB
Output Current	Source	V_{IN}^+ = 1V, V_{IN}^- = 0V, V^+ = 15V, V_O = 2V, T_A = 25°C		20	40		20	40		20	40		mA
	Sink	V_{IN}^- = 1V, V_{IN}^+ = 0V, V^+ = 15V, V_O = 2V, T_A = 25°C		10	20		10	20		10	20		
		V_{IN}^- = 1V, V_{IN}^+ = 0V, V^+ = 15V, V_O = 200 mV, T_A = 25°C		12	50		12	50		12	50		µA
Short Circuit to Ground		(Note 5) V^+ = 15V, T_A = 25°C			40	60		40	60		40	60	mA
Input Offset Voltage		(Note 8)				4			4			5	mV
V_{OS} Drift		R_S = 0Ω			7	20		7	20		7	30	µV/°C
Input Offset Current		$I_{IN(+)} - I_{IN(-)}$, V_{CM} = 0V				30			30			75	nA
I_{OS} Drift		R_S = 0Ω			10	200		10	200		10	300	pA/°C
Input Bias Current		$I_{IN(+)}$ or $I_{IN(-)}$			40	100		40	100		40	200	nA
Input Common-Mode Voltage Range (Note 10)		V^+ = +30V (LM2902, V^+ = 26V)		0		V^+-2	0		V^+-2	0		V^+-2	V
Large Signal Voltage Gain		V^+ = +15V (V_O Swing = 1 to 11V) $R_L \geq 2$ kΩ		25			25			15			V/mV
Output Voltage Swing	V_{OH}	V^+ = 30V (LM2902, V^+ = 26V)	R_L = 2 kΩ	26			26			26			V
			R_L = 10 kΩ	27	28		27	28		27	28		
	V_{OL}	V^+ = 5V, R_L = 10 kΩ			5	20		5	20		5	20	mV
Output Current	Source	V_O = 2V	V_{IN}^+ = +1V, V_{IN}^- = 0V, V^+ = 15V	10	20		10	20		10	20		mA
	Sink		V_{IN}^- = +1V, V_{IN}^+ = 0V, V^+ = 15V	10	15		5	8		5	8		

Electrical Characteristics

$V^+ = +5.0V$, (Note 7), unless otherwise stated

Parameter	Conditions	LM124/LM224			LM324			LM2902			Units
		Min	Typ	Max	Min	Typ	Max	Min	Typ	Max	
Input Offset Voltage	(Note 8) T_A = 25°C		2	5		2	7		2	7	mV
Input Bias Current (Note 9)	$I_{IN(+)}$ or $I_{IN(-)}$, V_{CM} = 0V, T_A = 25°C		45	150		45	250		45	250	nA
Input Offset Current	$I_{IN(+)}$ or $I_{IN(-)}$, V_{CM} = 0V, T_A = 25°C		3	30		5	50		5	50	nA
Input Common-Mode Voltage Range (Note 10)	V^+ = 30V, (LM2902, V^+ = 26V), T_A = 25°C	0		V^+-1.5	0		V^+-1.5	0		V^+-1.5	V
Supply Current	Over Full Temperature Range $R_L = \infty$ On All Op Amps V^+ = 30V (LM2902 V^+ = 26V)		1.5	3		1.5	3		1.5	3	mA
	V^+ = 5V		0.7	1.2		0.7	1.2		0.7	1.2	
Large Signal Voltage Gain	V^+ = 15V, $R_L \geq$ 2kΩ, (V_O = 1V to 11V), T_A = 25°C	50	100		25	100		25	100		V/mV
Common-Mode Rejection Ratio	DC, V_{CM} = 0V to V^+ - 1.5V, T_A = 25°C	70	85		65	85		50	70		dB
Power Supply Rejection Ratio	V^+ = 5V to 30V (LM2902, V^+ = 5V to 26V),	65	100		65	100		50	100		dB

Electrical Characteristics (Continued)

V^+ = +5.0V, (Note 7), unless otherwise stated

Parameter		Conditions	LM124/LM224			LM324			LM2902			Units
			Min	Typ	Max	Min	Typ	Max	Min	Typ	Max	
		T_A = 25°C										
Amplifier-to-Amplifier Coupling (Note 11)		f = 1 kHz to 20 kHz, T_A = 25°C (Input Referred)		−120			−120			−120		dB
Output Current	Source	V_{IN}^+ = 1V, V_{IN}^- = 0V, V^+ = 15V, V_O = 2V, T_A = 25°C	20	40		20	40		20	40		mA
	Sink	V_{IN}^- = 1V, V_{IN}^+ = 0V, V^+ = 15V, V_O = 2V, T_A = 25°C	10	20		10	20		10	20		
		V_{IN}^- = 1V, V_{IN}^+ = 0V, V^+ = 15V, V_O = 200 mV, T_A = 25°C	12	50		12	50		12	50		μA
Short Circuit to Ground		(Note 5) V^+ = 15V, T_A = 25°C		40	60		40	60		40	60	mA
Input Offset Voltage		(Note 8)			7			9			10	m'
V_{OS} Drift		R_S = 0Ω		7			7			7		μV/°C
Input Offset Current		$I_{IN(+)} - I_{IN(-)}$, V_{CM} = 0V			100			150		45	200	nA
I_{OS} Drift		R_S = 0Ω		10			10			10		pA/°C
Input Bias Current		$I_{IN(+)}$ or $I_{IN(-)}$		40	300		40	500		40	500	nA
Input Common-Mode Voltage Range (Note 10)		V^+ = +30V (LM2902, V^+ = 26V)	0		V^+−2	0		V^+−2	0		V^+−2	V
Large Signal Voltage Gain		V^+ = +15V (V_OSwing = 1V to 11V) $R_L \geq$ 2 kΩ	25			15			15			V/mV
Output Voltage Swing	V_{OH}	V^+ = 30V (LM2902, V^+ = 26V)	R_L = 2 kΩ	26			26			22			V
			R_L = 10 kΩ	27	28		27	28		23	24		
	V_{OL}	V^+ = 5V, R_L = 10 kΩ		5	20		5	20		5	100	mV
Output Current	Source	V_O = 2V	V_{IN}^+ = +1V, V_{IN}^- = 0V, V^+ = 15V	10	20		10	20		10	20		mA
	Sink		V_{IN}^- = +1V, V_{IN}^+ = 0V, V^+ = 15V	5	8		5	8		5	8		

Note 4: For operating at high temperatures, the LM324/LM324A/LM2902 must be derated based on a +125°C maximum junction temperature and a thermal resistance of 88°C/W which applies for the device soldered in a printed circuit board, operating in a still air ambient. The LM224/LM224A and LM124/LM124A can be derated based on a +150°C maximum junction temperature. The dissipation is the total of all four amplifiers—use external resistors, where possible, to allow the amplifier to saturate or to reduce the power which is dissipated in the integrated circuit.

Note 5: Short circuits from the output to V^+ can cause excessive heating and eventual destruction. When considering short circuits to ground, the maximum output current is approximately 40 mA independent of the magnitude of V^+. At values of supply voltage in excess of +15V, continuous short-circuits can exceed the power dissipation ratings and cause eventual destruction. Destructive dissipation can result from simultaneous shorts on all amplifiers.

Note 6: This input current will only exist when the voltage at any of the input leads is driven negative. It is due to the collector-base junction of the input PNP transistors becoming forward biased and thereby acting as input diode clamps. In addition to this diode action, there is also lateral NPN parasitic transistor action on the IC chip. This transistor action can cause the output voltages of the op amps to go to the V^+voltage level (or to ground for a large overdrive) for the time duration that an input is driven negative. This is not destructive and normal output states will re-establish when the input voltage, which was negative, again returns to a value greater than −0.3V (at 25°C).

Note 7: These specifications are limited to −55°C ≤ T_A ≤ +125°C for the LM124/LM124A. With the LM224/LM224A, all temperature specifications are limited to −25°C ≤ T_A ≤ +85°C, the LM324/LM324A temperature specifications are limited to 0°C ≤ T_A ≤ +70°C, and the LM2902 specifications are limited to −40°C ≤ T_A ≤ +85°C.

Note 8: V_O ≈ 1.4V, R_S = 0Ω with V^+ from 5V to 30V; and over the full input common-mode range (0V to V^+ − 1.5V) for LM2902, V^+ from 5V to 26V.

Note 9: The direction of the input current is out of the IC due to the PNP input stage. This current is essentially constant, independent of the state of the output so no loading change exists on the input lines.

Note 10: The input common-mode voltage of either input signal voltage should not be allowed to go negative by more than 0.3V (at 25°C). The upper end of the common-mode voltage range is V^+ − 1.5V (at 25°C), but either or both inputs can go to +32V without damage (+26V for LM2902), independent of the magnitude of V^+.

Note 11: Due to proximity of external components, insure that coupling is not originating via stray capacitance between these external parts. This typically can be detected as this type of capacitance increases at higher frequencies.

Note 12: Refer to RETS124AX for LM124A military specifications and refer to RETS124X for LM124 military specifications.

Note 13: Human body model, 1.5 kΩ in series with 100 pF.

LM124/LM224/LM324/LM2902

Typical Performance Characteristics

Input Voltage Range

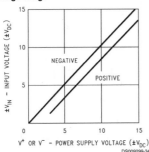

Input Current

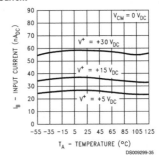

Supply Current

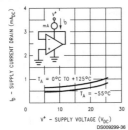

Voltage Gain

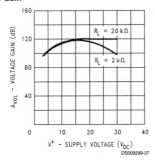

Open Loop Frequency Response

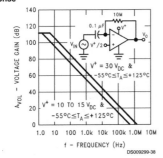

Common Mode Rejection Ratio

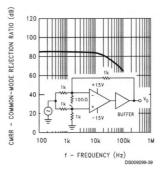

LM124/LM224/LM324/LM2902

Typical Performance Characteristics (Continued)

**Voltage Follower Pulse
Response**

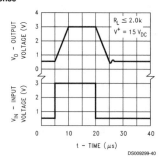

DS009299-40

**Voltage Follower Pulse
Response (Small Signal)**

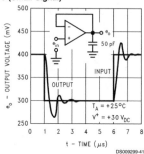

DS009299-41

**Large Signal Frequency
Response**

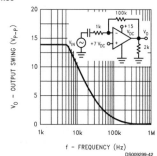

DS009299-42

**Output Characteristics
Current Sourcing**

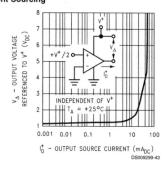

DS009299-43

**Output Characteristics
Current Sinking**

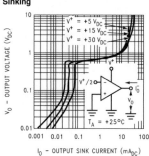

DS009299-44

Current Limiting

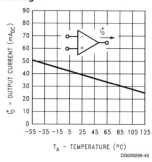

DS009299-45

LM124/LM224/LM324/LM2902

Typical Performance Characteristics (Continued)

Input Current (LM2902 only)

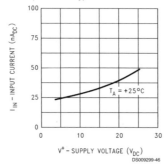

DS009299-46

Voltage Gain (LM2902 only)

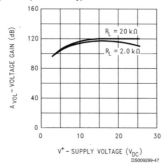

DS009299-47

Application Hints

The LM124 series are op amps which operate with only a single power supply voltage, have true-differential inputs, and remain in the linear mode with an input common-mode voltage of 0 V_{DC}. These amplifiers operate over a wide range of power supply voltage with little change in performance characteristics. At 25 C amplifier operation is possible down to a minimum supply voltage of 2.3 V_{DC}.

The pinouts of the package have been designed to simplify PC board layouts. Inverting inputs are adjacent to outputs for all of the amplifiers and the outputs have also been placed at the corners of the package (pins 1, 7, 8, and 14).

Precautions should be taken to insure that the power supply for the integrated circuit never becomes reversed in polarity or that the unit is not inadvertently installed backwards in a test socket as an unlimited current surge through the resulting forward diode within the IC could cause fusing of the internal conductors and result in a destroyed unit.

Large differential input voltages can be easily accommodated and, as input differential voltage protection diodes are not needed, no large input currents result from large differential input voltages. The differential input voltage may be larger than V+ without damaging the device. Protection should be provided to prevent the input voltages from going negative more than ,0.3 V_{DC} (at 25 C). An input clamp diode with a resistor to the IC input terminal can be used.

To reduce the power supply drain, the amplifiers have a class A output stage for small signal levels which converts to class B in a large signal mode. This allows the amplifiers to both source and sink large output currents. Therefore both NPN and PNP external current boost transistors can be used to extend the power capability of the basic amplifiers. The output voltage needs to raise approximately 1 diode drop above ground to bias the on-chip vertical PNP transistor for output current sinking applications.

For ac applications, where the load is capacitively coupled to the output of the amplifier, a resistor should be used, from the output of the amplifier to ground to increase the class A bias current and prevent crossover distortion.

Where the load is directly coupled, as in dc applications, there is no crossover distortion.

Capacitive loads which are applied directly to the output of the amplifier reduce the loop stability margin. Values of 50 pF can be accommodated using the worst-case non-inverting unity gain connection. Large closed loop gains or resistive isolation should be used if larger load capacitance must be driven by the amplifier.

The bias network of the LM124 establishes a drain current which is independent of the magnitude of the power supply voltage over the range of from 3 V_{DC} to 30 V_{DC}.

Output short circuits either to ground or to the positive power supply should be of short time duration. Units can be destroyed, not as a result of the short circuit current causing metal fusing, but rather due to the large increase in IC chip dissipation which will cause eventual failure due to excessive junction temperatures. Putting direct short-circuits on more than one amplifier at a time will increase the total IC power dissipation to destructive levels, if not properly protected with external dissipation limiting resistors in series with the output leads of the amplifiers. The larger value of output source current which is available at 25 C provides a larger output current capability at elevated temperatures (see typical performance characteristics) than a standard IC op amp.

The circuits presented in the section on typical applications emphasize operation on only a single power supply voltage. If complementary power supplies are available, all of the standard op amp circuits can be used. In general, introducing a pseudo-ground (a bias voltage reference of V+/2) will allow operation above and below this value in single power supply systems. Many application circuits are shown which take advantage of the wide input common-mode voltage range which includes ground. In most cases, input biasing is not required and input voltages which range to ground can easily be accommodated.

LM124/LM224/LM324/LM2902

Typical Single-Supply Applications (V⁺ = 5.0 V_DC)

Non-Inverting DC Gain (0V Input = 0V Output)

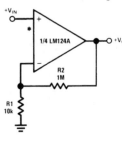

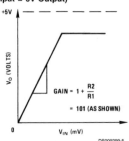

$$GAIN = 1 + \frac{R2}{R1}$$

$$= 101 \text{ (AS SHOWN)}$$

DS009299-5

*R not needed due to temperature independent I_{IN}

DC Summing Amplifier
($V_{IN'S} \geq 0$ V_{DC} and $V_O \geq V_{DC}$)

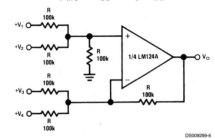

DS009299-6

Where: $V_0 = V_1 + V_2 , V_3 , V_4$
$(V_1 + V_2) \geq (V_3 + V_4)$ to keep $V_O > 0$ V_{DC}

Power Amplifier

$V_0 = 0$ V_{DC} for $V_{IN} = 0$ V_{DC}
$A_V = 10$

DS009299-7

Typical Single-Supply Applications (V$^+$ = 5.0 V$_{DC}$) (Continued)

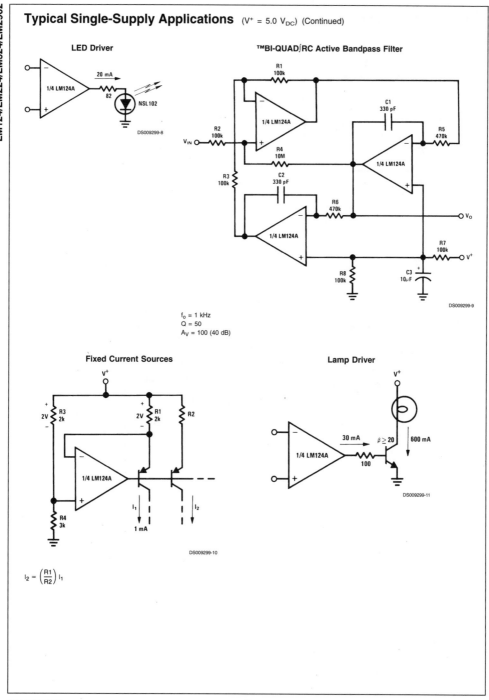

LED Driver

20 mA

82

NSL 102

1/4 LM124A

DS009299-8

™BI-QUAD∫RC Active Bandpass Filter

R1 100k

C1 330 pF

R5 470k

R2 100k

1/4 LM124A

V$_{IN}$

R4 10M

1/4 LM124A

R3 100k

C2 330 pF

R6 470k

V$_O$

1/4 LM124A

R7 100k

V$^+$

R8 100k

C3 10μF

DS009299-9

f$_o$ = 1 kHz
Q = 50
A$_V$ = 100 (40 dB)

Fixed Current Sources

V$^+$

2V R3 2k

2V R1 2k   R2

1/4 LM124A

I$_1$

I$_2$

1 mA

R4 3k

DS009299-10

$$I_2 = \left(\frac{R1}{R2}\right) I_1$$

Lamp Driver

V$^+$

1/4 LM124A

30 mA   β ≥ 20   600 mA

100

DS009299-11

LM124/LM224/LM324/LM2902

Typical Single-Supply Applications (V$^+$ = 5.0 V$_{DC}$) (Continued)

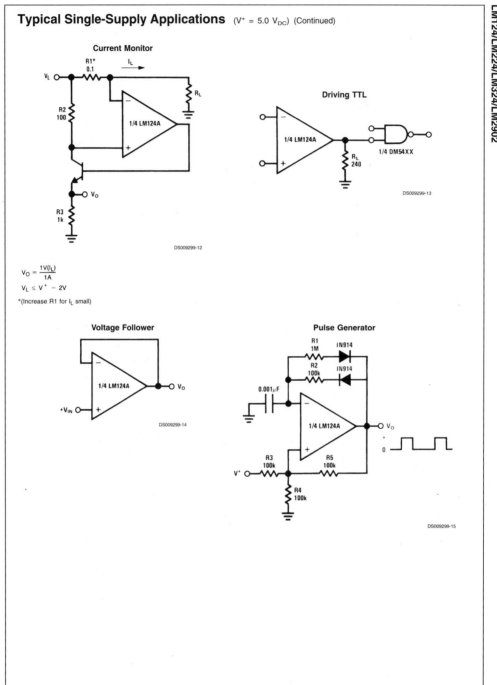

Current Monitor

DS009299-12

$$V_O = \frac{1V(I_L)}{1A}$$

$$V_L \leq V^+ - 2V$$

*(Increase R1 for I$_L$ small)

Driving TTL

DS009299-13

Voltage Follower

DS009299-14

Pulse Generator

DS009299-15

LM124/LM224/LM324/LM2902

Typical Single-Supply Applications (V⁺ = 5.0 V_DC) (Continued)

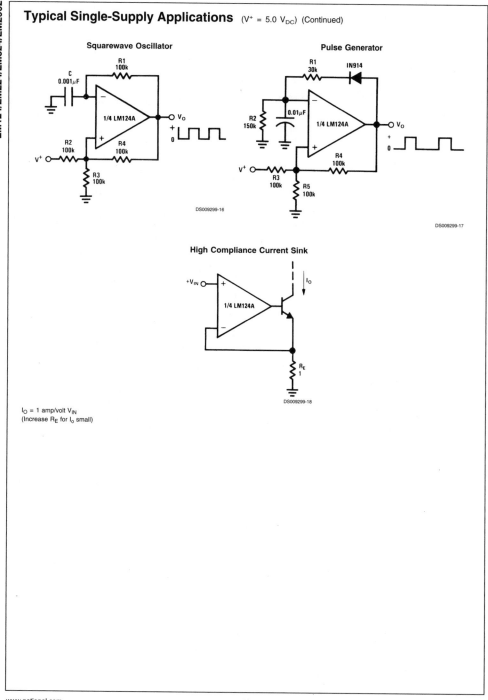

Squarewave Oscillator

DS009299-16

Pulse Generator

DS009299-17

High Compliance Current Sink

DS009299-18

I_O = 1 amp/volt V_{IN}
(Increase R_E for I_o small)

LM124/LM224/LM324/LM2902

Typical Single-Supply Applications (V⁺ = 5.0 V_DC) (Continued)

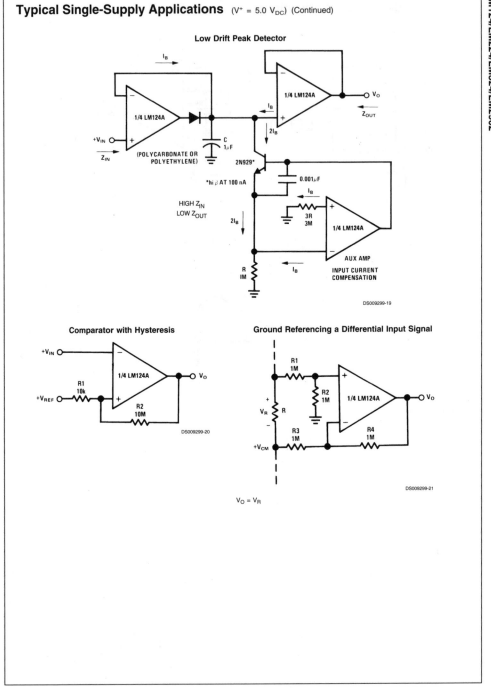

Low Drift Peak Detector

DS009299-19

Comparator with Hysteresis

DS009299-20

Ground Referencing a Differential Input Signal

$V_O = V_R$

DS009299-21

Typical Single-Supply Applications (V⁺ = 5.0 V_DC) (Continued)

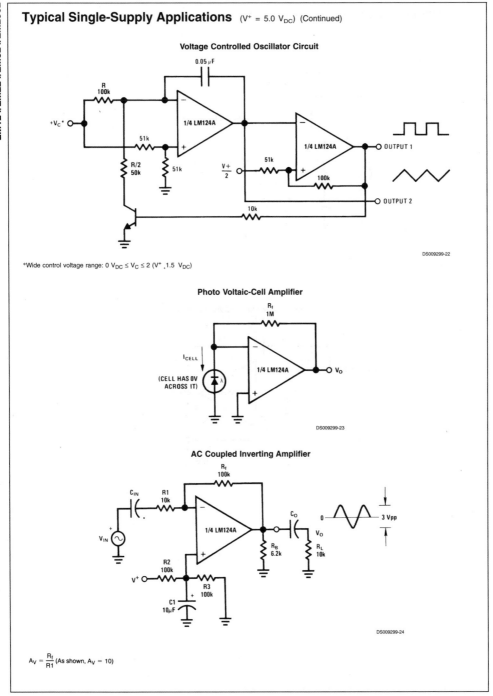

Voltage Controlled Oscillator Circuit

*Wide control voltage range: 0 V_DC ≤ V_C ≤ 2 (V⁺ .1.5 V_DC)

DS009299-22

Photo Voltaic-Cell Amplifier

DS009299-23

AC Coupled Inverting Amplifier

DS009299-24

$$A_V = \frac{R_f}{R1} \text{ (As shown, } A_V = 10\text{)}$$

Typical Single-Supply Applications (V$^+$ = 5.0 V$_{DC}$) (Continued)

AC Coupled Non-Inverting Amplifier

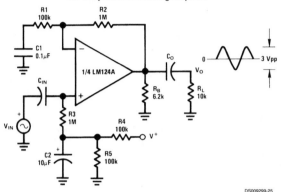

DS009299-25

$$A_V = 1 + \frac{R2}{R1}$$

$A_V = 11$ (As shown)

DC Coupled Low-Pass RC Active Filter

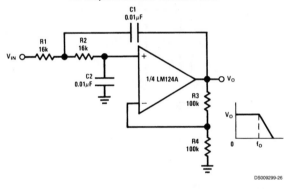

DS009299-26

$f_O = 1$ kHz
$Q = 1$
$A_V = 2$

LM124/LM224/LM324/LM2902

Typical Single-Supply Applications $(V^+ = 5.0\ V_{DC})$ (Continued)

High Input Z, DC Differential Amplifier

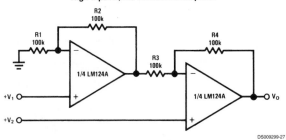

DS009299-27

For $\dfrac{R1}{R2} = \dfrac{R4}{R3}$ (CMRR depends on this resistor ratio match)

$V_O = 1 + \dfrac{R4}{R3}(V_2 - V_1)$

As shown: $V_O = 2(V_2 - V_1)$

**High Input Z Adjustable-Gain
DC Instrumentation Amplifier**

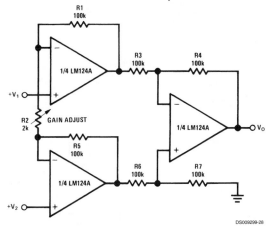

DS009299-28

If R1 = R5 & R3 = R4 = R6 = R7 (CMRR depends on match)

$V_O = 1 + \dfrac{2R1}{R2}(V_2 - V_1)$

As shown $V_O = 101\ (V_2 - V_1)$

LM124/LM224/LM324/LM2902

Typical Single-Supply Applications (V$^+$ = 5.0 V$_{DC}$) (Continued)

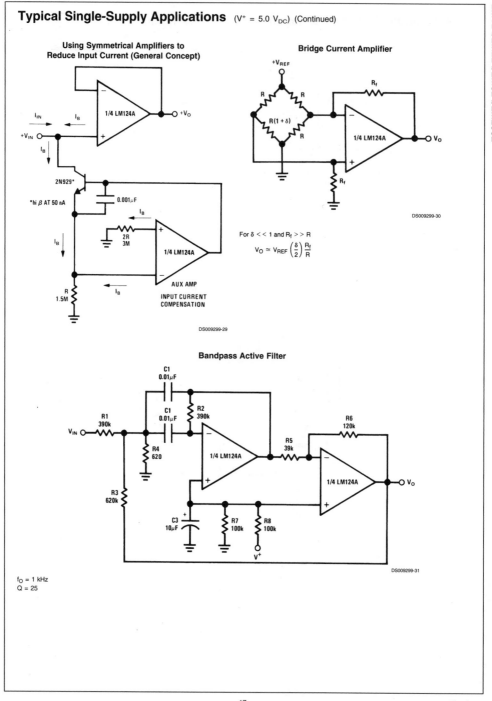

Using Symmetrical Amplifiers to Reduce Input Current (General Concept)

1/4 LM124A

+V$_O$

I$_{IN}$ I$_B$

+V$_{IN}$

I$_B$

2N929*

*hi β AT 50 nA

0.001μF

I$_B$

2R
3M

1/4 LM124A

I$_B$

I$_B$

R
1.5M

AUX AMP

INPUT CURRENT
COMPENSATION

DS009299-29

Bridge Current Amplifier

+V$_{REF}$

R   R

R(1 + δ)

R

R$_f$

1/4 LM124A

V$_O$

R$_f$

DS009299-30

For $\delta \ll 1$ and R$_f \gg$ R

$$V_O \approx V_{REF} \left(\frac{\delta}{2} \right) \frac{R_f}{R}$$

Bandpass Active Filter

C1
0.01μF

R1
390k

V$_{IN}$

C1
0.01μF

R2
390k

R4
620

1/4 LM124A

R5
39k

R6
120k

1/4 LM124A

V$_O$

R3
620k

C3
10μF

R7
100k

R8
100k

V$^+$

DS009299-31

f$_O$ = 1 kHz
Q = 25

LM124/LM224/LM324/LM2902

Physical Dimensions inches (millimeters) unless otherwise noted

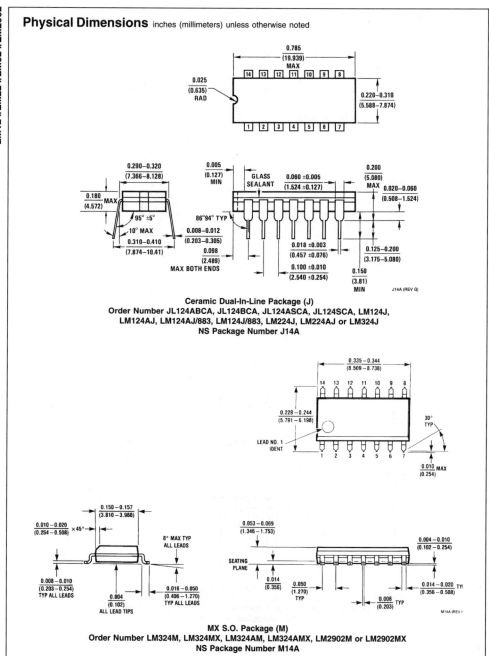

Ceramic Dual-In-Line Package (J)
Order Number JL124ABCA, JL124BCA, JL124ASCA, JL124SCA, LM124J,
LM124AJ, LM124AJ/883, LM124J/883, LM224J, LM224AJ or LM324J
NS Package Number J14A

MX S.O. Package (M)
Order Number LM324M, LM324MX, LM324AM, LM324AMX, LM2902M or LM2902MX
NS Package Number M14A

LM124/LM224/LM324/LM2902

Physical Dimensions inches (millimeters) unless otherwise noted (Continued)

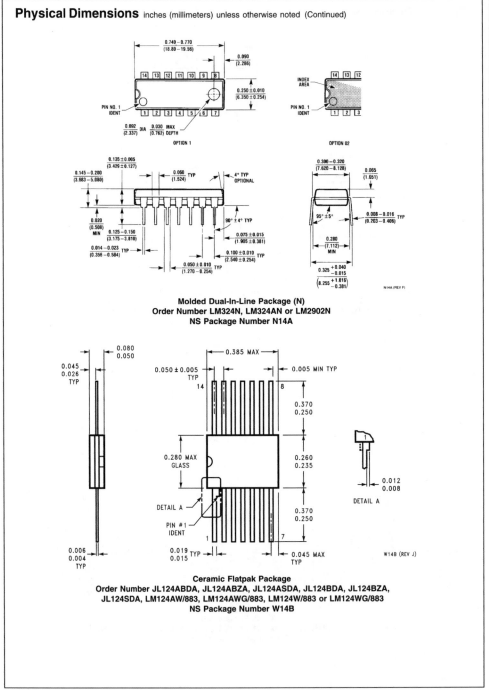

Molded Dual-In-Line Package (N)
Order Number LM324N, LM324AN or LM2902N
NS Package Number N14A

Ceramic Flatpak Package
Order Number JL124ABDA, JL124ABZA, JL124ASDA, JL124BDA, JL124BZA,
JL124SDA, LM124AW/883, LM124AWG/883, LM124W/883 or LM124WG/883
NS Package Number W14B

Physical Dimensions inches (millimeters) unless otherwise noted (Continued)

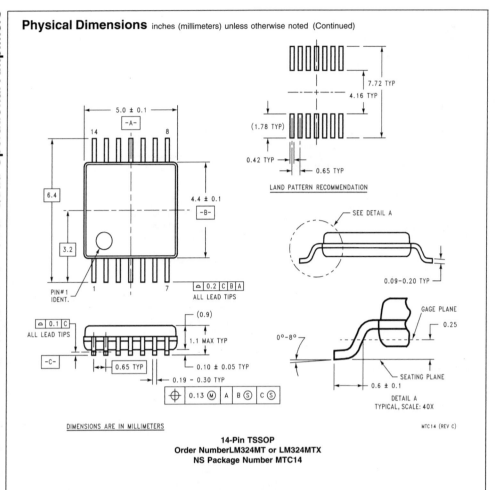

LAND PATTERN RECOMMENDATION

DIMENSIONS ARE IN MILLIMETERS

MTC14 (REV C)

14-Pin TSSOP
Order NumberLM324MT or LM324MTX
NS Package Number MTC14

LIFE SUPPORT POLICY

NATIONAL'S PRODUCTS ARE NOT AUTHORIZED FOR USE AS CRITICAL COMPONENTS IN LIFE SUPPORT DEVICES OR SYSTEMS WITHOUT THE EXPRESS WRITTEN APPROVAL OF THE PRESIDENT AND GENERAL COUNSEL OF NATIONAL SEMICONDUCTOR CORPORATION. As used herein:

1. Life support devices or systems are devices or systems which, (a) are intended for surgical implant into the body, or (b) support or sustain life, and whose failure to perform when properly used in accordance with instructions for use provided in the labeling, can be reasonably expected to result in a significant injury to the user.

2. A critical component is any component of a life support device or system whose failure to perform can be reasonably expected to cause the failure of the life support device or system, or to affect its safety or effectiveness.

 National Semiconductor Corporation
Americas
Tel: 1-800-272-9959
Fax: 1-800-737-7018
Email: support@nsc.com
www.national.com

National Semiconductor Europe
Fax: +49 (0) 180-530 85 86
Email: europe.support@nsc.com
Deutsch Tel: +49 (0) 69 9508 6208
English Tel: +44 (0) 870 24 0 2171
Fran'ais Tel: +33 (0) 1 41 91 8790

National Semiconductor Asia Pacific Customer Response Group
Tel: 65-2544466
Fax: 65-2504466
Email: ap.support@nsc.com

National Semiconductor Japan Ltd.
Tel: 81-3-5639-7560
Fax: 81-3-5639-7507

LM555 Timer

National Semiconductor

February 2000

LM555
Timer

General Description

The LM555 is a highly stable device for generating accurate time delays or oscillation. Additional terminals are provided for triggering or resetting if desired. In the time delay mode of operation, the time is precisely controlled by one external resistor and capacitor. For astable operation as an oscillator, the free running frequency and duty cycle are accurately controlled with two external resistors and one capacitor. The circuit may be triggered and reset on falling waveforms, and the output circuit can source or sink up to 200mA or drive TTL circuits.

Features

- Direct replacement for SE555/NE555
- Timing from microseconds through hours
- Operates in both astable and monostable modes
- Adjustable duty cycle
- Output can source or sink 200 mA
- Output and supply TTL compatible
- Temperature stability better than 0.005% per °C
- Normally on and normally off output
- Available in 8-pin MSOP package

Applications

- Precision timing
- Pulse generation
- Sequential timing
- Time delay generation
- Pulse width modulation
- Pulse position modulation
- Linear ramp generator

Schematic Diagram

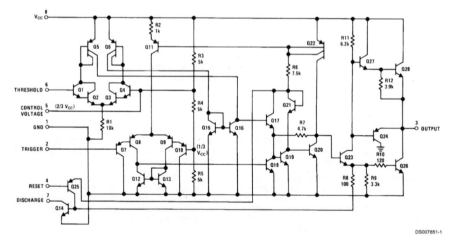

DS007851-1

Connection Diagram

**Dual-In-Line, Small Outline
and Molded Mini Small Outline Packages**

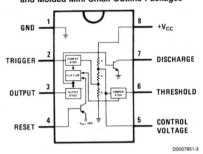

Top View

DS007851-3

Ordering Information

Package	Part Number	Package Marking	Media Transport	NSC Drawing
8-Pin SOIC	LM555CM	LM555CM	Rails	M08A
	LM555CMX	LM555CM	2.5k Units Tape and Reel	
8-Pin MSOP	LM555CMM	Z55	1k Units Tape and Reel	MUA08A
	LM555CMMX	Z55	3.5k Units Tape and Reel	
8-Pin MDIP	LM555CN	LM555CN	Rails	N08E

Electrical Characteristics (Notes 1, 2) (Continued)

(T_A = 25 C, V_{CC} = +5V to +15V, unless othewise specified)

Parameter	Conditions	Limits LM555C			Units
		Min	Typ	Max	
Output Voltage Drop (Low)	V_{CC} = 15V				
	I_{SINK} = 10mA		0.1	0.25	V
	I_{SINK} = 50mA		0.4	0.75	V
	I_{SINK} = 100mA		2	2.5	V
	I_{SINK} = 200mA		2.5		V
	V_{CC} = 5V				
	I_{SINK} = 8mA				V
	I_{SINK} = 5mA		0.25	0.35	V
Output Voltage Drop (High)	I_{SOURCE} = 200mA, V_{CC} = 15V		12.5		V
	I_{SOURCE} = 100mA, V_{CC} = 15V	12.75	13.3		V
	V_{CC} = 5V	2.75	3.3		V
Rise Time of Output			100		ns
Fall Time of Output			100		ns

Note 1: All voltages are measured with respect to the ground pin, unless otherwise specified.

Note 2: Absolute Maximum Ratings indicate limits beyond which damage to the device may occur. Operating Ratings indicate conditions for which the device is functional, but do not guarantee specific performance limits. Electrical Characteristics state DC and AC electrical specifications under particular test conditions which guarantee specific performance limits. This assumes that the device is within the Operating Ratings. Specifications are not guaranteed for parameters where no limit is given, however, the typical value is a good indication of device performance.

Note 3: For operating at elevated temperatures the device must be derated above 25 C based on a +150 C maximum junction temperature and a thermal resistance of 106 C/W (DIP), 170 C/W (S0-8), and 204 C/W (MSOP) junction to ambient.

Note 4: Supply current when output high typically 1 mA less at V_{CC} = 5V.

Note 5: Tested at V_{CC} = 5V and V_{CC} = 15V.

Note 6: This will determine the maximum value of $R_A + R_B$ for 15V operation. The maximum total ($R_A + R_B$) is 20MΩ.

Note 7: No protection against excessive pin 7 current is necessary providing the package dissipation rating will not be exceeded.

Note 8: Refer to RETS555X drawing of military LM555H and LM555J versions for specifications.

LM555

Typical Performance Characteristics

**Minimuim Pulse Width
Required for Triggering**

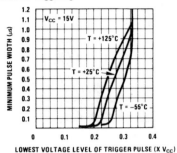

DS007851-4

**Supply Current vs.
Supply Voltage**

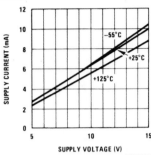

DS007851-19

**High Output Voltage vs.
Output Source Current**

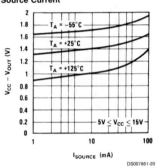

DS007851-20

**Low Output Voltage vs.
Output Sink Current**

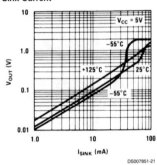

DS007851-21

**Low Output Voltage vs.
Output Sink Current**

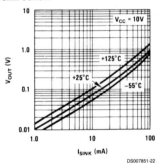

DS007851-22

**Low Output Voltage vs.
Output Sink Current**

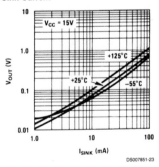

DS007851-23

Typical Performance Characteristics (Continued)

Output Propagation Delay vs.
Voltage Level of Trigger Pulse

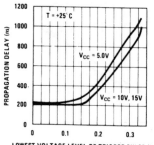

DS007851-24

Output Propagation Delay vs.
Voltage Level of Trigger Pulse

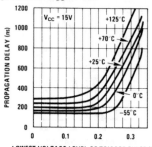

DS007851-25

Discharge Transistor (Pin 7)
Voltage vs. Sink Current

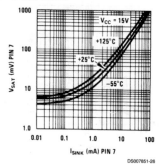

DS007851-26

Discharge Transistor (Pin 7)
Voltage vs. Sink Current

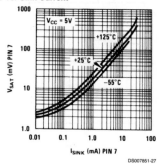

DS007851-27

Applications Information

MONOSTABLE OPERATION

In this mode of operation, the timer functions as a one-shot (*Figure 1*). The external capacitor is initially held discharged by a transistor inside the timer. Upon application of a negative trigger pulse of less than 1/3 V_{CC} to pin 2, the flip-flop is set which both releases the short circuit across the capacitor and drives the output high.

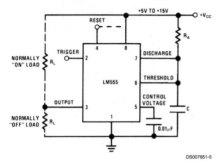

FIGURE 1. Monostable

The voltage across the capacitor then increases exponentially for a period of t = 1.1 R_A C, at the end of which time the voltage equals 2/3 V_{CC}. The comparator then resets the flip-flop which in turn discharges the capacitor and drives the output to its low state. *Figure 2* shows the waveforms generated in this mode of operation. Since the charge and the threshold level of the comparator are both directly proportional to supply voltage, the timing internal is independent of supply.

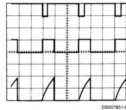

DS007851-6

V_{CC} = 5V   Top Trace: Input 5V/Div.
TIME = 0.1 ms/DIV.   Middle Trace: Output 5V/Div.
R_A = 9.1kΩ   Bottom Trace: Capacitor Voltage 2V/Div.
C = 0.01 F

FIGURE 2. Monostable Waveforms

During the timing cycle when the output is high, the further application of a trigger pulse will not effect the circuit so long as the trigger input is returned high at least 10 s before the end of the timing interval. However the circuit can be reset during this time by the application of a negative pulse to the reset terminal (pin 4). The output will then remain in the low state until a trigger pulse is again applied.

When the reset function is not in use, it is recommended that it be connected to V_{CC} to avoid any possibility of false triggering.

Figure 3 is a nomograph for easy determination of R, C values for various time delays.

NOTE: In monostable operation, the trigger should be driven high before the end of timing cycle.

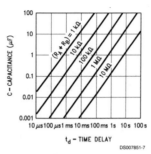

t_d – TIME DELAY

DS007851-7

FIGURE 3. Time Delay

ASTABLE OPERATION

If the circuit is connected as shown in *Figure 4* (pins 2 and 6 connected) it will trigger itself and free run as a multivibrator. The external capacitor charges through R_A + R_B and discharges through R_B. Thus the duty cycle may be precisely set by the ratio of these two resistors.

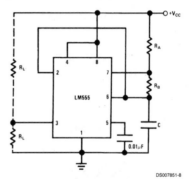

DS007851-8

FIGURE 4. Astable

In this mode of operation, the capacitor charges and discharges between 1/3 V_{CC} and 2/3 V_{CC}. As in the triggered mode, the charge and discharge times, and therefore the frequency are independent of the supply voltage.

LM555

Applications Information (Continued)

Figure 5 shows the waveforms generated in this mode of operation.

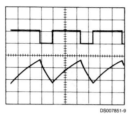

V_CC = 5V   Top Trace: Output 5V/Div.
TIME = 20 s/DIV.   Bottom Trace: Capacitor Voltage 1V/Div.
R_A = 3.9kΩ
R_B = 3kΩ
C = 0.01 F

FIGURE 5. Astable Waveforms

The charge time (output high) is given by:

$$t_1 = 0.693 \, (R_A + R_B) \, C$$

And the discharge time (output low) by:

$$t_2 = 0.693 \, (R_B) \, C$$

Thus the total period is:

$$T = t_1 + t_2 = 0.693 \, (R_A + 2R_B) \, C$$

The frequency of oscillation is:

$$f = \frac{1}{T} = \frac{1.44}{(R_A + 2\,R_B)\,C}$$

Figure 6 may be used for quick determination of these RC values.

The duty cycle is:

$$D = \frac{R_B}{R_A + 2R_B}$$

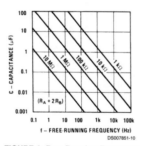

FIGURE 6. Free Running Frequency

FREQUENCY DIVIDER

The monostable circuit of *Figure 1* can be used as a frequency divider by adjusting the length of the timing cycle. *Figure 7* shows the waveforms generated in a divide by three circuit.

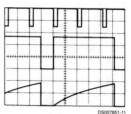

V_CC = 5V   Top Trace: Input 4V/Div.
TIME = 20 s/DIV.   Middle Trace: Output 2V/Div.
R_A = 9.1kΩ   Bottom Trace: Capacitor 2V/Div.
C = 0.01 F

FIGURE 7. Frequency Divider

PULSE WIDTH MODULATOR

When the timer is connected in the monostable mode and triggered with a continuous pulse train, the output pulse width can be modulated by a signal applied to pin 5. *Figure 8* shows the circuit, and in *Figure 9* are some waveform examples.

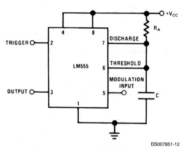

FIGURE 8. Pulse Width Modulator

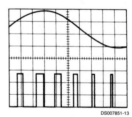

V_CC = 5V   Top Trace: Modulation 1V/Div.
TIME = 0.2 ms/DIV.   Bottom Trace: Output Voltage 2V/Div.
R_A = 9.1kΩ
C = 0.01 F

FIGURE 9. Pulse Width Modulator

Applications Information (Continued)

PULSE POSITION MODULATOR

This application uses the timer connected for astable operation, as in *Figure 10*, with a modulating signal again applied
to the control voltage terminal. The pulse position varies with
the modulating signal, since the threshold voltage and hence
the time delay is varied. *Figure 11* shows the waveforms
generated for a triangle wave modulation signal.

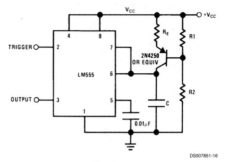

FIGURE 12.

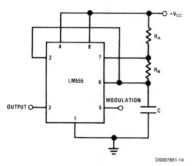

FIGURE 10. Pulse Position Modulator

Figure 13 shows waveforms generated by the linear ramp.
The time interval is given by:

$$T = \frac{2/3\ V_{CC}\ R_E\ (R_1 + R_2)\ C}{R_1\ V_{CC} - V_{BE}\ (R_1 + R_2)}$$

$$V_{BE} \cong 0.6V$$

$$V_{BE}\ .\ 0.6V$$

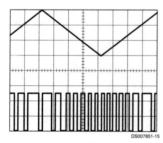

$V_{CC} = 5V$ Top Trace: Modulation Input 1V/Div.
TIME = 0.1 ms/DIV.   Bottom Trace: Output 2V/Div.
$R_A = 3.9k\Omega$
$R_B = 3k\Omega$
C = 0.01 F

FIGURE 11. Pulse Position Modulator

LINEAR RAMP

When the pullup resistor, R_A, in the monostable circuit is replaced by a constant current source, a linear ramp is generated. *Figure 12* shows a circuit configuration that will perform
this function.

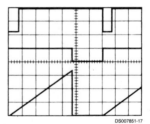

$V_{CC} = 5V$ Top Trace: Input 3V/Div.
TIME = 20 s/DIV .   Middle Trace: Output 5V/Div.
$R_1 = 47k\Omega$ Bottom Trace: Capacitor Voltage 1V/Div.
$R_2 = 100k\Omega$
$R_E = 2.7\ k\Omega$
C = 0.01 F

FIGURE 13. Linear Ramp

LM555

Applications Information (Continued)

50% DUTY CYCLE OSCILLATOR

For a 50% duty cycle, the resistors R_A and R_B may be connected as in *Figure 14*. The time period for the output high is the same as previous, $t_1 = 0.693\ R_A\ C$. For the output low it is $t_2 =$

$$\left[(R_A R_B)/(R_A + R_B)\right] C\ ln\left[\frac{R_B - 2R_A}{2R_B - R_A}\right]$$

Thus the frequency of oscillation is

$$f = \frac{1}{t_1 + t_2}$$

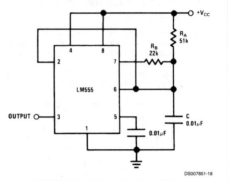

DS007851-18

FIGURE 14. 50% Duty Cycle Oscillator

Note that this circuit will not oscillate if R_B is greater than 1/2 R_A because the junction of R_A and R_B cannot bring pin 2 down to 1/3 V_{CC} and trigger the lower comparator.

ADDITIONAL INFORMATION

Adequate power supply bypassing is necessary to protect associated circuitry. Minimum recommended is 0.1 F in parallel with 1 F electrolytic.

Lower comparator storage time can be as long as 10 s when pin 2 is driven fully to ground for triggering. This limits the monostable pulse width to 10 s minimum.

Delay time reset to output is 0.47 s typical. Minimum reset pulse width must be 0.3 s, typical.

Pin 7 current switches within 30ns of the output (pin 3) voltage.

LT1013, LT1013A, LT1013D
DUAL PRECISION OPERATIONAL AMPLIFIERS

SLOS018E – MAY 1988 REVISED SEPTEMBER 2001

- **Single-Supply Operation**
  - Input Voltage Range Extends to Ground
  - Output Swings to Ground While Sinking Current
- **Input Offset Voltage**
  - 150 μV Max at 25°C for LT1013A
- **Offset Voltage Temperature Coefficient**
  - 2.5 μV/°C Max for LT1013A
- **Input Offset Current**
  - 0.8 nA Max at 25°C for LT1013A
- **High Gain . . . 1.5 V/μV Min (R_L = 2 kΩ), 0.8 V/μV Min (R_L = 600 kΩ) for LT1013A**
- **Low Supply Current . . . 0.5 mA Max at T_A = 25°C for LT1013A**
- **Low Peak-to-Peak Noise Voltage . . . 0.55 μV Typ**
- **Low Current Noise . . . 0.07 pA/√Hz Typ**

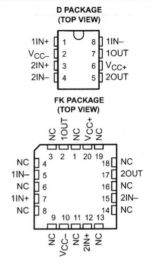

D PACKAGE
(TOP VIEW)

```
1IN+  [ 1      8 ]  1IN–
VCC–  [ 2      7 ]  1OUT
2IN+  [ 3      6 ]  VCC+
2IN–  [ 4      5 ]  2OUT
```

FK PACKAGE
(TOP VIEW)

```
         NC  1OUT  NC  VCC+  NC
          3   2    1   20   19
NC   [ 4               18 ]  NC
1IN– [ 5               17 ]  2OUT
NC   [ 6               16 ]  NC
1IN+ [ 7               15 ]  2IN–
NC   [ 8               14 ]  NC
          9  10  11  12  13
         NC VCC– NC 2IN+ NC
```

NC – No internal connection

JG OR P PACKAGE
(TOP VIEW)

```
1OUT  [ 1      8 ]  VCC+
1IN–  [ 2      7 ]  2OUT
1IN+  [ 3      6 ]  2IN–
VCC–  [ 4      5 ]  2IN+
```

description

The LT1013 devices are dual precision operational amplifiers, featuring high gain, low supply current, low noise, and low-offset-voltage temperature coefficient.

The LT1013 devices can be operated from a single 5-V power supply; the common-mode input voltage range includes ground, and the output can also swing to within a few millivolts of ground. Crossover distortion is eliminated. The LT1013 can be operated with both dual ±15-V and single 5-V supplies.

The LT1013C and LT1013AC, and LT1013D are characterized for operation from 0°C to 70°C. The LT1013I and LT1013AI, and LT1013DI are characterized for operation from –40°C to 105°C. The LT1013M and LT1013AM, and LT1013DM are characterized for operation over the full military temperature range of –55°C to 125°C.

AVAILABLE OPTIONS

T_A	V_{IO}max AT 25°C (μV)	PACKAGED DEVICES			
		SMALL OUTLINE (D)	CHIP CARRIER (FK)	CERAMIC DIP (JG)	PLASTIC DIP (P)
0°C to 70°C	150	—	—	—	LT1013ACP
	300	—	—	—	LT1013CP
	800	LT1013DD	—	—	LT1013DP
–40°C to 105°C	150	—	—	—	LT1013AIP
	300	—	—	—	LT1013IP
	800	LT1013DID	—	—	LT1013DIP
–55°C to 125°C	150	—	LT1013AMFK	LT1013AMJG	—
	300	—	—	LT1013MJG	—
	800	LT1013DMD	—	LT1013DMJG	—

The D package is available taped and reeled. Add the suffix R to the device type (e.g., LT1013DDR).

 Please be aware that an important notice concerning availability, standard warranty, and use in critical applications of Texas Instruments semiconductor products and disclaimers thereto appears at the end of this data sheet.

TEXAS INSTRUMENTS

POST OFFICE BOX 655303 ● DALLAS, TEXAS 75265

Template Release Date: 7–11–94

LT1013, LT1013A, LT1013D
DUAL PRECISION OPERATIONAL AMPLIFIERS

SLOS018E — MAY 1988 — REVISED SEPTEMBER 2001

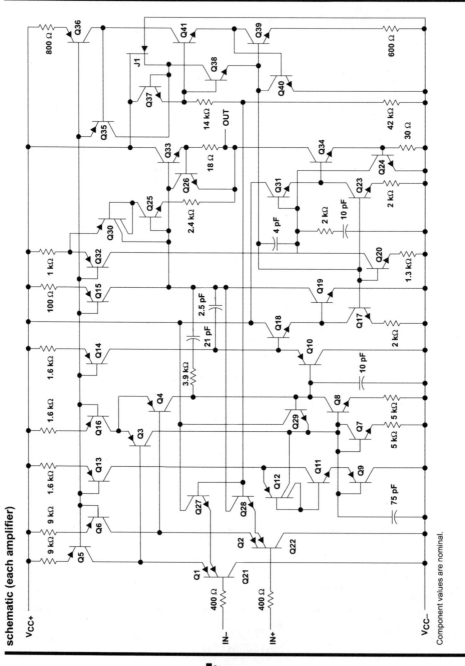

schematic (each amplifier)

Component values are nominal.

LT1013, LT1013A, LT1013D
DUAL PRECISION OPERATIONAL AMPLIFIERS

SLOS018E – MAY 1988 – REVISED SEPTEMBER 2001

absolute maximum ratings over operating free-air temperature range (unless otherwise noted †

Supply voltage (see Note 1): V_{CC+} .... 22 V
V_{CC-} .... −22 V
Input voltage range, V_I (any input, see Note 1) .... V_{CC-} −5 V to V_{CC+}
Differential input voltage (see Note 2) .... ±30 V
Duration of short-circuit current at (or below) 25°C (see Note 3) .... Unlimited
Package thermal impedance, θ_{JA} (see Note 4): D package .... 97°C/W
P package .... 85°C/W
Case temperature for 60 seconds: FK package .... 260°C
Lead temperature 1,6 mm (1/16 inch) from case for 10 seconds: D or P package .... 260°C
JG package .... 300°C
Storage temperature range, T_{stg} .... −65°C to 150°C

† Stresses beyond those listed under "absolute maximum ratings" may cause permanent damage to the device. These are stress ratings only, and
  functional operation of the device at these or any other conditions beyond those indicated under "recommended operating conditions" is not
  implied. Exposure to absolute-maximum-rated conditions for extended periods may affect device reliability.
NOTES: 1. All voltage values, except differential voltages, are with respect to the midpoint between V_{CC+} and V_{CC-}.
        2. Differential voltages are at IN+ with respect to IN−.
        3. The output may be shorted to either supply.
        4. The package thermal impedance is calculated in accordance with JESD 51-7.

TEXAS
INSTRUMENTS
POST OFFICE BOX 655303 ● DALLAS, TEXAS 75265

3

Template Release Date: 7–11–94

LT1013, LT1013A, LT1013D
DUAL PRECISION OPERATIONAL AMPLIFIERS

SLOS018E — MAY 1988 — REVISED SEPTEMBER 2001

electrical characteristics at specified free-air temperature, $V_{CC\pm} = \pm15$ V, $V_{IC} = 0$ (unless otherwise noted)

	PARAMETER	TEST CONDITIONS	T_A†	LT1013C			LT1013AC			LT1013DC			UNIT
				MIN	TYP‡	MAX	MIN	TYP‡	MAX	MIN	TYP‡	MAX	
V_{IO}	Input offset voltage	$R_S = 50\ \Omega$	25°C		60	300		40	150		200	800	μV
			Full range			400			240			1000	μV
$\alpha_{V_{IO}}$	Temperature coefficient of input offset voltage		Full range		0.4	2.5		0.3	2		0.7	5	μV/°C
	Long-term drift of input offset voltage		25°C		0.5			0.4			0.5		μV/mo
I_{IO}	Input offset current		25°C		0.2	1.5		0.15	0.8		0.2	1.5	nA
			Full range			2.8			1.5			2.8	nA
I_{IB}	Input bias current		25°C		-15	-30		-12	-20		-15	-30	nA
			Full range			-38			-25			-38	nA
V_{ICR}	Common-mode input voltage range		25°C	-15 to 13.5	-15.3 to 13.8		-15 to 13.5	-15.3 to 13.8		-15 to 13.5	-15.3 to 13.8		V
			Full range	-15 to 13			-15 to 13			-15 to 13			V
V_{OM}	Maximum peak output voltage swing	$R_L = 2\ K\Omega$	25°C	±12.5	±14		±13	±14		±12.5	±14		V
			Full range	±12			±12.5			±12			V
A_{VD}	Large-signal differential voltage amplification	$V_O = \pm10$ V, $R_L = 600\ \Omega$	25°C	0.5	2		0.8	2.5		0.5	2		V/μV
		$V_O = \pm10$ V, $R_L = 2$ kΩ	25°C	1.2	7		1.5	8		1.2	7		V/μV
			Full range	0.7			1			0.7			V/μV
CMRR	Common-mode rejection ratio	$V_{IC} = -15$ V to 13.5 V	25°C	97	114		100	117		97	114		dB
		$V_{IC} = -14.9$ V to 13 V	Full range	94			98			94			dB
k_{SVR}	Supply-voltage rejection ratio ($\Delta V_{CC}/\Delta V_{IO}$)	$V_{CC+} = \pm2$ V to ±18 V	25°C	100	117		103	120		100	117		dB
			Full range	97			101			97			dB
	Channel separation	$V_O = \pm10$ V, $R_L = 2$ kΩ	25°C	120	137		123	140		120	137		dB
r_{id}	Differential input resistance		25°C	70	300		100	400		70	300		MΩ
r_{ic}	Common-mode input resistance		25°C		4			5			4		GΩ
I_{CC}	Supply current per amplifier		25°C		0.35	0.55		0.35	0.5		0.35	0.55	mA
			Full range			0.7			0.55			0.6	mA

† Full range is 0°C to 70°C.
‡ All typical values are at T_A = 25°C.

LT1013, LT1013A, LT1013D
DUAL PRECISION OPERATIONAL AMPLIFIERS

SLOS018E — MAY 1988 — REVISED SEPTEMBER 2001

electrical characteristics at specified free-air temperature, V_{CC+} = 5 V, V_{CC-} = 0, V_O = 1.4 V, V_{IC} = 0 (unless otherwise noted)

PARAMETER		TEST CONDITIONS	T_A†	LT1013C MIN	LT1013C TYP‡	LT1013C MAX	LT1013AC MIN	LT1013AC TYP‡	LT1013AC MAX	LT1013DC MIN	LT1013DC TYP‡	LT1013DC MAX	UNIT
V_{IO}	Input offset voltage	R_S = 50 Ω	25°C		90	450		60	250		250	950	μV
			Full range			570			350			1200	
I_{IO}	Input offset current		25°C		0.3	2		0.2	1.3		0.3	2	nA
			Full range			6			3.5			6	
I_{IB}	Input bias current		25°C		−18	−50		−15	−35		−18	−50	nA
			Full range			−90			−55			−90	
V_{ICR}	Common-mode input voltage range		25°C	0 to 3.5	−0.3 to 3.8		0 to 3.5	−0.3 to 3.8		0 to 3.5	−0.3 to 3.8		V
			Full range	0 to 3			0 to 3			0 to 3			
V_{OM}	Maximum peak output voltage swing	Output low, No load	25°C		15	25		15	25		15	25	mV
		Output low, R_L = 600 Ω to GND, I_{sink} = 1 mA	25°C		5	10		5	10		5	10	
			Full range			13			13			13	
		Output high, No load	25°C	4	4.4		4	4.4		4	4.4		V
		Output high, R_L = 600 Ω to GND	25°C	3.4	4		3.4	4		3.4	4		
			Full range	3.2			3.3			3.2			
A_{VD}	Large-signal differential voltage amplification	V_O = 5 mV to 4 V, R_L = 500 Ω	25°C		1			1			1		V/μV
I_{CC}	Supply current per amplifier		25°C		0.32	0.5		0.31	0.45		0.32	0.5	mA
			Full range			0.55			0.5			0.55	

† Full range is 0°C to 70°C.
‡ All typical values are at T_A = 25°C.

operating characteristics, $V_{CC\pm}$ = ±15 V, V_{IC} = 0, T_A = 25°C

PARAMETER		TEST CONDITIONS	MIN	TYP	MAX	UNIT
SR	Slew rate		0.2	0.4		V/μs
V_n	Equivalent input noise voltage	f = 10 Hz		24		nV/√Hz
		f = 1 kHz		22		nV/√Hz
$V_{N(PP)}$	Peak-to-peak equivalent input noise voltage	f = 0.1 Hz to 10 Hz		0.55		μV
I_n	Equivalent input noise current	f = 10 Hz		0.07		pA/√Hz

TEXAS
INSTRUMENTS
POST OFFICE BOX 655303• DALLAS, TEXAS 75265

Template Release Date: 7–11–94

LT1013, LT1013A, LT1013D
DUAL PRECISION OPERATIONAL AMPLIFIERS

SLOS018E — MAY 1988 — REVISED SEPTEMBER 2001

electrical characteristics at specified free-air temperature, $V_{CC\pm} = \pm15$ V, $V_{IC} = 0$ (unless otherwise noted)

PARAMETER	TEST CONDITIONS	T_A†	LT1013I MIN	LT1013I TYP‡	LT1013I MAX	LT1013AI MIN	LT1013AI TYP‡	LT1013AI MAX	LT1013DI MIN	LT1013DI TYP‡	LT1013DI MAX	UNIT
V_{IO} Input offset voltage	$R_S = 50\ \Omega$	25°C		60	300		40	150		200	800	µV
		Full range			550			300			1000	
$\alpha_{V_{IO}}$ Temperature coefficient of input offset voltage		Full range		0.4	2.5		0.3	2		0.7	5	µV/°C
Long-term drift of input offset voltage		25°C		0.5			0.4			0.5		µV/mo
I_{IO} Input offset current		25°C		0.2	1.5		0.15	0.8		0.2	1.5	nA
		Full range			2.8			1.5			2.8	
I_{IB} Input bias current		25°C		-15	-30		-12	-20		-15	-30	nA
		Full range			-38			-25			-38	
V_{ICR} Common-mode input voltage range		25°C	-15 to 13.5	-15.3 to 13.8		-15 to 13.5	-15.3 to 13.8		-15 to 13.5	-15.3 to 13.8		V
		Full range	-15 to 13			-15 to 13			-15 to 13			
V_{OM} Maximum peak output voltage swing	$R_L = 2\ k\Omega$	25°C	±12.5	±14		±13	±14		±12.5	±14		V
		Full range	±12			±12.5			±12			
A_{VD} Large-signal differential voltage amplification	$V_O = \pm10$ V, $R_L = 600\ \Omega$	25°C	0.5	0.2		0.8	2.5		0.5	2		V/µV
	$V_O = \pm10$ V, $R_L = 2\ k\Omega$	25°C	1.2	7		1.5	8		1.2	7		
		Full range	0.7			1			0.7			
$CMRR$ Common-mode rejection ratio	$V_{IC} = -15$ V to 13.5 V	25°C	97	114		100	117		97	114		dB
	$V_{IC} = -14.9$ V to 13 V	Full range	94			97			94			
k_{SVR} Supply-voltage rejection ratio ($\Delta V_{CC}/\Delta V_{IO}$)	$V_{CC\pm} = \pm2$ V to ±18 V	25°C	100	117		103	120		100	117		dB
		Full range	97			101			97			
Channel separation	$V_O = \pm10$ V, $R_L = 2\ k\Omega$	25°C	120	137		123	140		120	137		dB
r_{id} Differential input resistance		25°C	70	300		100	400		70	300		MΩ
r_{ic} Common-mode input resistance		25°C		4			5			4		GΩ
I_{CC} Supply current per amplifier		25°C		0.35	0.55		0.35	0.5		0.35	0.55	mA
		Full range			0.7			0.55			0.6	

† Full range is –40°C to 105°C.
‡ All typical values are at T_A = 25°C.

TEXAS INSTRUMENTS
POST OFFICE BOX 655303• DALLAS, TEXAS 75265

LT1013, LT1013A, LT1013D
DUAL PRECISION OPERATIONAL AMPLIFIERS

SLOS018E — MAY 1988 — REVISED SEPTEMBER 2001

electrical characteristics at specified free-air temperature, $V_{CC+} = 5\ V$, $V_{CC-} = 0$, $V_O = 1.4\ V$, $V_{IC} = 0$ (unless otherwise noted)

PARAMETER	TEST CONDITIONS	T_A†	LT1013I MIN	LT1013I TYP‡	LT1013I MAX	LT1013AI MIN	LT1013AI TYP‡	LT1013AI MAX	LT1013DI MIN	LT1013DI TYP‡	LT1013DI MAX	UNIT
V_{IO} Input offset voltage	$R_S = 50\ \Omega$	25°C		90	450		60	250		250	950	µV
		Full range			570			350			1200	
I_{IO} Input offset current		25°C		0.3	2		0.2	1.3		0.3	2	nA
		Full range			6			3.5			6	
I_{IB} Input bias current		25°C		−18	−50		−15	−35		−18	−50	nA
		Full range			−90			−55			−90	
V_{ICR} Common-mode input voltage range		25°C	0 to 3.5	−0.3 to 3.8		0 to 3.5	−0.3 to 3.8		0 to 3.5	−0.3 to 3.8		V
		Full range	0 to 3			0 to 3			0 to 3			
V_{OM} Maximum peak output voltage swing	Output low, No load	25°C		15	25		15	25		15	25	mV
	Output low, $R_L = 600\ \Omega$ to GND, $I_{sink} = 1\ mA$	25°C		5	10		5	10		5	10	
		Full range			13			13			13	
	Output low, No load	25°C		220	350		220	350		220	350	
	Output high,	25°C	4	4.4		4	4.4		4	4.4		V
	Output high, $R_L = 600\ \Omega$ to GND	25°C	3.4	4		3.4	4		3.4	4		
		Full range	3.2			3.3			3.2			
A_{VD} Large-signal differential voltage amplification	$V_O = 5\ mV$ to $4\ V$, $R_L = 500\ \Omega$	25°C		1			1			1		V/µV
I_{CC} Supply current per amplifier		25°C		0.32	0.5		0.31	0.45		0.32	0.5	mA
		Full range			0.55			0.5			0.55	

† Full range is −40°C to 105°C.
‡ All typical values are at $T_A = 25°C$.

operating characteristics, $V_{CC\pm} = \pm15\ V$, $V_{IC} = 0$, $T_A = 25°C$

PARAMETER	TEST CONDITIONS	MIN	TYP	MAX	UNIT
SR Slew rate		0.2	0.4		V/µs
V_n Equivalent input noise voltage	$f = 10\ Hz$		24		nV/$\sqrt{Hz}$
	$f = 1\ kHz$		22		
$V_{N(PP)}$ Peak-to-peak equivalent input noise voltage	$f = 0.1\ Hz$ to $10\ Hz$		0.55		µV
I_n Equivalent input noise current	$f = 10\ Hz$		0.07		pA/$\sqrt{Hz}$

TEXAS INSTRUMENTS

POST OFFICE BOX 655303• DALLAS, TEXAS 75265

Template Release Date: 7–11–94

LT1013, LT1013A, LT1013D
DUAL PRECISION OPERATIONAL AMPLIFIERS

SLOS018E — MAY 1988 — REVISED SEPTEMBER 2001

electrical characteristics at specified free-air temperature, $V_{CC\pm} = \pm15$ V, $V_{IC} = 0$ (unless otherwise noted)

PARAMETER	TEST CONDITIONS	TA†	LT1013M MIN	LT1013M TYP‡	LT1013M MAX	LT1013AM MIN	LT1013AM TYP‡	LT1013AM MAX	LT1013DM MIN	LT1013DM TYP‡	LT1013DM MAX	UNIT
V_{IO} Input offset voltage	$R_S = 50\ \Omega$	25°C		60	300		40	150		200	800	μV
		Full range			550			300			1000	
$\alpha_{V_{IO}}$ Temperature coefficient of input offset voltage		Full range		0.5	2.5*		0.4	2*		0.5	2.5*	μV/°C
Long-term drift of input offset voltage		25°C		0.5			0.4			0.5		μV/mo
I_{IO} Input offset current		25°C		0.2	1.5		0.15	0.8		0.2	1.5	nA
		Full range			5			2.5			5	
I_{IB} Input bias current		25°C		-15	-30		-12	-20		-15	-30	nA
		Full range			-45			-30			-45	
V_{ICR} Common-mode input voltage range		25°C	-15 to 13.5	-15.3 to 13.8		-15 to 13.5	-15.3 to 13.8		-15 to 13.5	-15.3 to 13.8		V
		Full range	-14.9 to 13			-14.9 to 13			-14.9 to 13			
V_{OM} Maximum peak output voltage swing	$R_L = 2\ k\Omega$	25°C	±12.5	±14		±13	±14		±12.5	±14		V
		Full range	±11.5			±12			±11.5			
A_{VD} Large-signal differential voltage amplification	$V_O = \pm10$ V, $R_L = 600\ \Omega$	25°C	0.5	2		0.8	2.5		0.5	2		V/μV
	$V_O = \pm10$ V, $R_L = 2\ k\Omega$	25°C	1.2	7		1.5	8		1.2	7		
		Full range	0.25			0.5			0.25			
$CMRR$ Common-mode rejection ratio	$V_{IC} = -15$ V to 13.5 V	25°C	97	117		100	117		97	114		dB
	$V_{IC} = -14.9$ V to 13 V	Full range	94			97			94			
k_{SVR} Supply-voltage rejection ratio ($\Delta V_{CC}/\Delta V_{IO}$)	$V_{CC\pm} = \pm2$ V to ±18 V	25°C	100	117		103	120		100	117		dB
		Full range	97			100			97			
Channel separation	$V_O = \pm10$ V, $R_L = 2\ k\Omega$	25°C	120	137		123	140		120	137		dB
r_{id} Differential input resistance		25°C	70	300		100	400		70	300		MΩ
r_{ic} Common-mode input resistance		25°C		4			5			4		GΩ
I_{CC} Supply current per amplifier		25°C		0.35	0.55		0.35	0.5		0.35	0.55	mA
		Full range			0.7			0.6			0.7	

* On products compliant to MIL-PRF-38535, Class B, this parameter is not production tested.
† Full range is –55°C to 125°C.
‡ All typical values are at T_A = 25°C.

TEXAS
INSTRUMENTS
POST OFFICE BOX 655303• DALLAS, TEXAS 75265

LT1013, LT1013A, LT1013D
DUAL PRECISION OPERATIONAL AMPLIFIERS

SLOS018E — MAY 1988 — REVISED SEPTEMBER 2001

electrical characteristics at specified free-air temperature, V_{CC+} = 5 V, V_{CC-} = 0, V_O = 1.4 V, V_{IC} = 0 (unless otherwise noted)

PARAMETER		TEST CONDITIONS	T_A†	LT1013M			LT1013AM			LT1013DM			UNIT
				MIN	TYP‡	MAX	MIN	TYP‡	MAX	MIN	TYP‡	MAX	
V_{IO}	Input offset voltage	R_S = 50 Ω	25°C		90	450		60	250		250	950	µV
		R_S = 50 Ω	Full range		400	1500		250	900		800	2000	
		R_S = 50 Ω, V_{IC} = 0.1 V	125°C		200	750		120	450		560	1200	
I_{IO}	Input offset current		25°C		0.3	2		0.2	1.3		0.3	2	nA
			Full range			10			6			10	
I_{IB}	Input bias current		25°C		−18	−50		−15	−35		−18	−50	nA
			Full range			−120			−80			−120	
V_{ICR}	Common-mode input voltage range		25°C	0 to 3.5	−0.3 to 3.8		0 to 3.5	−0.3 to 3.8		0 to 3.5	−0.3 to 3.8		V
			Full range	0 to 3			0 to 3			0 to 3			
V_{OM}	Maximum peak output voltage swing	Output low, No load	25°C		15	25		15	25		15	25	mV
		Output low, RL = 600 Ω to GND	25°C		5	10		5	10		5	10	
			Full range			18			15			18	
		Output low, I_{sink} = 1 mA	25°C		220	350		220	350		220	350	
		Output high, No load	25°C	4	4.4		4	4.4		4	4.4		V
		Output high, RL = 600 Ω to GND	25°C	3.4	4		3.4	4		3.4	4		
			Full range	3.1			3.2			3.1			
A_{VD}	Large-signal differential voltage amplification	V_O = 5 mV to 4 V, RL = 500 Ω	25°C		1			1			1		V/µV
I_{CC}	Supply current per amplifier		25°C		0.32	0.45		0.31	0.45		0.32	0.5	mA
			Full range			0.55			0.55			0.65	

† Full range is −55°C to 125°C.
‡ All typical values are at T_A = 25°C.

operating characteristics, $V_{CC\pm}$ = ±15 V, V_{IC} = 0, T_A = 25°C

	PARAMETER	TEST CONDITIONS	MIN	TYP	MAX	UNIT
SR	Slew rate		0.2	0.4		V/µs
V_n	Equivalent input noise voltage	f = 10 Hz		24		nV/√Hz
		f = 1 kHz		22		
$V_{N(PP)}$	Peak-to-peak equivalent input noise voltage	f = 0.1 Hz to 10 Hz		0.55		µV
I_n	Equivalent input noise current	f = 10 Hz		0.07		pA/√Hz

TEXAS
INSTRUMENTS
POST OFFICE BOX 655303 • DALLAS, TEXAS 75265

LT1013, LT1013A, LT1013D
DUAL PRECISION OPERATIONAL AMPLIFIERS

SLOS018E – MAY 1988 – REVISED SEPTEMBER 2001

TYPICAL CHARACTERISTICS

Table of Graphs

			FIGURE
V_{IO}	Input offset voltage	vs Supply voltage	1
		vs Temperature	2
ΔV_{IO}	Change in input offset voltage	vs Time	3
I_{IO}	Input offset current	vs Temperature	4
I_{IB}	Input bias current	vs Temperature	5
V_{IC}	Common-mode input voltage	vs Input bias current	6
A_{VD}	Differential voltage amplification	vs Load resistance	7, 8
		vs Frequency	9, 10
	Channel separation	vs Frequency	11
	Output saturation voltage	vs Temperature	12
CMRR	Common-mode rejection ratio	vs Frequency	13
k_{SVR}	Supply-voltage rejection ratio	vs Frequency	14
I_{CC}	Supply current	vs Temperature	15
I_{OS}	Short-circuit output current	vs Time	16
V_n	Equivalent input noise voltage	vs Frequency	17
I_n	Equivalent input noise current	vs Frequency	17
$V_{N(PP)}$	Peak-to-peak input noise voltage	vs Time	18
	Pulse response	Small signal	19, 21
		Large signal	20, 22, 23
	Phase shift	vs Frequency	9

TEXAS
INSTRUMENTS

POST OFFICE BOX 655303 ● DALLAS, TEXAS 75265

LT1013, LT1013A, LT1013D
DUAL PRECISION OPERATIONAL AMPLIFIERS

SLOS018E – MAY 1988 – REVISED SEPTEMBER 2001

TYPICAL CHARACTERISTICS†

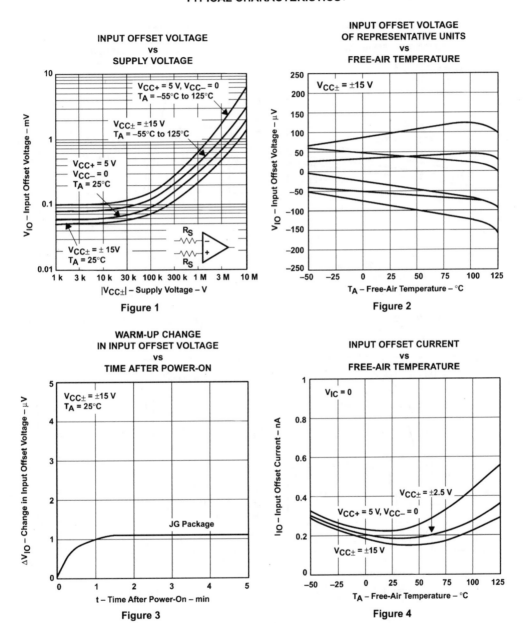

Figure 1

Figure 2

Figure 3

Figure 4

† Data at high and low temperatures are applicable only within the rated operating free-air temperature ranges of the various devices.

POST OFFICE BOX 655303 ● DALLAS, TEXAS 75265

11

LT1013, LT1013A, LT1013D
DUAL PRECISION OPERATIONAL AMPLIFIERS

SLOS018E – MAY 1988 – REVISED SEPTEMBER 2001

TYPICAL CHARACTERISTICS†

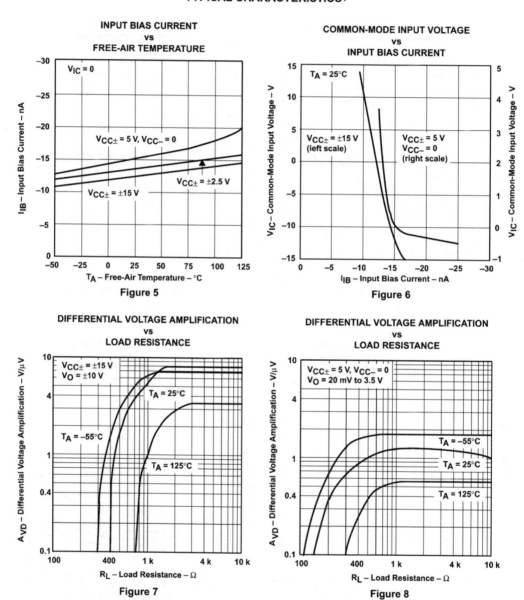

INPUT BIAS CURRENT
vs
FREE-AIR TEMPERATURE

Figure 5

COMMON-MODE INPUT VOLTAGE
vs
INPUT BIAS CURRENT

Figure 6

DIFFERENTIAL VOLTAGE AMPLIFICATION
vs
LOAD RESISTANCE

Figure 7

DIFFERENTIAL VOLTAGE AMPLIFICATION
vs
LOAD RESISTANCE

Figure 8

† Data at high and low temperatures are applicable only within the rated operating free-air temperature ranges of the various devices.

TEXAS
INSTRUMENTS

POST OFFICE BOX 655303 ● DALLAS, TEXAS 75265

LT1013, LT1013A, LT1013D
DUAL PRECISION OPERATIONAL AMPLIFIERS

SLOS018E – MAY 1988 – REVISED SEPTEMBER 2001

TYPICAL CHARACTERISTICS†

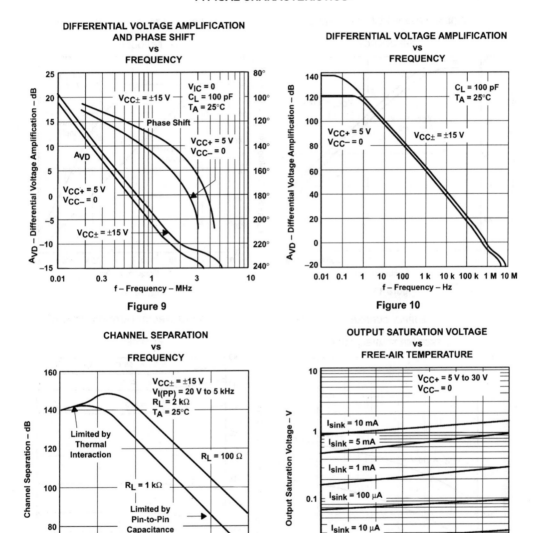

DIFFERENTIAL VOLTAGE AMPLIFICATION AND PHASE SHIFT vs FREQUENCY

Figure 9

DIFFERENTIAL VOLTAGE AMPLIFICATION vs FREQUENCY

Figure 10

CHANNEL SEPARATION vs FREQUENCY

Figure 11

OUTPUT SATURATION VOLTAGE vs FREE-AIR TEMPERATURE

Figure 12

† Data at high and low temperatures are applicable only within the rated operating free-air temperature ranges of the various devices.

LT1013, LT1013A, LT1013D
DUAL PRECISION OPERATIONAL AMPLIFIERS

SLOS018E – MAY 1988 – REVISED SEPTEMBER 2001

TYPICAL CHARACTERISTICS†

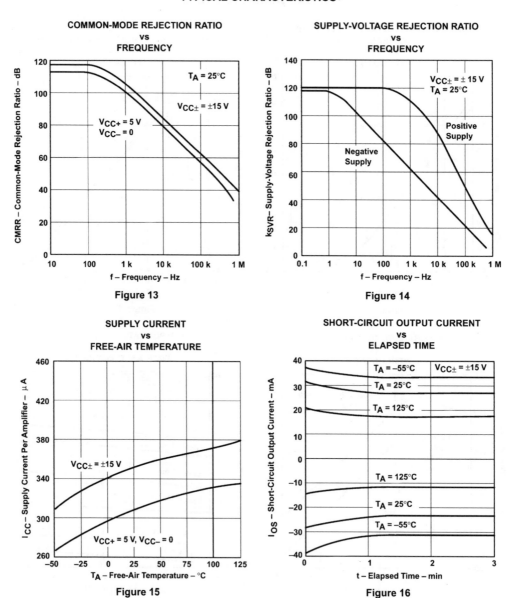

Figure 13

Figure 14

Figure 15

Figure 16

† Data at high and low temperatures are applicable only within the rated operating free-air temperature ranges of the various devices.

TEXAS
INSTRUMENTS
POST OFFICE BOX 655303 • DALLAS, TEXAS 75265

LT1013, LT1013A, LT1013D
DUAL PRECISION OPERATIONAL AMPLIFIERS

SLOS018E – MAY 1988 – REVISED SEPTEMBER 2001

TYPICAL CHARACTERISTICS

EQUIVALENT INPUT NOISE VOLTAGE
AND EQUIVALENT INPUT NOISE CURRENT
vs
FREQUENCY

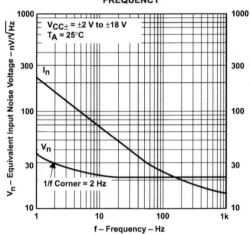

Figure 17

PEAK-TO-PEAK INPUT NOISE VOLTAGE
OVER A
10-SECOND PERIOD

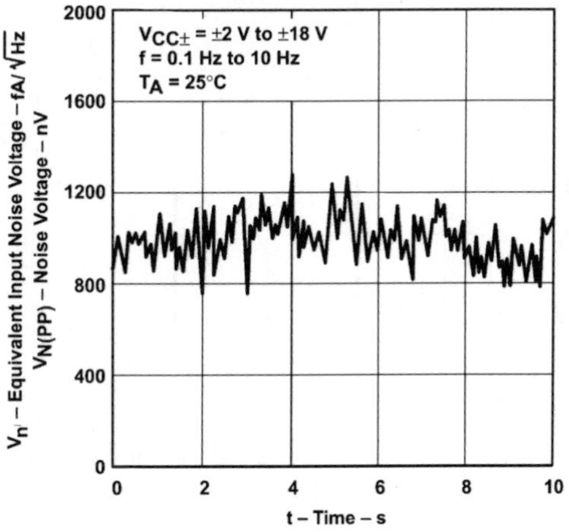

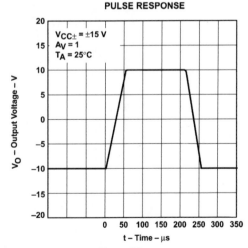

Figure 18

VOLTAGE-FOLLOWER
SMALL-SIGNAL
PULSE RESPONSE

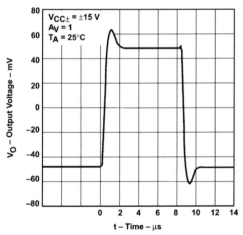

Figure 19

VOLTAGE-FOLLOWER
LARGE-SIGNAL
PULSE RESPONSE

Figure 20

![Texas Instruments logo]
TEXAS
INSTRUMENTS
POST OFFICE BOX 655303 ● DALLAS, TEXAS 75265

LT1013, LT1013A, LT1013D
DUAL PRECISION OPERATIONAL AMPLIFIERS

SLOS018E – MAY 1988 – REVISED SEPTEMBER 2001

TYPICAL CHARACTERISTICS

**VOLTAGE-FOLLOWER
SMALL-SIGNAL
PULSE RESPONSE**

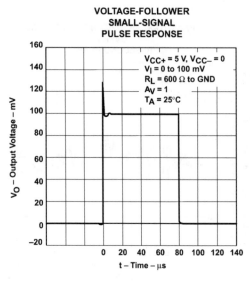

Figure 21

**VOLTAGE-FOLLOWER
LARGE-SIGNAL
PULSE RESPONSE**

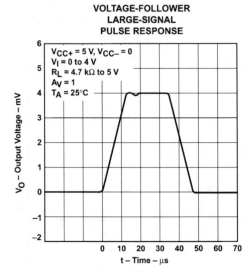

Figure 22

**VOLTAGE-FOLLOWER
LARGE-SIGNAL
PULSE RESPONSE**

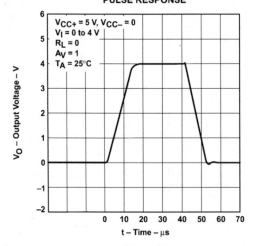

Figure 23

![TEXAS INSTRUMENTS]
POST OFFICE BOX 655303 ● DALLAS, TEXAS 75265

16

LT1013, LT1013A, LT1013D
DUAL PRECISION OPERATIONAL AMPLIFIERS

SLOS018E – MAY 1988 – REVISED SEPTEMBER 2001

APPLICATION INFORMATION

single-supply operation

The LT1013 is fully specified for single-supply operation (V_{CC-} = 0). The common-mode input voltage range includes ground, and the output swings to within a few millivolts of ground.

Furthermore, the LT1013 has specific circuitry that addresses the difficulties of single-supply operation, both at the input and at the output. At the input, the driving signal can fall below 0 V, either inadvertently or on a transient basis. If the input is more than a few hundred millivolts below ground, the LT1013 is designed to deal with the following two problems that can occur:

1. On many other operational amplifiers, when the input is more than a diode drop below ground, unlimited current flows from the substrate (V_{CC-} terminal) to the input, which can destroy the unit. On the LT1013, the 400-Ω resistors in series with the input [see *schematic (each amplifier)*] protect the device, even when the input is 5 V below ground.

2. When the input is more than 400 mV below ground (at T_A = 25°C), the input stage of similar type operational amplifiers saturates and phase reversal occurs at the output. This can cause lockup in servo systems. Because of a unique phase-reversal protection circuitry (Q21, Q22, Q27, and Q28), the LT1013 outputs do not reverse, even when the inputs are at –1.5 V (see Figure 24).

This phase-reversal protection circuitry does not function when the other operational amplifier on the LT1013 is driven hard into negative saturation at the output. Phase-reversal protection does not work on amplifier 1 when amplifier 2 output is in negative saturation nor on amplifier 2 when amplifier 1 output is in negative saturation.

At the output, other single-supply designs either cannot swing to within 600 mV of ground or cannot sink more than a few microamperes while swinging to ground. The all-npn output stage of the LT1013 maintains its low output resistance and high-gain characteristics until the output is saturated. In dual-supply operations, the output stage is free of crossover distortion.

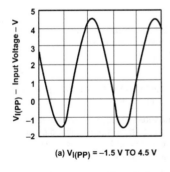

(a) $V_{I(PP)}$ = –1.5 V TO 4.5 V

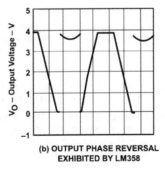

(b) OUTPUT PHASE REVERSAL
EXHIBITED BY LM358

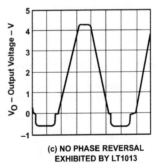

(c) NO PHASE REVERSAL
EXHIBITED BY LT1013

Figure 24. Voltage-Follower Response With Input Exceeding
the Negative Common-Mode Input Voltage Range

TEXAS INSTRUMENTS
POST OFFICE BOX 655303 ● DALLAS, TEXAS 75265

LT1013, LT1013A, LT1013D
DUAL PRECISION OPERATIONAL AMPLIFIERS

SLOS018E – MAY 1988 – REVISED SEPTEMBER 2001

APPLICATION INFORMATION

comparator applications

The single-supply operation of the LT1013 lends itself for use as a precision comparator with TTL-compatible output. In systems using both operational amplifiers and comparators, the LT1013 can perform multiple duties (see Figures 25 and 26).

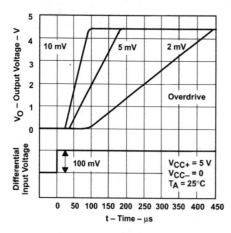

Figure 25. Low-to-High-Level Output Response for Various Input Overdrives   **Figure 26. High-to-Low-Level Output Response for Various Input Overdrives**

low-supply operation

The minimum supply voltage for proper operation of the LT1013 is 3.4 V (three NiCad batteries). Typical supply current at this voltage is 290 µA; therefore, power dissipation is only 1 mW per amplifier.

offset voltage and noise testing

The test circuit for measuring input offset voltage and its temperature coefficient is shown in Figure 30. This circuit, with supply voltages increased to ±20 V, also is used as the burn-in configuration.

The peak-to-peak equivalent input noise voltage of the LT1013 is measured using the test circuit shown in Figure 27. The frequency response of the noise tester indicates that the 0.1-Hz corner is defined by only one zero. The test time to measure 0.1-Hz to 10-Hz noise should not exceed 10 seconds, as this time limit acts as an additional zero to eliminate noise contribution from the frequency band below 0.1 Hz.

An input noise voltage test is recommended when measuring the noise of a large number of units. A 10-Hz input noise voltage measurement correlates well with a 0.1-Hz peak-to-peak noise reading because both results are determined by the white noise and the location of the 1/f corner frequency.

Current noise is measured by the circuit and formula shown in Figure 28. The noise of the source resistors is subtracted.

LT1013, LT1013A, LT1013D
DUAL PRECISION OPERATIONAL AMPLIFIERS

SLOS018E – MAY 1988 – REVISED SEPTEMBER 2001

APPLICATION INFORMATION

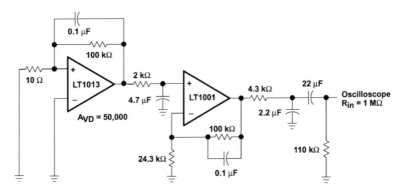

NOTE A: All capacitor values are for nonpolarized capacitors only.

Figure 27. 0.1-Hz to 10-Hz Peak-to-Peak Noise Test Circuit

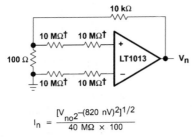

$$I_n = \frac{[V_{no}2 - (820 \text{ nV})^2]^{1/2}}{40 \text{ M}\Omega \times 100}$$

† Metal-film resistor

**Figure 28. Noise-Current Test Circuit
and Formula**

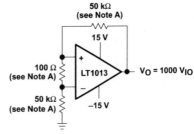

NOTE A: Resistors must have low thermoelectric potential.

Figure 29. Test Circuit for V_{IO} and $\alpha_{V_{IO}}$

**TEXAS
INSTRUMENTS**
POST OFFICE BOX 655303 ● DALLAS, TEXAS 75265

LT1013, LT1013A, LT1013D
DUAL PRECISION OPERATIONAL AMPLIFIERS

SLOS018E – MAY 1988 – REVISED SEPTEMBER 2001

APPLICATION INFORMATION

typical applications

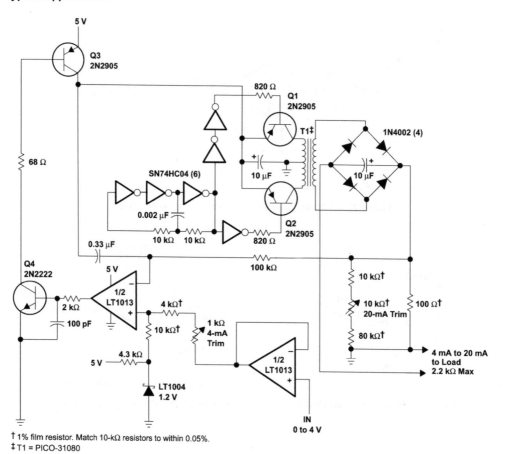

† 1% film resistor. Match 10-kΩ resistors to within 0.05%.
‡ T1 = PICO-31080

Figure 30. 5-V 4-mA to 20-mA Current Loop Transmitter With 12-Bit Accuracy

LT1013, LT1013A, LT1013D
DUAL PRECISION OPERATIONAL AMPLIFIERS

SLOS018E – MAY 1988 – REVISED SEPTEMBER 2001

APPLICATION INFORMATION

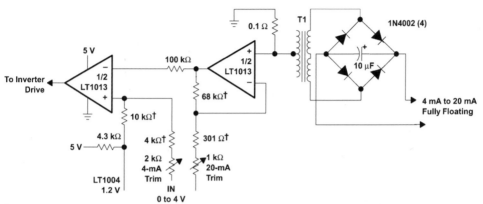

† 1% film resistor

Figure 31. Fully Floating Modification to 4-mA to 20-mA Current Loop Transmitter With 8-Bit Accuracy

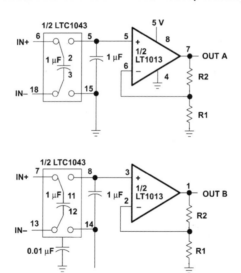

NOTE A: V_{IO} = 150 µV, A_{VD} = (R1/R2) + 1, CMRR = 120 dB, V_{ICR} = 0 to 5 V

Figure 32. 5-V Single-Supply Dual Instrumentation Amplifier

TEXAS INSTRUMENTS
POST OFFICE BOX 655303 ● DALLAS, TEXAS 75265

LT1013, LT1013A, LT1013D
DUAL PRECISION OPERATIONAL AMPLIFIERS

SLOS018E – MAY 1988 – REVISED SEPTEMBER 2001

APPLICATION INFORMATION

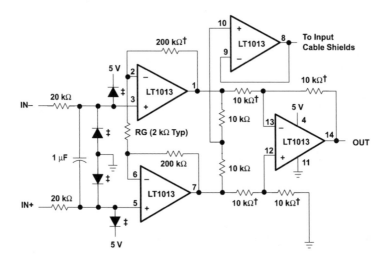

† 1% film resistor. Match 10-kΩ resistors to within 0.05%.
‡ For high source impedances, use 2N2222 diodes.
NOTE A: $A_{VD} = (400{,}000/RG) + 1$

Figure 33. 5-V Precision Instrumentation Amplifier

IMPORTANT NOTICE

B&B electronics
MANUFACTURING COMPANY

Technical Article 10   Page 1/2

Unfortunately, in most serial communications systems,
surge suppression is not the best choice.

DATA LINE ISOLATION THEORY

When it comes time to protect data lines from electrical transients, surge suppression is often the first thing that leaps to mind. The concept of surge suppression is intuitive and there are a large variety of devices on the market to choose from. Models are available to protect everything from your computer to answering machine as well as those serial devices found in RS-232, RS-422 and RS-485 systems.

Unfortunately, in most serial communications systems, surge suppression is not the best choice. The result of most storm and inductively induced surges is to cause a difference in ground potential between points in a communications system. The more physical area covered by the system, the more likely those differences in ground potential will exist.

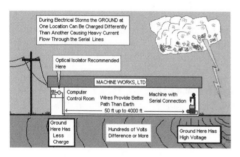

How Ground Changes During An Electrical Storm

The water analogy helps explain this phenomenon as well as any. Instead of water in a pipe, we'll think a little bigger and use waves on the ocean. Ask anyone what the elevation of the ocean is, and you will get an answer of zero – so common that we call it sea level. While the average ocean elevation is zero, we know that tides and waves can cause large short-term changes in the actual height of the water. This is very similar to earth ground. The effect of a large amount of current dumped into the earth (such as a lightning strike) can be visualized in the same way, as a wave propagating outwards from the origin. Until this energy dissipates, the voltage level of the earth will vary greatly between two locations.

Adding a twist to the ocean analogy, what is the best way to protect a boat from high waves? We could lash the boat to a fixed dock, forcing the boat to remain at one elevation. This will protect against small waves, but this solution obviously has limitations. While a little rough, this comparison isn't far off from what a typical surge suppressor is trying to accomplish. Attempting to clamp a surge of energy to a level safe for the local equipment requires that the clamping device be able to completely absorb or redirect transient energy.

Instead of lashing the boat to a fixed dock, let's let the dock float. Now the boat can rise and fall with the ocean swells (until we hit the end of our floating dock's posts). Instead of fighting nature, we're simply moving along with it. This is our data line isolation solution.

Isolation is not a new idea. It has always been implemented in telephone and ethernet equipment. For asynchronous data applications such as many RS-232, RS-422 and RS-485 systems, optical isolators are the most common isolation element. With isolation, two different grounds (better thought of as reference voltages) can exist on opposite sides of the isolation element without any current flowing through the element. With an optical isolator, this is performed with an LED and a photosensitive transistor. Only light passes between the two elements.

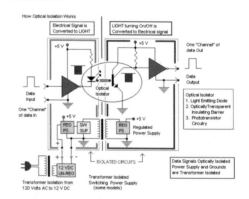

How Optical Isolation Works

B&B Electronics Mfg. Co.
707 Dayton Road - P.O. Box 1040 - Ottawa, IL 61350 USA
Phone: (815) 433-5100 - Fax: (815) 433-5105
Home Page: www.bb-elec.com
Sales e-mail: orders@bb-elec.com - Fax: (815) 433-5109
Technical Support e-mail: support@bb-elec.com - Fax: (815) 433-5104

B&B Electronics Ltd
Westlink Commercial Park - Oranmore, Co. Galway - Ireland
Phone: +353 91 792444 - Fax: +353 91 792445
Home Page: www.bb-europe.com
Sales e-mail: orders@bb-europe.com
Technical Support e-mail: support@bb-europe.com

B&B electronics
MANUFACTURING COMPANY

Technical Article 10   Page 2/2

(Data Line Isolation Theory – cont'd.)

Another benefit of optical isolation is that it is not dependent on installation quality. Typical surge suppressors used in data line protection use special diodes to shunt excess energy to ground. The installer must provide an extremely low impedance ground connection to handle this energy, which can be thousands of amps at frequencies into the tens of megahertz. A small impedance in the ground connection, such as in 1.8 meters (6 feet) of 18 gauge wire, can cause a voltage drop of hundreds of volts – enough voltage to damage most equipment. Isolation, on the other hand, does not require an additional ground connection, making it insensitive to installation quality.

Isolation is not a perfect solution. An additional isolated power supply is required to support the circuitry. This supply may be built-in as an isolated DC-DC converter or external. Simple surge suppressors require no power source. Isolation voltages are limited as well, usually ranging from 500V to 4000V. In some cases, applying both surge suppression and isolation is an effective solution.

When choosing data line protection for a system, it is important to consider all available options. There are pros and cons to both surge suppression and optical isolation. However, isolation is a more effective solution for most systems. If in doubt, choose isolation.

© B&B Electronics — November 1998

B&B Electronics Mfg.Co.
707 Dayton Road - P.O. Box 1040 - Ottawa, IL 61350 USA
Phone: (815) 433-5100 - Fax: (815) 433-5105
Home Page: www.bb-elec.com
Sales e-mail: orders@bb-elec.com - Fax: (815) 433-5109
Technical Support e-mail: support@bb-elec.com - Fax: (815) 433-5104

B&B Electronics Ltd
Westlink Commercial Park - Oranmore, Co. Galway - Ireland
Phone: +353 91 792444 - Fax: +353 91 792445
Home Page: www.bb-europe.com
Sales e-mail: orders@bb-europe.com
Technical Support e-mail: support@bb-europe.com

B&B electronics
MANUFACTURING COMPANY

Technical Article 10   Page 2/2

(Data Line Isolation Theory – cont'd.)

Another benefit of optical isolation is that it is not dependent on installation quality. Typical surge suppressors used in data line protection use special diodes to shunt excess energy to ground. The installer must provide an extremely low impedance ground connection to handle this energy, which can be thousands of amps at frequencies into the tens of megahertz. A small impedance in the ground connection, such as in 1.8 meters (6 feet) of 18 gauge wire, can cause a voltage drop of hundreds of volts – enough voltage to damage most equipment. Isolation, on the other hand, does not require an additional ground connection, making it insensitive to installation quality.

Isolation is not a perfect solution. An additional isolated power supply is required to support the circuitry. This supply may be built-in as an isolated DC-DC converter or external. Simple surge suppressors require no power source. Isolation voltages are limited as well, usually ranging from 500V to 4000V. In some cases, applying both surge suppression and isolation is an effective solution.

When choosing data line protection for a system, it is important to consider all available options. There are pros and cons to both surge suppression and optical isolation. However, isolation is a more effective solution for most systems. If in doubt, choose isolation.

© B&B Electronics — November 1998

B&B Electronics Mfg. Co.
707 Dayton Road - P.O. Box 1040 - Ottawa, IL 61350 USA
Phone: (815) 433-5100 - Fax: (815) 433-5105
Home Page: www.bb-elec.com
Sales e-mail: orders@bb-elec.com - Fax: (815) 433-5109
Technical Support e-mail: support@bb-elec.com - Fax: (815) 433-5104

B&B Electronics Ltd
Westlink Commercial Park - Oranmore, Co. Galway - Ireland
Phone: +353 91 792444 - Fax: +353 91 792445
Home Page: www.bb-europe.com
Sales e-mail: orders@bb-europe.com
Technical Support e-mail: support@bb-europe.com

Appendix F

Setting Up the Origin Examples

From the website http://www.pharmalabauto.com, copy all the files in "\Origin\Fitting" to Origin's fitting functions directory "C:\Program Files\OriginLab\Origin61\FitFunc". Edit the nlsf.ini file located in Origin's main directory. Change the pharmacology section of the nlsf.ini file to look like this:

```
[Pharmacology]

Default Function=

BiPhasic=BIPHASIC

DoseResp=DRESP

OneSiteBind=BIND1

OneSiteComp=COMP1

TwoSiteBind=BIND2

TwoSiteComp=COMP2

monopKa=monopKa

bispKa=bispKa
```

Copy the file TestCNF.cnf in "\Origin\cnf" to "C:\Program Files\OriginLab\Origin61". In the [Config] section of the Origin .ini file modify the following statement from:

```
File1=Macros FullMen
```

to:

```
File1=Macros FullMenu TestCNF
```

Copy the file SimpSamp.ogs in "\Origin\Scripts" to "C:\Program Files\OriginLab\Origin61". The files located in "\Origin\Projects" and "\Origin\Worksheets" are example worksheets and projects used within the text. These files should be copied over to the user's hard drive.

Keep in mind that copied files from the website may have the "read only" attribute set, which will prevent any modification to these files unless this attribute is disabled. This attribute can be modified by selecting the filename, left clicking and selecting properties, and, from the general tab unchecking the "read-only" attribute. This attribute can also be removed from many files in "batch mode" by using the MS-DOS command at the command prompt "attrib -r path/s". For example, to remove the read-only attribute from all the files residing in and under the C:\TestFiles directory, type "attrib -r c:\testfiles/s" at the command prompt in DOS. In Windows 2000 and XP, the read only attribute can be removed from all files contained with a folder by right clicking the folder, unchecking the read only attribute, and when prompted selecting "apply changes to this folder, subfolders, and files".

Appendix G

Setting Up the Agilent Chemstation Macros

From the website http://www.pharmalabauto.com, copy all the HPLC macro example files (*.mac) files in "\AgilentMacros" to the "HPChem\Core" directory.

Appendix H

HPLC Chemstation Macro Hook Information

(From *Macro HOOK Information CE, GC, LC and A/D ChemStation Revisions: A.05.0x- A.08.0x*, Agilent Technologies, Palo Alto, CA, http:// www.chem.agilent.com. With permission.)

Agilent Technologies

Innovating the HP Way

Macro HOOK Information
CE, GC, LC and A/D ChemStation
Revisions: A.05.0x- A.08.0x

Dynamic Data Exchange

Situation: The user wants to modify the operation of the standard ChemStation by adding own macros that should be executed in certain places, so-called *hooks*.

Problem: It is not desired to force the user to modify the standard macros other than in those files that are delivered as .mac files anyhow. And even in .mac files there is no complete freedom to change at will.

Goal: It should be possible that the standard ChemStation can be updated to a new version without disrupting the user macro hooks, provided that the user macros are still compatible.

The Hook Concept

At certain points in the standard ChemStation a (named) hook is defined. For each hook it is described what data is available at this point and what the hook macro might do here.

If in the following it is mentioned that a hook is called during Run Method, this also implies Run Sequence. If the called hook macro needs to differentiate between the two, the variable _SequenceOn can be checked.

Some of the hooks are called both during Run Method (and Run Sequence) and also interactively from various menu items. If the called hook macro needs to differentiate between automatic and interactive mode, the variable _MethodOn can be checked.

The Hooks

PreViewMenu

Called during setting up the menu immediately before the View menu is added. Can be used to define a new menu in the current view. Note that the current view is stored as table header item CurrentView in the Window table. This can be used to find more info, esp. the MenuFile, in the MenusForViews table. By comparison of the MenuFile table header item with the ShortMenuFile and FullMenuFile column names in there the short/full mode can be determined.

LoadMethod

Called at the end of loading a method, either interactively from the menu item or automatically during a sequence. At this point _MethPath$ and _MethFile$ are updated to the new method, and all parameter files within that .m directory are loaded. Can be used to load additional files from the .m directory.

SaveMethod

Called at the end of interactively saving a method, either from the Save or Save As menu items. At this point _MethPath$ and _MethFile$ are updated to the new method, and all parameter files are saved to that .m directory. Can be used to save additional files to the .m directory.

EditMethod

Called at the end of Edit Entire Method, after all other method sections are displayed. Can be used to edit additional parameters. Note that the hook is called even if (through Cancel) the rest of the originally selected method sections are skipped.

PrintMethod

Called at the end of printing a method, either interactively from the menu item or automatically as part of a sequence summary report. At this point the output destination port #5 is still open. With the DevNum command it can be checked whether this port is connected to the printer or to a .txt file. Can be used to print additional parameters.

PreRun

Called during Run Method immediately before the PreRun macro is done, regardless whether this is enabled by the user or not. Can be used to modify the Runtime Checklist's system variables.

AcqRes

Called during Run Method immediately after data acquisition is finished in an online system, while the acqres.reg file is being generated. At this time the register AcqRes exists. This will afterwards be saved to the file acqres.reg. The hook can be used to add information to the Data File. The first object of acqres.reg will become the basis of the ChromRes register in Data Analysis. Therefore the names of object header items and tables must not coincide

User Defined Hooks
The Hooks

	with those added by Data Analysis later on. To avoid naming conflicts it is strongly suggested that names generated by the user, e.g. for info generated in the PreRun macro, start with User, Usr, or Cust. Names generated by HP should use prefixes like Acq, Mod, Inst, or Dev.
PostAcq	Called during Run Method immediately after data acquisition is finished in an online system. At this time all raw data files exist. Can be used to do instrument-specific control.
PreLoadFile	Called during Run Method before standard Data Analysis loads the rawdata file. Can be used to change the file name and several other variables that control the standard Data Analysis.
PreInteg	Called during Run Method before standard Data Analysis does the integration. Also called during interactive Load Signal or Overlay Signal or Subtract Blank Run from the menu in the Data Analysis view. At all these times all required signals in the ChromReg register are loaded, ChromRes register is setup. Can be used to modify the loaded signals, e.g., by scaling or shifting, subtracting a blank run, etc. It is advisable to put some kind of flag, for example, a new object header item, into each modified object. This enables the hook macro to check before the modification whether the modification has already been done, especially in the case of Overlay Signal or Subtract Blank Run where the ChromReg register is merely enhanced by several objects, but not generated completely anew.
PostInteg	Called during Run Method after standard Data Analysis has done the integration, but before peak identification. Also called during interactive Integrate or Auto Integrate from the menu in the Data Analysis view. At all these times the signals are integrated. Can be used to modify the integration results, for example, through manual integration, re-integration with optimized parameters, or similar things.
PreQuant	Called during Run Method before standard Data Analysis does the quantification or automatic recalibration (in case of calibration runs). Also called during interactive Print Report from the menu in the Data Analysis view. At both these times the peak identification is finished. Can be used to modify the identification results in ChromRes.
PostQuant	Called during Run Method after standard Data Analysis does the quantification or automatic recalibration (in case of calibration runs). Also called during interactive Print Report from the menu in the Data Analysis view. At both these times the quantification results in ChromRes are fully available. Can be used to modify the Amount values in ChromRes through special calculation. Also any additional result calculation should be done

User Defined Hooks
The Hooks

here and added to ChromRes because this is the latest point before report formatting starts.

PostDA

Called during Run Method at the end of processing the data file but before the PostRun macro. At this time both standard and custom Data Analysis is finished, all results have been calculated and printed, all result files have been generated and the Save GLP Data has been done (that is, the glpsave.reg file has been generated). Can be used to transfer the result files from the .D directory to some other place. The difference between this hook and the PostRun macro is that this hook is done on a data file basis, that is, it might be delayed in case of bracketing, whereas the PostRun macro is done on an instrument-run basis, that is, in bracketing it might be done before the final quantification of the data file takes place. Another difference is that in the GC with two injectors the PostDA hook is executed once per data file, whereas the PostRun macro is called just once.

PostRun

Called during Run Method immediately after the PostRun macro is done, regardless whether this is enabled by the user or not. Can be used to wrap up the instrument run in connection with other external devices or a server. Note that this hook will not be called if a sequence is running which has Acquisition Only selected in its Sequence Parameter screen. Note also that in case of a bracketing sequence this hook might be called before the calculation and report of the sample run is done because that is delayed until the calibration run of the closing bracket is performed.

PreSeq

Called during Run Sequence before the actual sequence is started. Can be used for initialization, for example, opening a data base or establishing contact to a host computer for data transfers during a sequence.

PostSeq

Called during Run Sequence when the sequence table is finished and the user-specified Post-Sequence macro has been executed, and the sequence summary report macro is finished. Can be used to close a data base or instruct a host computer to start some processing before closing the communication links.

CloseApp

Called during the Exit or Close menu item. Can be used to close open files or save registers.

Initialization of a Hook

For each hook that the user wants to use, a macro has to be written that is executed when the hook point is reached. This macro can either be put into user.mac directly, or, especially if several such macros have to be written, put into a separate .mac file with the loading Macro command in user.mac.

Next, the hook macro has to be connected to the hook itself. For this a macro called SetHook is available. It has two string parameters: the name of the hook, as defined in the previous chapter, and the name of the macro to be executed.

Hook Example

In the following an example is shown which can be used as a structure for specific hooks. It shows the principle of the hook concept:

Example

```
name MyPreIntegHook
 ! this will be called as PreInteg hook
 ! As the hook will be called each time (it is not possible
 ! to remove the hook again) it should be possible to
 ! enable or disable the modifications done by the hook.
 ! For this it is necessary to check whether we really need
 ! to do something in this macro.
 ! This could be done by checking a global variable, like:
 !   if Check (Variable, USR_MyPreInteg) = 1 then
 !     if USR_MyPreInteg = 1 then
 !       < do the real work here >
 !     endif
 !   endif
 !
 ! or by checking a header item in the method, like:
 !   if ObjHdrType (_DAMethod[1], "USR_MyPreInteg") <> 99 then
 !     if ObjHdrVal (_DAMethod[1], "USR_MyPreInteg") = 1 then
 !       < do the real work here >
 !     endif
 !   endif
endmacro
SetHook "PreInteg", "MyPreIntegHook"
```

Appendix I

Further Reading

VISUAL BASIC

Eidahl, L.D., *Using Excel Visual Basic for Applications*, Platinum Edition, QUE, Indianapolis, IN, 1999.
Eidahl, L.D., *Using Visual Basic 5.0*, Platinum Edition, QUE, Indianapolis, IN, 1999.
Eidahl, L.D., *Using Visual Basic 6.0*, Platinum Edition, QUE, Indianapolis, IN, 1999.

NUMERICAL METHODS

Binstock, A. and Rex, J., *Practical Algorithms for Programmers*, Addison-Wesley, Reading, MA, 1995.
Broesch, J.D., *DSP Demystified*, LLH Publishing, Eagle Rock, VA, 1998.
Lyons, R.G., *Understanding Digital Signal Processing*, Addison Wesley, Reading, MA, 1996.
Mathews, J.H., *Numerical Methods for Mathematics, Science, and Engineering*, Prentice Hall, New York, 1992.

C++

Hollingworth, J., Butterfield, D., Swart, B., and Allsop, J., *C++ Builder 5 Developer's Guide*, SAMS, Indianapolis, IN, 2001.
Schildt, H., *Borland C++ Builder*, Osborne Press, Emeryville, CA, 2001.
Schildt, H., *C++ from the Ground Up*, Osborne Press, Emeryville, CA, 1994.
Schildt, H., *C/C++ Programmer's Reference*, Osborne Press, Emeryville, CA, 1997.

ELECTRONICS AND ELECTRICAL ENGINEERING

Carr, J.J., *Integrated Electronics*, Harcourt Brace Jovanovich, New York, 1990.
Cathey, J.J. and Nasar, S.A., *Basic Electrical Engineering*, McGraw-Hill, New York, 1984.
Horowitz, P. and Hill, W., *The Art of Electronics*, Cambridge University Press, Cambridge, U.K., 1989.
Mano, M.M., *Computer Engineering and Hardware Design*, Prentice Hall, New York, 1998.
Mims, F.M., *The Forrest Mims Circuit Scrapbook*, LLH Publishing, Eagle Rock, VA, 2000.
Mims, F.M., *The Forrest Mims Engineer's Notebook*, LLH Publishing, Eagle Rock, VA, 1993.
Pease, R.A., *Troubleshooting Analog Circuits*, Butterworth-Heinemann, Woburn, MA, 1993.
Williams, J., *The Art and Science of Analog Circuit Design*, Newnes, division of Butterworth-Heinemann, Woburn, MA, 1998.

MISCELLANEOUS

Abbey, M. and Corey, M., *Oracle, A Beginner's Guide*, Oracle Press (Osborne), Berkeley, CA, 1995.
HP Chemstation, Macro Programming Guide, Hewlett Packard, Palo Alto, CA, 1994.
Schildt, H., *STL Programming from the Ground Up*, Osborne Press, Emeryville, CA, 1999.

References

50300 Hardware User's Manual, Appendices A–C, Kloehn Ltd., Las Vegas, NV, 1996.

Data Line Isolation Theory, Technical article #10, B&B Electronics, Ottawa, IL, www.bb-elec.com, November 1998.

Fundamentals of RS–232 Serial Communications, Application note 83, Dallas Semiconductor, Sunnyvale, CA, March 1998.

Goldie, J., *Ten ways to bulletproof RS-485 interfaces*, Application note 1057, National Semiconductor Corporation, Santa Clara, CA, www.national.com, October 1996.

Grounding and RS-422/485, RobustDC Application note #5, Robust DataComm Inc., Addison, TX, www.robustdc.com, 1997.

HP Chemstation Macro Programming Guide, Hewlett-Packard, Palo Alto, CA, 1995.

Labtalk Manual, Version 6, Origin Lab Corporation, Northampton, MA, 1999.

LM124/LM224/LM324/LM2902: Low Power Quad Operational Amplifiers, National Semiconductor Corporation, Santa Clara, CA, www.national.com, August 2000.

LM555: Timer, National Semiconductor Corporation, Santa Clara, CA, www.national.com, February 2000.

LT1013, LT1013A, LT1013D: Dual Precision Operational Amplifiers, Texas Instruments Corporation, Dallas, TX, www.ti.com, September 2001.

Macro HOOK Information CE, GC, LC and A/D ChemStation Revisions: A.05.0x- A.08.0x, Agilent Technologies, Palo Alto, CA, http:// www.chem.agilent.com.

Martin, A., *Decoupling: basics*, AVX Corporation, Myrtle Beach, SC, 1994.

McConnell, E., Choosing a data acquisition method, *Electron. Design Mag.*, June, 1995.

Metrabus User's Guide, Revision F, Keithley Metrabyte Division, Keithley Instruments, Inc., Cleveland, OH, July 1994.

Microsoft Visual Basic 6.0 Help File, Microsoft Corporation, Redmond, WA, 2000.

Nelson, T., *The practical limits of RS-485*, Application note 979, National Semiconductor Corporation, Santa Clara, CA, www.national.com, March 1995.

PortI/O DLL, Shareware, Scientific Software Tools, Inc., Media, PA, 1998–2002.

Savant, C.J., Roden, M.S., and Carpenter, G.L., *Electronic Design*, 2nd ed., Discovery Press, Los Angeles, CA, 1991.

The SpreadSheet Page, menumakr.xls utility, JWalk & Associates, La Jolla, CA, http://www.j-walk.com, 2002.

Index